Enzyme Technology

Enzyme Technology

D. Vijaya Lakshmi

SBW PUBLISHERS
DELHI (INDIA)

Published by:
SBW Publishers
4831/24, Prahlad Gali, Ansari Road, Darya Ganj
New Delhi-110002 (INDIA)
E-mail: sbwpublishers@gmail.com
website: www.sbwpublishers.com

Enzyme Technology

ISBN : 978-81-85708-45-4

Published by: SBW Publishers.
Printed at: Thomson Press, Delhi-110092.
Lasertypset by Gaurav Graphices & Design.

Preface

Enzyme Technology has made a major impact on society and is the central factor around which most biotechnological science revolves. The aim of the present book is to provide a deeper insight into the technology of enzymes. It especially stresses on the following interrelationships: A thorough understanding of enzymes as biocatalysts requires the integration of natural science knowledge: biology, biochemistry, cell and molecular biology; physicochemical aspects of catalysis and molecular interactions in solutions, heterogeneous systems and interphase boundaries; the physics of mass transfer processes. The same applies for the interrelations between Enzyme Technology and chemical and process engineering, which are based on the above natural sciences.

In less than a century since the start of industrial enzyme production, enzyme biotechnology and its products have continuously gained importance. For the industrial production of goods to meet demands in human need for everyday life, enzymes play an important - often hardly known role. Their application spans from the production of processed foods, such as bread, cheese, juice beer; pharmaceuticals; fine chemicals; to the processing of leather and textiles, and their application as process aids in detergents or environmental engineering.

To meet the demand for new products, such as new or pure pharmaceuticals and fine chemicals, is an important incentive for the further development of biocatalysts and enzyme technology. Of similar importance is the development of new sustainable production processes for existing products.

Enzymes as catalysts are of key importance in biotechnology, similar to the role of nucleic acids as carriers of genetic information. Their application as isolated catalysts justifies the detailed treatment of the fundamentals of enzymes as biocatalyst in chapter of this book. They can be analyzed on a molecular level and their kinetics can be described by mathematics. This is essential for an analytical description and the rational design of enzyme processes. Enzymes can catalyze a reaction in both directions. This is applied in Enzyme Biotechnology, in order to reach the desired endpoint of a reaction rapidly and with high product yield. The thermodynamics of the catalyzed reaction must also be considered, as well as the properties of the enzyme. The enzyme amount required for a given conversion of substrate per unit time must be calculated in order to estimate the enzyme cost and the economic feasibility of the process. Therefore the quantitative treatment of biocatalysis is stressed in the chapter of this book.

Author

Contents

Chapter 1

Introduction to Enzymes

The use of enzymes in the diagnosis of disease is one of the important benefits derived from the intensive research in biochemistry since the 1940's. Enzymes have provided the basis for the field of clinical chemistry. It is, however, only within the recent past few decades that interest in diagnostic enzymology has multiplied. Many methods currently on record in the literature are not in wide use, and there are still large areas of medical research in which the diagnostic potential of enzyme reactions has not been explored at all. Enzymes are biological catalysts responsible for supporting almost all of the chemical reactions that maintain animal homeostasis. Because of their role in maintaining life processes, the assay and pharmacological regulation of enzymes have become key elements in clinical diagnosis and therapeutics.

The macromolecular components of almost all enzymes are composed of protein, except for a class of RNA modifying catalysts known as ribozymes. Ribozymes are molecules of ribonucleic acid that catalyze reactions on the phosphodiester bond of other RNAs. Enzymes are found in all tissues and fluids of the body. Intracellular enzymes catalyze the reactions of metabolic pathways. Plasma membrane enzymes regulate catalysis within cells in response to extracellular signals, and enzymes of the circulatory system are responsible for regulating the clotting of blood. Almost every significant life process is dependent on enzyme activity.

ENZYME CLASSIFICATIONS

Traditionally, enzymes were simply assigned names by the investigator who discovered the enzyme. As knowledge expanded, systems of enzyme classification became more comprehensive and complex. Currently enzymes are grouped into six functional classes by the International Union of Biochemists (I.U.B.).

Number	Classification	Biochemical Properties
1	Oxidoreductases	Act on many chemical groupings to add or remove hydrogen atoms.

2	Transferases	Transfer functional groups between donor and acceptor molecules. Kinases are specialized transferases that regulate metabolism by transferring phosphate from ATP to other molecules.
3	Hydrolases	Add water across a bond, hydrolyzing it.
4.	Lyases	Add water, ammonia or carbon dioxide across double bonds, or remove these elements to produce double bonds.
5	Isomerases	Carry out many kinds of isomerization: L to D isomerizations, mutase reactions (shifts of chemical groups) and others.
6	Ligases	Catalyze reactions in which two chemical groups are joined (or ligated) with the use of energy from ATP.

These rules give each enzyme a unique number. The I.U.B. system also specifies a textual name for each enzyme. The enzyme's name is comprised of the names of the substrate(s), the product(s) and the enzyme's functional class. Because many enzymes, such as alcohol dehydrogenase, are widely known in the scientific community by their common names, the change to I.U.B.-approved nomenclature has been slow. In everyday usage, most enzymes are still called by their common name.

Enzymes are also classified on the basis of their composition. Enzymes composed wholly of protein are known as simple enzymes in contrast to complex enzymes, which are composed of protein plus a relatively small organic molecule. Complex enzymes are also known as holoenzymes. In this terminology the protein component is known as the apoenzyme, while the non-protein component is known as the coenzyme or prosthetic group where prosthetic complex in which the small organic molecule is bound to the apoenzyme by covalent bonds; when the binding between the apoenzyme and non-protein components is non-covalent, the small organic molecule is called a coenzyme. Many prosthetic groups and coenzymes are water-soluble derivatives of vitamins. It should be noted that the main clinical symptoms of dietary vitamin insufficiency generally arise from the malfunction of enzymes, which lack sufficient cofactors derived from vitamins to maintain homeostasis.

The non-protein component of an enzyme may be as simple as a metal ion or as complex as a small non-protein organic molecule. Enzymes that require a metal in their composition are known as metalloenzymes if they bind and retain their metal atom(s) under all conditions with very high affinity. Those which have a lower affinity for metal ion, but still require the metal ion for activity, are known as metal-activated enzymes.

ENZYMES AND LIFE PROCESSES

The living cell is the site of tremendous biochemical activity called metabolism. This is the process of chemical and physical change which goes on continually in the living organism. Build-up of new tissue, replacement of old tissue, conversion of food to energy, disposal of waste materials, reproduction - all the activities that we characterize as "life." This building up and tearing down takes place in the face of an apparent paradox. The greatest majority of these biochemical reactions do not take place spontaneously. The phenomenon of catalysis makes possible biochemical reactions necessary for all life processes. Catalysis is defined as the acceleration of a chemical reaction by some substance which itself undergoes no permanent chemical change. The catalysts of biochemical reactions are enzymes and are responsible for bringing about almost all of the chemical reactions in living organisms. Without enzymes, these reactions take place at a rate far too slow for the pace of metabolism.

The oxidation of a fatty acid to carbon dioxide and water is not a gentle process in a test tube - extremes of pH, high temperatures and corrosive chemicals are required. Yet in the body, such a reaction takes place smoothly and rapidly within a narrow range of pH and temperature. In the laboratory, the average protein must be boiled for about 24 hours in a 20% HCl solution to achieve a complete breakdown. In the body, the breakdown takes place in four hours or less under conditions of mild physiological temperature and pH.

CHEMICAL NATURE OF ENZYMES

All known enzymes are proteins. They are high molecular weight compounds made up principally of chains of amino acids linked together by peptide bonds.

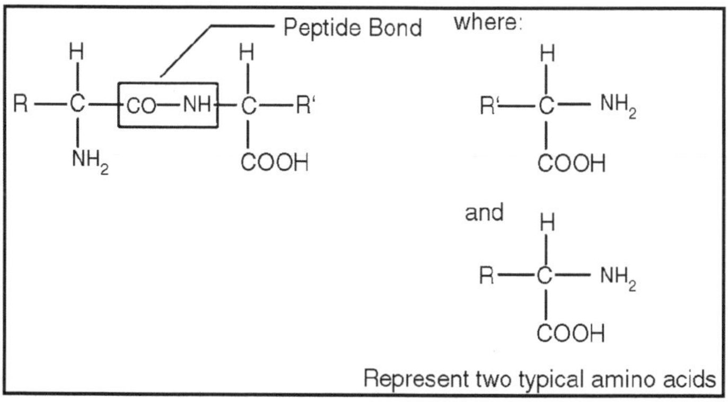

Fig. Typical Protein Structure - Two Amino Acids Joined by a Peptide Bond.

Enzymes can be denatured and precipitated with salts, solvents and other reagents. They have molecular weights ranging from 10,000 to 2,000,000. Many enzymes require the presence of other compounds - cofactors - before their catalytic activity can be exerted. This entire active complex is referred to as the holoenzyme; *i.e.*, apoenzyme (protein portion) plus the cofactor (coenzyme, prosthetic group or metal-ion-activator) is called the holoenzyme.

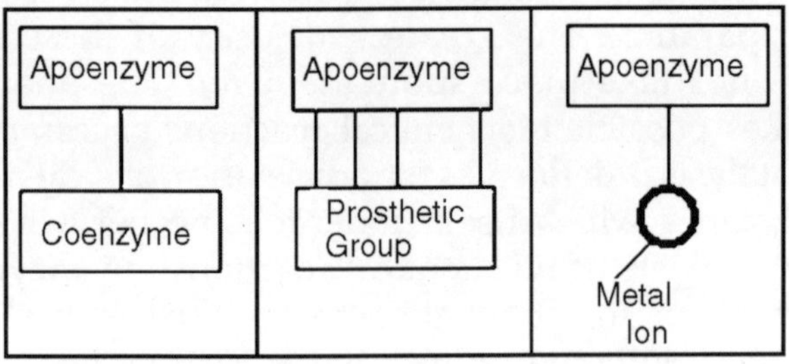

Fig. Holoenzymes- Apoenzymes Plus Various Types of Cofactors.

Apoenzyme + Cofactor = Holoenzyme

According to Holum, the cofactor may be:

- *A coenzyme*: A non-protein organic substance which is dialyzable, thermostable and loosely attached to the protein part.
- *A prosthetic group*: An organic substance which is dialyzable and thermostable which is firmly attached to the protein or apoenzyme portion.
- *A metal-ion-activator*: These include K^+, Fe^{++}, Fe^{+++}, Cu^{++}, Co^{++}, Zn^{++}, Mn^{++}, Mg^{++}, Ca^{++}, and Mo^{+++}.

Specificity of Enzymes

One of the properties of enzymes that makes them so important as diagnostic and research tools is the specificity they exhibit relative to the reactions they catalyze. A few enzymes exhibit absolute specificity; that is, they will catalyze only one particular reaction. Other enzymes will be specific for a particular type of chemical bond or functional group.

In general, there are four distinct types of specificity:

1. *Absolute specificity*: The enzyme will catalyze only one reaction.
2. *Group specificity*: The enzyme will act only on molecules that have specific functional groups, such as amino, phosphate and methyl groups.
3. *Linkage specificity*: The enzyme will act on a particular type of chemical bond regardless of the rest of the molecular structure.
4. *Stereochemical specificity*: The enzyme will act on a particular steric or optical isomer.

Though enzymes exhibit great degrees of specificity, cofactors may serve many apoenzymes. For example, nicotinamide adenine dinucleotide (NAD) is a coenzyme for a great number of dehydrogenase reactions in which it acts as a hydrogen acceptor. Among them are the alcohol dehydrogenase, malate dehydrogenase and lactate dehydrogenase reactions.

Naming and Classification

Except for some of the original enzymes such as pepsin, rennin, and trypsin, most enzyme names end in "ase". The International Union of Biochemistry (I.U.B.) initiated standards of enzyme nomenclature which recommend that enzyme names indicate both the substrate acted upon and the type of reaction catalyzed. Under this system, the enzyme uricase is called urate: O_2 oxidoreductase, while the enzyme glutamic oxaloacetic transaminase (GOT) is called L-aspartate: 2-oxoglutarate aminotransferase.

Enzymes can be classified by the kind of chemical reaction catalyzed:

- Addition or removal of water
  - *Hydrolases*: These include esterases, carbohydrases, nucleases, deaminases, amidases, and proteases
  - Hydrases such as fumarase, enolase, aconitase and carbonic anhydrase
- Transfer of electrons
  - Oxidases
  - Dehydrogenases
- Transfer of a radical
  - *Transglycosidases*: Of monosaccharides
  - *Transphosphorylases and phosphomutases*: Of a phosphate group
  - *Transaminases*: Of amino group
  - *Transmethylases*: Of a methyl group
  - *Transacetylases*: Of an acetyl group
- Splitting or forming a C-C bond
  - Desmolases
- Changing geometry or structure of a molecule
  - Isomerases
- Joining two molecules through hydrolysis of pyrophosphate bond in ATP or other triphosphate
  - Ligases

ROLE OF COENZYMES

The functional role of coenzymes is to act as transporters of chemical groups from one reactant to another. The chemical groups carried can be as simple as the hydride ion ($H^+ + 2e^-$) carried by NAD or the mole of hydrogen carried by FAD; or they can be even more complex than the amine ($-NH_2$) carried by pyridoxal phosphate.

Since coenzymes are chemically changed as a consequence of enzyme action, it is often useful to consider coenzymes to be a special class of substrates, or second substrates, which are common to many different holoenzymes. In all cases, the coenzymes donate the carried chemical grouping to an acceptor molecule and are thus regenerated to their original form. This regeneration of coenzyme and holoenzyme fulfills the definition of an enzyme as a chemical catalyst, since (unlike the usual substrates, which are used up during the course of a reaction) coenzymes are generally regenerated.

ENZYME RELATIVE TO SUBSTRATE TYPE

Although enzymes are highly specific for the kind of reaction they catalyze, the same is not always true of substrates they attack. For example, while succinic dehydrogenase (SDH) always catalyzes an oxidation-reduction reaction and its substrate is invariably succinic acid, alcohol dehydrogenase (ADH) always catalyzes oxidation-reduction reactions but attacks a number of different alcohols, ranging from methanol to butanol. Generally, enzymes having broad substrate specificity are most active against one particular substrate. In the case of ADH, ethanol is the preferred substrate.

Enzymes also are generally specific for a particular steric configuration (optical isomer) of a substrate. Enzymes that attack D sugars will not attack the corresponding L isomer. Enzymes that act on L amino acids will not employ the corresponding D optical isomer as a substrate. The enzymes known as racemases provide a striking exception to these generalities; in fact, the role of racemases is to convert D isomers to L isomers and vice versa. Thus racemases attack both D and L forms of their substrate.

As enzymes have a more or less broad range of substrate specificity, it follows that a given substrate may be acted on by a number of different enzymes, each of which uses the same substrate(s) and produces the same product(s). The individual members of a set of enzymes sharing such characteristics are known as isozymes. These are the products of genes that vary only slightly; often, various isozymes of a group are expressed in different tissues of the body. The best studied set of isozymes is the lactate dehydrogenase (LDH) system. LDH is a tetrameric enzyme composed of all possible arrangements of two different protein subunits; the subunits are known as H (for heart) and M (for skeletal muscle). These subunits combine in various combinations leading to 5 distinct isozymes. The all H isozyme is characteristic of that from heart tissue, and the all M isozyme is typically found in skeletal muscle and liver. These isozymes all catalyze the same chemical reaction, but they exhibit differing degrees of efficiency. The detection of specific LDH isozymes in the blood is highly diagnostic of tissue damage such as occurs during cardiac infarct.

ENZYME-SUBSTRATE INTERACTIONS

The favoured model of enzyme substrate interaction is known as the induced fit model. This model proposes that the initial interaction between enzyme and substrate is relatively weak, but that these weak interactions rapidly induce conformational changes in the enzyme that strengthen binding and bring catalytic sites close to substrate bonds to be altered. After binding takes place, one or more mechanisms of catalysis generates transition- state complexes and reaction products.

The possible mechanisms of catalysis are four in number:

1. *Catalysis by Bond Strain*: In this form of catalysis, the induced structural rearrangements that take place with the binding of substrate and enzyme ultimately produce strained substrate bonds, which more easily attain the transition state. The new conformation often forces substrate atoms and bulky catalytic groups, such as aspartate and glutamate, into conformations that strain existing substrate bonds.
2. *Catalysis by Proximity and Orientation*: Enzyme-substrate interactions orient reactive groups and bring them into proximity with one another. In addition to inducing strain, groups such as aspartate are frequently chemically reactive as well, and their proximity and orientation towards the substrate thus favours their participation in catalysis.
3. *Catalysis Involving Proton Donors (Acids) and Acceptors (Bases)*: Other mechanisms also contribute significantly to the completion of catalytic events initiated by a strain mechanism, for example, the use of glutamate as a general acid catalyst (proton donor).
4. *Covalent Catalysis*: In catalysis that takes place by covalent mechanisms, the substrate is oriented to active sites on the enzymes in such a way that a covalent intermediate forms between the enzyme or coenzyme and the substrate. One of the best-known examples of this mechanism is that involving proteolysis by serine proteases, which include both digestive enzymes (trypsin, chymotrypsin, and elastase) and several

enzymes of the blood clotting cascade. These proteases contain an active site serine whose R group hydroxyl forms a covalent bond with a carbonyl carbon of a peptide bond, thereby causing hydrolysis of the peptide bond.

Chemical Reactions and Rates

According to the conventions of biochemistry, the rate of a chemical reaction by the number of molecules of reactant(s) that are converted into product(s) in a specified time period. Reaction rate is always dependent on the concentration of the chemicals involved in the process and on rate constants that are characteristic of the reaction.

For example, the reaction in which A is converted to B is written as follows:

$$A \rightarrow B$$

The rate of this reaction is expressed algebraically as either a decrease in the concentration of reactant A:

$$-[A] = k[B]$$

or an increase in the concentration of product B:

$$[B] = k[A]$$

In the second equation (of the 3 above) the negative sign signifies a decrease in concentration of A as the reaction progresses, brackets define concentration in molarity and the *k* is known as a rate constant. Rate constants are simply proportionality constants that provide a quantitative connection between chemical concentrations and reaction rates.

Each chemical reaction has characteristic values for its rate constants; these in turn directly relate to the equilibrium constant for that reaction. Thus, reaction can be rewritten as an equilibrium expression in order to show the relationship between reaction rates, rate constants and the equilibrium constant for this simple case. The rate constant for the forward reaction is defined as k_{+1} and the reverse as k_{-1}.

At equilibrium the rate (v) of the forward reaction ($A \rightarrow B$) is, by definition, equal to that of the reverse or back reaction ($B \rightarrow A$), a relationship which is algebraically symbolized as:

$$v_{forward} = v_{reverse}$$

where, for the forward reaction:

$$v_{forward} = k_{+1}[A]$$

and for the reverse reaction:

$$v_{reverse} = k_{-1}[B]$$

In the above equations, k_{+1} and k_{-1} represent rate constants for the forward and reverse reactions, respectively. The negative subscript refers only to a reverse reaction, not to an actual negative value for the constant. To put the relationships of the two equations into words, we state that the rate of the forward reaction [$v_{forward}$] is equal to the product of the forward rate constant k_{+1} and the molar concentration of A. The rate of the reverse reaction is equal to the product of the reverse rate constant k_{-1} and the molar concentration of B.

At equilibrium, the rate of the forward reaction is equal to the rate of the reverse reaction leading to the equilibrium constant of the reaction and is expressed by:

$$K_{eq} = \frac{K_{+1}}{K_{-1}} = \frac{[B]}{[A]}$$

This equation demonstrates that the equilibrium constant for a chemical reaction is not only equal to the equilibrium ratio of product and reactant concentrations, but is also equal to the ratio of the characteristic rate constants of the reaction.

Chemical Reaction Order

Reaction order refers to the number of molecules involved in forming a reaction complex that is competent to proceed to product(s). Empirically, order is easily determined by summing the exponents of each concentration term in the rate equation for a reaction. A reaction characterized by the conversion of one molecule of A to one molecule of B with no influence from any other reactant or solvent is a first-order reaction. The exponent on the substrate concentration in the rate equation for this type of reaction is 1. A reaction with two substrates forming two products would a second-order reaction. However, the reactants in second– and higher– order reactions need not be different chemical species. An example of a second order reaction is the formation of ATP through the condensation of ADP with orthophosphate:

$$ADP + H_2PO_4^- \leftrightarrow ATP + H_2O$$

For this reaction the forward reaction rate would be written as:

$$v_{forward} = k_1[ADP][H_2PO_4^-]$$

FACTORS AFFECTING ENZYME ACTIVITY

Knowledge of basic enzyme kinetic theory is important in enzyme analysis in order both to understand the basic enzymatic mechanism and to select a method for enzyme analysis. The conditions selected to measure the activity of an enzyme would not be the same as those selected to measure the concentration of its substrate. Several factors affect the rate at which enzymatic reactions proceed - temperature, pH, enzyme concentration, substrate concentration, and the presence of any inhibitors or activators.

Enzyme Concentration

In order to study the effect of increasing the enzyme concentration upon the reaction rate, the substrate must be present in an excess amount; *i.e.*, the reaction must be independent of the substrate concentration. Any change in the amount of product formed over a specified period of time will be dependent upon the level of enzyme present. Graphically this can be represented as:

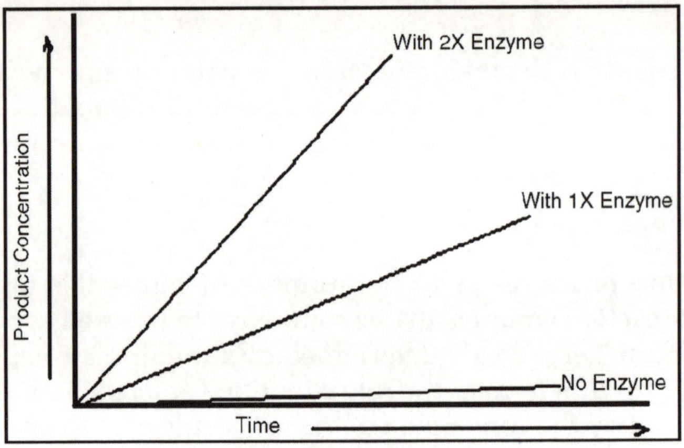

Fig. "Zero Order" Reaction rate is Independent of Substrate Concentration.

These reactions are said to be "zero order" because the rates are independent of substrate concentration, and are equal to some constant k. The formation of product proceeds at a rate which is linear with time. The addition of more substrate does not serve to increase the rate. In zero order kinetics, allowing the assay to run for double time results in double the amount of product.

Table. Reaction Orders with Respect to Substrate Concentration

Order	Rate Equation	Comments
Zero	rate = k	Rate is independent of substrate concentration
First	rate = k[S]	Rate is proportional to the first power of substrate concentration
Second	rate = $k[S][S]=k[S]^2$	Rate is proportional to the square of the substrate concentration
Second	rate = $k[S_1][S_2]$	Rate is proportional to the first power of each of two reactants

The amount of enzyme present in a reaction is measured by the activity it catalyzes. The relationship between activity and concentration is affected by many factors such as temperature, pH, etc. An enzyme assay must be designed so that the observed activity is proportional to the amount of enzyme present in order that the enzyme concentration is the only limiting factor. It is satisfied only when the reaction is zero order.

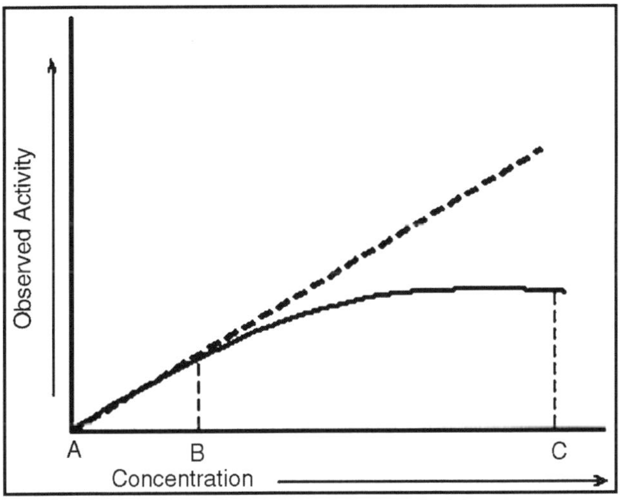

Fig. Activity vs. Concentration.

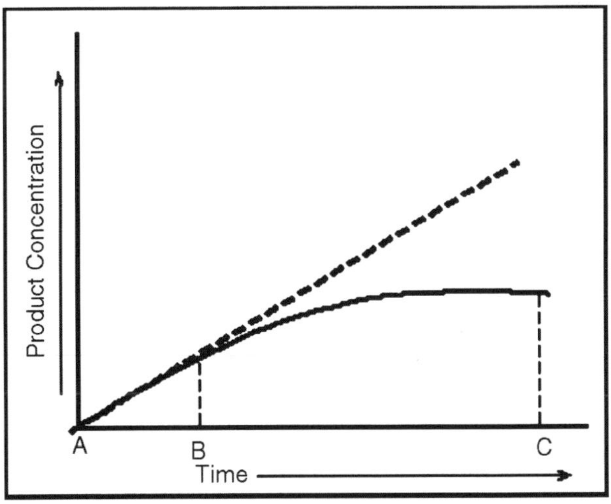

Fig. Reaction Rate Limited by Substrate Concentration.

When the concentration of the product of an enzymatic reaction is plotted against time, a similar curve results.

Between A and B, the curve represents a zero order reaction; that is, one in which the rate is constant with time. As substrate is used up, the enzyme's active sites are no longer saturated, substrate concentration becomes rate limiting, and the reaction becomes first order between B and C. To measure enzyme activity ideally, the measurements must be made in that portion of the curve where the reaction is zero order. A reaction is most likely to be zero order initially since substrate concentration is then highest. To be certain that a reaction is zero order, multiple measurements of product (or substrate) concentration must be made. Three types of reactions which might be encountered in enzyme assays and shows the problems which might be enountered if only single measurements are made.

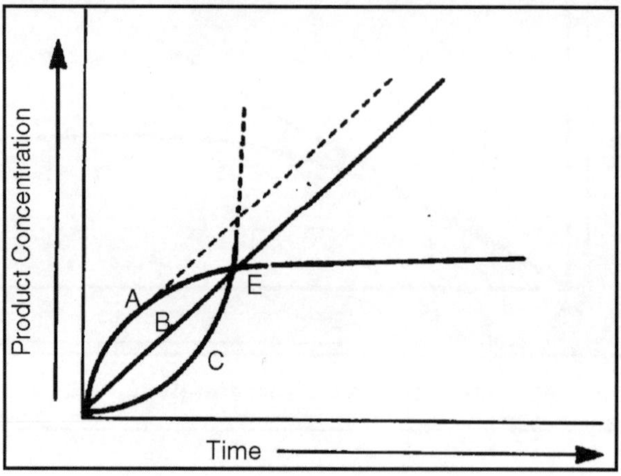

Fig. Time vs Concentration.

B is a straight line representing a zero order reaction which permits accurate determination of enzyme activity for part or all of the reaction time. This reaction is zero order initially and then slows, presumably due to substrate exhaustion or product inhibition. This type of reaction is sometimes referred to as a "leading" reaction. True "potential" activity is represented by the dotted line. Curve C represents a reaction with an initial "lag" phase. Again the dotted line represents the potentially measurable activity. Multiple determinations of product concentration enable each curve to be plotted and true activity determined. A single end point determination at E would lead to the false conclusion that all three samples had identical enzyme concentration.

Substrate Concentration

It has been shown experimentally that if the amount of the enzyme is kept constant and the substrate concentration is then gradually increased, the reaction velocity will increase until it reaches a maximum. After this point, increases in substrate concentration will not increase the velocity ($\Delta A/\Delta T$).

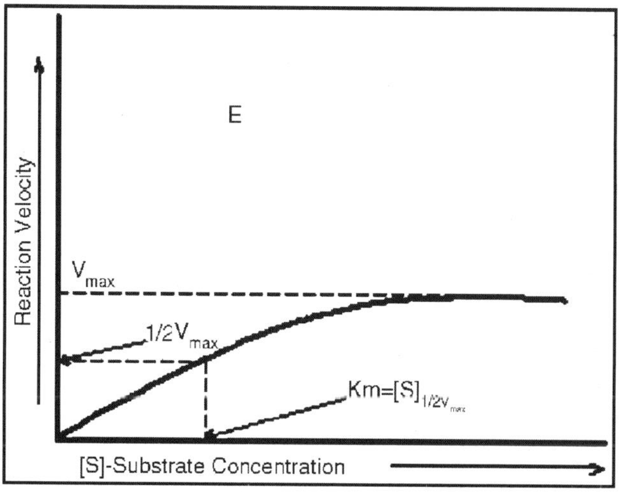

Fig. Effect of Substrate Concentration.

It is theorized that when this maximum velocity had been reached, all of the available enzyme has been converted to ES, the enzyme substrate complex. This point on the graph is designated Vmax. Using this maximum velocity and equation, Michaelis developed a set of mathematical expressions to calculate enzyme activity in terms of reaction speed from measurable laboratory data.

$$E + S \underset{K-1}{\overset{K+1}{\rightleftarrows}} ES \underset{K-2}{\overset{K+2}{\rightleftarrows}} P + E$$

The Michaelis constant Km is defined as the substrate concentration at 1/2 the maximum velocity. Using this constant and the fact that Km can also be defined as:

$$K_m = K_{-1} + K_{2/}K_{+1}$$

K^{+1}, K^{-1} and K^{+2} being the rate constants from equation. Michaelis developed the following,

$$V_1 = \frac{V_{max}[s]}{K_m + [S]}$$

where,

$V_{\uparrow}$	=	the velocity at any time
[s]	=	the substrate concentration at this time
V_{max}	=	the highest under this set of experimental conditions (pH, temperature, etc.)
K_m	=	the Michaelis constant for the particular enzyme being investigated

Michaelis constants have been determined for many of the commonly used enzymes.

The size of K_{m3} tells us several things about a particular enzyme:

- A small K_{m3} indicates that the enzyme requires only a small amount of substrate to become saturated. Hence, the maximum velocity is reached at relatively low substrate concentrations.
- A large K_{m3} indicates the need for high substrate concentrations to achieve maximum reaction velocity.
- The substrate with the lowest K_{m3} upon which the enzyme acts as a catalyst is frequently assumed to be enzyme's natural substrate, though this is not true for all enzymes.

ENZYMES AS BIOLOGICAL CATALYSTS

In cells and organisms most reactions are catalyzed by enzymes, which are regenerated during the course of a reaction. These biological catalysts are physiologically important because they speed up the rates of reactions that would otherwise be too slow to support life. Enzymes increase reaction rates, sometimes by as much as one million-fold, but more typically by about one thousand fold. Catalysts speed up the forward and reverse reactions proportionately so that, although the magnitude of the rate constants of the forward and reverse reactions is are increased, the ratio of the rate constants remains the same in the presence or absence of enzyme. Since the equilibrium constant is equal to a ratio of rate constants, it is apparent that enzymes and other catalysts have no effect on the equilibrium constant of the reactions they catalyze.

Enzymes increase reaction rates by decreasing the amount of energy required to form a complex of reactants that is competent to produce reaction products. This complex is known as the activated state or transition state complex for the reaction. Enzymes and other catalysts accelerate reactions by lowering the energy of the transition state. The free energy required to form an activated complex is much lower in the catalyzed reaction. The amount of energy required to achieve the transition state is lowered; consequently, at any instant a greater proportion of the molecules in the population can achieve the transition state. The result is that the reaction rate is increased.

Michaelis-Menten Kinetics

In typical enzyme-catalyzed reactions, reactant and product concentrations are usually hundreds or thousands of times greater than the enzyme concentration. Consequently, each enzyme molecule catalyzes the conversion to product of many reactant molecules. In biochemical reactions, reactants are commonly known as substrates. The catalytic event that converts substrate to product involves the formation of a transition state, and it occurs most easily at a specific binding site on the enzyme. This site, called the catalytic site of the enzyme, has been evolutionarily structured to provide specific, high-affinity binding of substrate(s) and to provide an environment that favours the catalytic events. The complex that forms, when substrate(s) and enzyme combine, is called the enzyme substrate (ES) complex. Reaction products arise when the ES complex breaks down releasing free enzyme.

Between the binding of substrate to enzyme, and the reappearance of free enzyme and product, a series of complex events must take place. At a minimum an ES complex must be formed; this complex must pass to the transition state (ES*); and the transition state complex must advance to an enzyme product complex (EP). The latter is finally competent to dissociate to product and free enzyme.

The series of events can be shown thus:

$$E + S \leftrightarrow ES \leftrightarrow ES^* \leftrightarrow EP \leftrightarrow E + P$$

The kinetics of simple reactions like that above were first characterized by biochemists Michaelis and Menten. The concepts underlying their analysis of enzyme kinetics continue to provide the cornerstone for understanding metabolism today, and for the development and clinical use of drugs aimed at selectively altering rate constants and interfering with the progress of disease states. The Michaelis-Menten equation is a quantitative description of the relationship among the rate of an enzyme- catalyzed reaction [v_1], the concentration of substrate [S] and two constants, V_{max} and K_m (which are set by the particular equation). The symbols used in the Michaelis-Menten equation refer to the reaction rate [v_1], maximum reaction rate (V_{max}), substrate concentration [S] and the Michaelis-Menten constant (K_m).

$$V_1 = \frac{V_{max}[s]}{\{K_m + [S]\}}$$

The Michaelis-Menten equation can be used to demonstrate that at the substrate concentration that produces exactly half of the maximum reaction rate, *i.e.* ½ V_{max}, the substrate concentration is numerically equal to K_m. This fact provides a simple yet powerful bioanalytical tool that has been used to characterize both normal and altered enzymes, such as those that produce the symptoms of genetic diseases. Rearranging the Michaelis-Menten equation leads to:

$$K_m = [S]\left\{\left[\frac{V_{max}}{V_1}\right] - 1\right\}$$

From this equation it should be apparent that when the substrate concentration is half that required to support the maximum rate of reaction, the observed rate, v_1, will, be equal to V_{max} divided by 2; in other words, $v_1 = [V_{max}/2]$. At this substrate concentration V_{max}/v_1 will be exactly equal to 2, with the result that:

$$[S](1) = K_m$$

The latter is an algebraic statement of the fact that, for enzymes of the Michaelis-Menten type, when the observed reaction rate is half of the maximum possible reaction rate, the substrate concentration is numerically equal to the Michaelis-Menten constant. In this derivation, the units of K_m are those used to specify the concentration of S, usually Molarity.

The Michaelis-Menten equation has the same form as the equation for a rectangular hyperbola; graphical analysis of reaction rate (v) versus substrate concentration [S] produces a hyperbolic rate plot.

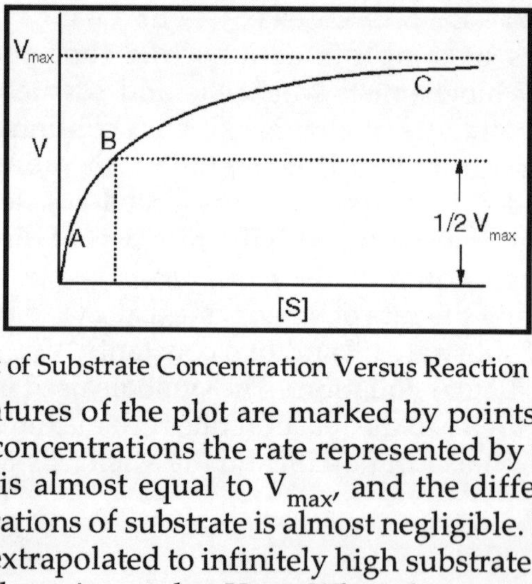

Fig. Plot of Substrate Concentration Versus Reaction Velocity

The key features of the plot are marked by points A, B and C. At high substrate concentrations the rate represented by point C the rate of the reaction is almost equal to V_{max}, and the difference in rate at nearby concentrations of substrate is almost negligible. If the Michaelis-Menten plot is extrapolated to infinitely high substrate concentrations, the extrapolated rate is equal to V_{max}. When the reaction rate becomes independent of substrate concentration, or nearly so, the rate is said to be zero order. (Note that the reaction is zero order only with respect to this substrate. If the reaction has two substrates, it may or may not be zero order with respect to the second substrate).

The very small differences in reaction velocity at substrate concentrations around point C (near V_{max}) reflect the fact that at these concentrations almost all of the enzyme molecules are bound to substrate and the rate is virtually independent of substrate, hence zero order. At lower substrate concentrations, such as at points A and B, the lower reaction velocities indicate that at any moment only a portion of the enzyme molecules are bound to the substrate. In fact, at the substrate concentration denoted by point B, exactly half the enzyme molecules are in an ES complex at any instant and the rate is exactly one half of V_{max}. At substrate concentrations near point A the rate appears to be directly proportional to substrate concentration, and the reaction rate is said to be first order.

Inhibition of Enzyme Catalyzed Reactions

To avoid dealing with curvilinear plots of enzyme catalyzed reactions, biochemists Lineweaver and Burk introduced an analysis of enzyme kinetics based on the following rearrangement of the Michaelis-Menten equation:

$$\frac{1}{V} = \left[\frac{K_m (1)}{V_{max} (S)} + \frac{1}{V_{max}}\right]$$

Plots of 1/v versus 1/[S] yield straight lines having a slope of K_m/V_{max} and an intercept on the ordinate at $1/V_{max}$.

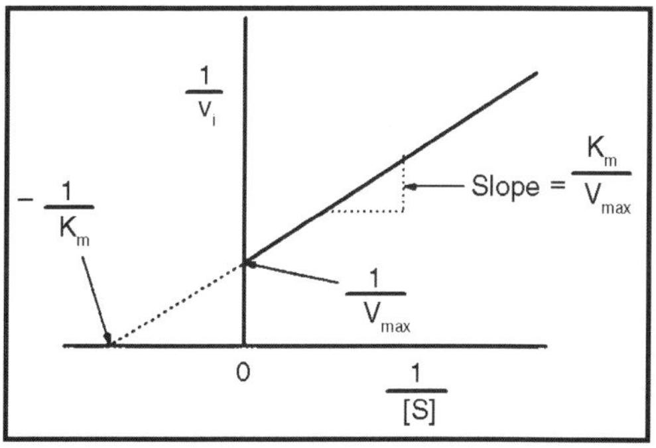

Fig. A Lineweaver-Burk Plot

An alternative linear transformation of the Michaelis-Menten equation is the Eadie-Hofstee transformation:

$$\frac{v}{[S]} = -v\left[\frac{1}{K_m} + \right] + \left[\frac{V_{max}}{K_m}\right]$$

and when v/[S] is plotted on the y-axis versus v on the x-axis, the result is a linear plot with a slope of $-1/K_m$ and the value V_{max}/K_m as the intercept on the y-axis and V_{max} as the intercept on the x-axis.

Both the Lineweaver-Burk and Eadie-Hofstee transformation of the Michaelis-Menton equation are useful in the analysis of enzyme inhibition. Well- known examples of such therapy include the use of methotrexate in cancer chemotherapy to semi-selectively inhibit DNA synthesis of malignant cells, the use of aspirin to inhibit the synthesis of prostaglandins which are at least partly responsible for the aches and pains of arthritis, and the use of sulfa drugs to inhibit the folic acid synthesis that is essential for the metabolism and growth of disease-causing bacteria. In addition, many poisons, such as cyanide, carbon monoxide and polychlorinated biphenols (PCBs) produce their life-threatening effects by means of enzyme inhibition.

Enzyme inhibitors fall into two broad classes: those causing irreversible inactivation of enzymes and those whose inhibitory effects can be reversed. Inhibitors of the first class usually cause an inactivating, covalent modification of enzyme structure. Cyanide is a classic example of an irreversible enzyme inhibitor: by covalently binding mitochondrial cytochrome oxidase, it inhibits all the reactions associated with electron transport.

The kinetic effect of irreversible inhibitors is to decrease the concentration of active enzyme, thus decreasing the maximum possible concentration of ES complex. Since the limiting enzyme reaction rate is often k_2[ES], it is clear that under these circumstances the reduction of enzyme concentration will lead to decreased reaction rates. Note that when enzymes in cells are only partially inhibited by irreversible inhibitors, the remaining unmodified enzyme molecules are not distinguishable from those in untreated cells; in particular, they have the same turnover number and the same K_m. Turnover number, related to V_{max}, is defined as the maximum number of moles of substrate that can be converted to product per mole of catalytic site per second. Irreversible inhibitors are usually considered to be poisons and are generally unsuitable for therapeutic purposes.

Reversible inhibitors can be divided into two main categories; competitive inhibitors and noncompetitive inhibitors, with a third category, uncompetitive inhibitors, rarely encountered.

Inhibitor Type	Binding Site on Enzyme	Kinetic effect
Competitive Inhibitor	Specifically at the catalytic site, where it competes with substrate for binding in a dynamic equilibrium-like process. Inhibition is reversible by substrate.	V_{max} is unchanged; K_m, as defined by [S] required for ½ maximal activity, is increased.
Noncompetitive Inhibitor	Binds E or ES complex other than at the catalytic site. Substrate binding unaltered, but ESI complex	K_m appears unaltered; V_{max} is decreased

	cannot form products. Inhibition cannot be reversed by substrate.	proportionately to inhibitor concentration.
Uncompetitive Inhibitor	Binds only to ES complexes at locations other than the catalytic site. Substrate binding modifies enzyme structure, making inhibitor- binding site available. Inhibition cannot be reversed by substrate.	Apparent V_{max} decreased; K_m, as defined by [S] required for ½ maximal activity, is decreased.

The hallmark of all the reversible inhibitors is that when the inhibitor concentration drops, enzyme activity is regenerated. Usually these inhibitors bind to enzymes by non-covalent forces and the inhibitor maintains a reversible equilibrium with the enzyme. The equilibrium constant for the dissociation of enzyme inhibitor complexes is known as K_i:

$$K_i = \frac{[E][I]}{[EI]}$$

The importance of K_I is that in all enzyme reactions where substrate, inhibitor and enzyme interact, the normal K_m and or V_{max} for substrate enzyme interaction appear to be altered. These changes are a consequence of the influence of K_i on the overall rate equation for the reaction. The effects of K_i are best observed in Lineweaver-Burk plots.

Probably the best known reversible inhibitors are competitive inhibitors, which always bind at the catalytic or active site of the enzyme. Most drugs that alter enzyme activity are of this type. Competitive inhibitors are especially attractive as clinical modulators of enzyme activity because they offer two routes for the reversal of enzyme inhibition, while other reversible inhibitors offer only one. First, as with all kinds of reversible inhibitors, a decreasing concentration of the inhibitor reverses the equilibrium regenerating active free enzyme. Second, since substrate and competitive inhibitors both bind at the same site they compete with one another for binding

Raising the concentration of substrate (S), while holding the concentration of inhibitor constant, provides the second route for reversal of competitive inhibition. The greater the proportion of substrate, the greater the proportion of enzyme present in competent ES complexes.

As noted earlier, high concentrations of substrate can displace virtually all competitive inhibitor bound to active sites. Thus, it is apparent that V_{max} should be unchanged by competitive inhibitors. This characteristic of competitive inhibitors is reflected in the identical vertical-axis intercepts of Lineweaver-Burk plots, with and without inhibitor.

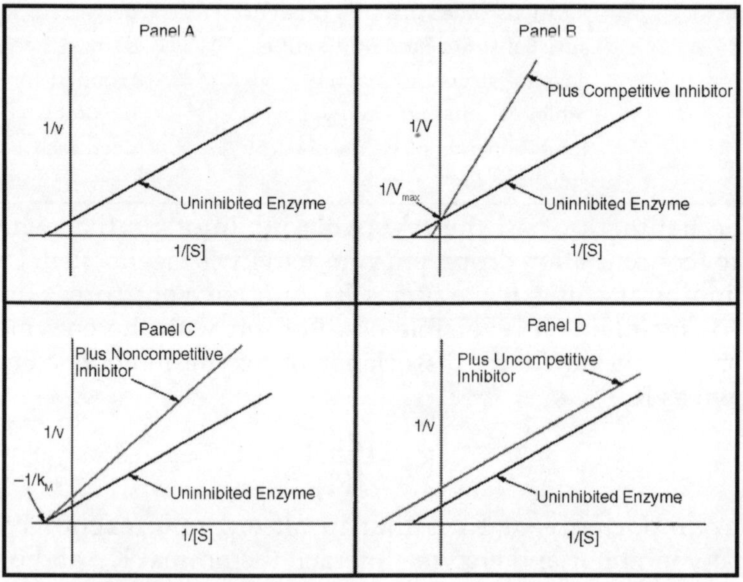

Fig. Lineweaver-Burk Plots of Inhibited Enzymes

Since attaining V_{max} requires appreciably higher substrate concentrations in the presence of competitive inhibitor, K_m (the substrate concentration at half maximal velocity) is also higher, as demonstrated by the differing negative intercepts on the horizontal axis in panel B.

Analogously, panel C illustrates that noncompetitive inhibitors appear to have no effect on the intercept at the x-axis implying that noncompetitive inhibitors have no effect on the K_m of the enzymes they inhibit. Since noncompetitive inhibitors do not interfere in the equilibration of enzyme, substrate and ES complexes, the K_m's of Michaelis-Menten type enzymes are not expected to be affected by noncompetitive inhibitors, as demonstrated by x-axis intercepts in panel C. However, because complexes that contain inhibitor (ESI) are incapable of progressing to reaction products, the effect of a noncompetitive inhibitor is to reduce the concentration of ES complexes that can advance to product. Since $V_{max} = k_2[E_{total}]$, and the concentration of competent E_{total} is diminished by the amount of ESI formed, noncompetitive inhibitors are expected to decrease V_{max}, as illustrated by the y-axis intercepts in panel C.

A corresponding analysis of uncompetitive inhibition leads to the expectation that these inhibitors should change the apparent values of K_m as well as V_{max}. Changing both constants leads to double reciprocal plots, in which intercepts on the x and y axes are proportionately changed; this leads to the production of parallel lines in inhibited and uninhibited reactions.

REGULATION OF ENZYME ACTIVITY

While it is clear that enzymes are responsible for the catalysis of almost all biochemical reactions, it is important to also recognize that rarely, if ever, do enzymatic reactions proceed in isolation. The most common scenario is that enzymes catalyze individual steps of multi-step metabolic pathways, as is the case with glycolysis, gluconeogenesis or the synthesis of fatty acids. As a consequence of these lock- step sequences of reactions, any given enzyme is dependent on the activity of preceding reaction steps for its substrate.

In humans, substrate concentration is dependent on food supply and is not usually a physiologically important mechanism for the routine regulation of enzyme activity. Enzyme concentration, by contrast, is continually modulated in response to physiological needs.

Three principal mechanisms are known to regulate the concentration of active enzyme in tissues:

1. Regulation of gene expression controls the quantity and rate of enzyme synthesis.
2. Proteolytic enzyme activity determines the rate of enzyme degradation.
3. Covalent modification of preexisting pools of inactive proenzymes produces active enzymes.

Enzyme synthesis and proteolytic degradation are comparatively slow mechanisms for regulating enzyme concentration, with response times of hours, days or even weeks. Proenzyme activation is a more rapid method of increasing enzyme activity but, as a regulatory mechanism, it has the disadvantage of not being a reversible process. Proenzymes are generally synthesized in abundance, stored in secretory granules and covalently activated upon release from their storage sites. Examples of important proenzymes include pepsinogen, trypsinogen and chymotrypsinogen, which give rise to the proteolytic digestive enzymes. Likewise, many of the proteins involved in the cascade of chemical reactions responsible for blood clotting are synthesized as proenzymes. Other important proteins, such as peptide hormones and collagen, are also derived by covalent modification of precursors.

Another mechanism of regulating enzyme activity is to sequester enzymes in compartments where access to their substrates is limited. For example, the proteolysis of cell proteins and glycolipids by enzymes responsible for their degradation is controlled by sequestering these enzymes within the lysosome.

In contrast to regulatory mechanisms that alter enzyme concentration, there is an important group of regulatory mechanisms that do not affect enzyme concentration, are reversible and rapid in action, and actually carry out most of the moment to moment physiological regulation of enzyme activity. These mechanisms include allosteric regulation, regulation by reversible covalent modification and regulation by control proteins such as calmodulin. Reversible covalent modification is a major mechanism for the rapid and transient regulation of enzyme activity.

The best examples, again, come from studies on the regulation of glycogen metabolism where phosphorylation of glycogen synthase and glycogen phosphorylase kinase results in the stimulation of glycogen degradation while glycogen synthesis is coordinately inhibited. Numerous other enzymes of intermediary metabolism are affected by phosphorylation, either positively or negatively. These covalent phosphorylations can be reversed by a separate sub-subclass of enzymes known as phosphatases. Recent research has indicated that the aberrant phosphorylation of growth factor and hormone receptors, as well as of proteins that regulate cell division, often leads to unregulated cell growth or cancer. The usual sites for phosphate addition to proteins are the serine, threonine and tyrosine R group hydroxyl residues.

ALLOSTERIC ENZYMES

In addition to simple enzymes that interact only with substrates and inhibitors, there is a class of enzymes that bind small, physiologically important molecules and modulate activity. These are known as allosteric enzymes; the small regulatory molecules to which they bind are known as effectors. Allosteric effectors bring about catalytic modification by binding to the enzyme at distinct allosteric sites, well removed from the catalytic site, and causing conformational changes that are transmitted through the bulk of the protein to the catalytically active site(s).

The hallmark of effectors is that when they bind to enzymes, they alter the catalytic properties of an enzyme's active site. Those that increase catalytic activity are known as positive effectors. Effectors that reduce or inhibit catalytic activity are negative effectors.

Most allosteric enzymes are oligomeric (consisting of multiple subunits); generally they are located at or near branch points in metabolic pathways, where they are influential in directing substrates along one or another of the available metabolic paths. The effectors that modulate the activity of these allosteric enzymes are of two types. Those activating and inhibiting effectors that bind at allosteric sites are called heterotropic effectors. (Thus there exist both positive and negative heterotropic effectors.) These effectors can assume a vast diversity of chemical forms, ranging from simple inorganic molecules to complex nucleotides such as cyclic adenosine monophosphate (cAMP). Their single defining feature is that they are not identical to the substrate. In many cases the substrate itself induces distant allosteric effects when it binds to the catalytic site. Substrates acting as effectors are said to be homotropic effectors. When the substrate is the effector, it can act as such, either by binding to the substrate-binding site, or to an allosteric effector site. When the substrate binds to the catalytic site it transmits an activity-modulating effect to other subunits of the molecule. Often used as the model of a homotropic effector is hemoglobin, although it is not a branch-point enzyme and thus does not fit the definition on all counts.

There are two ways that enzymatic activity can be altered by effectors: the V_{max} can be increased or decreased, or the K_m can be raised or lowered. Enzymes whose K_m is altered by effectors are said to be K-type enzymes and the effector a K-type effector. If V_{max} is altered, the enzyme and effector are said to be V-type. Many allosteric enzymes respond to multiple effectors with V-type and K-type behaviour. Here again, hemoglobin is often used as a model to study allosteric interactions, although it is not strictly an enzyme.

In the preceding discussion we assumed that allosteric sites and catalytic sites were homogeneously present on every subunit of an allosteric enzyme. While this is often the case, there is another class of allosteric enzymes that are comprised of separate catalytic and regulatory subunits. The archetype of this class of enzymes is cAMP-dependent protein kinase (PKA), whose mechanism of activation. The enzyme is tetrameric, containing two catalytic subunits and two regulatory subunits, and enzymatically inactive. When intracellular cAMP levels rise, one molecule of cAMP binds to each regulatory subunit, causing the tetramer to dissociate into one regulatory dimer and two catalytic monomers. In the dissociated form, the catalytic subunits are fully active; they catalyze the phosphorylation of a number of other enzymes, such as those involved in regulating glycogen metabolism. The regulatory subunits have no catalytic activity.

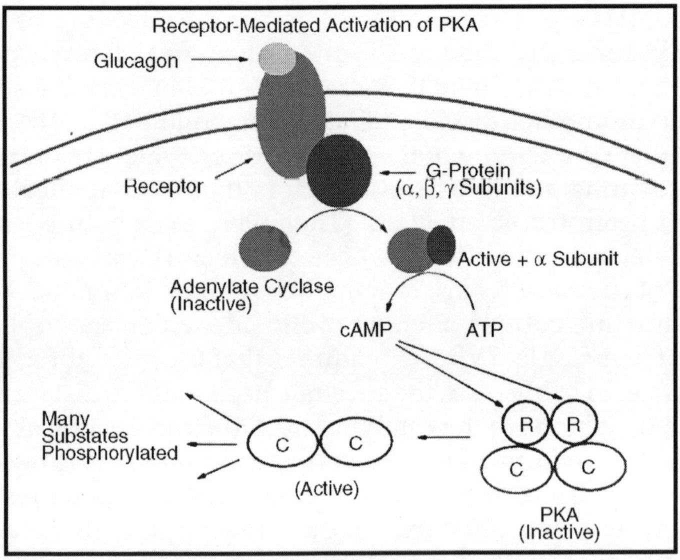

Fig. Activation of cAMP-dependent Protein Kinase (PKA).

ENZYMES IN THE DIAGNOSIS OF PATHOLOGY

The measurement of the serum levels of numerous enzymes has been shown to be of diagnostic significance. This is because the presence of these enzymes in the serum indicates that tissue or cellular damage has occurred resulting in the release of intracellular components into the blood. Hence, when a physician indicates that he/she is going to assay for liver enzymes, the purpose is to ascertain the potential for liver cell damage. Commonly assayed enzymes are the amino transferases: alanine transaminase, ALT (sometimes still referred to as serum glutamate-pyruvate aminotransferase, SGPT) and aspartate aminotransferase, AST (also referred to as serum glutamate-oxaloacetate aminotransferase, SGOT); lactate dehydrogenase, LDH; creatine kinase, CK (also called creatine phosphokinase, CPK); gamma-glutamyl transpeptidase, GGT. Other enzymes are assayed under a variety of different clinical situations but they will not be covered here.

The typical liver enzymes measured are AST and ALT. ALT is particularly diagnostic of liver involvement as this enzyme is found predominantly in hepatocytes. When assaying for both ALT and AST the ratio of the level of these two enzymes can also be diagnostic. Normally in liver disease or damage that is not of viral origin the ratio of ALT/AST is less than 1. However, with viral hepatitis the ALT/AST ratio will be greater than 1. Measurement of AST is useful not only for liver involvement but also for heart disease or damage. The level of

AST elevation in the serum is directly proportional to the number of cells involved as well as on the time following injury that the AST assay was performed. Following injury, levels of AST rise within 8 hours and peak 24-36 hours later. Within 3-7 days the level of AST should return to pre-injury levels, provided a continuous insult is not present or further injury occurs. Although measurement of AST is not, in and of itself, diagnostic for myocardial infarction, taken together with LDH and CK measurements the level of AST is useful for timing of the infarct.

The measurement of LDH is especially diagnostic for myocardial infarction because this enzyme exist in 5 closely related, but slightly different forms (isozymes).

The 5 types and their normal distribution and levels in non-disease/injury are listed below:

1. *LDH* 1: Found in heart and red-blood cells and is 17% - 27% of the normal serum total.
2. *LDH* 2: Found in heart and red-blood cells and is 27% - 37% of the normal serum total.
3. *LDH* 3: Found in a variety of organs and is 18% - 25% of the normal serum total.
4. *LDH* 4: Found in a variety of organs and is 3% - 8% of the normal serum total.
5. *LDH* 5: Found in liver and skeletal muscle and is 0% - 5% of the normal serum total.

Following a myocardial infarct the serum levels of LDH rise within 24-48 hours reaching a peak by 2-3 days and return to normal in 5-10 days. Especially diagnostic is a comparison of the LDH-1/LDH-2 ratio. Normally, this ration is less than 1. A reversal of this ration is referred to as a "flipped LDH". Following an acute myocardial infarct the flipped LDH ratio will appear in 12-24 hours and is definitely present by 48 hours in over 80% of patients. Also important is the fact that persons suffering chest pain due to angina only will not likely have altered LDH levels.

CPK is found primarily in heart and skeletal muscle as well as the brain. Therefore, measurement of serum CPK levels is a good diagnostic for injury to these tissues. The levels of CPK will rise within 6 hours of injury and peak by around 18 hours. If the injury is not persistent the level of CK returns to normal within 2-3 days.

Like LDH, there are tissue-specific isozymes of CPK and there designations:

- CPK3 (CPK-MM) is the predominant isozyme in muscle and is 100% of the normal serum total.

- CPK2 (CPK-MB) accounts for about 35% of the CPK activity in cardiac muscle, but less than 5% in skeletal muscle and is 0% of the normal serum total.
- CPK1 (CPK-BB) is the characteristic isozyme in brain and is in significant amounts in smooth muscle and is 0% of the normal serum total.

Since most of the released CPK after a myocardial infarction is CPK-MB, an increased ratio of CPK-MB to total CPK may help in diagnosis of an acute infarction, but an increase of total CPK in itself may not. CPK-MB levels rise 3–6 hours after a myocardial infarct and peak 12–24 hours later if no further damage occurs and returns to normal 12–48 hours after the infarct.

Chapter 2

Enzyme Catalysis

Catalysts increase the rate of otherwise slow or imperceptible reactions, without undergoing any net change in their structure. The early development of the concept of catalysis in the nineteenth century went hand in hand with the discovery of powerful catalysts from biological sources. These were called enzymes and were later found to be proteins. They mediate all synthetic and degradative reactions carried out by living organisms. They are very efficient catalysts, often far superior to conventional chemical catalysts, for which reason they are being employed increasingly in today's high-technological society, as a highly significant part of the biotechnological expansion. Their utilization has created a billion dollar business, including a wide diversity of industrial processes, consumer products, and the field of biosensors.

Enzymes have a number of distinct advantages over chemical catalysts:

- *Enzymes are very efficient catalysts.* Typically the rates of enzymemediated processes are accelerated, compared to those of the corresponding non-enzymatic reactions, by a factor of 10^8-10^{10}, which is far above the values that chemical catalysts are capable of achieving.
- *Enzymes are environmentally acceptable.* Unlike heavy metals, biocatalysts are environmentally benign reagents since they are completely degradable.
- *Enzymes act under mild conditions.* Enzymes act in a pH range of about 5-8, typically around 7, and in a temperature range of 20–40°C, preferably at around 30°C. This minimizes problems of undesired sidereactions such as decomposition, isomerization, racemization and rearrangement, which often plague traditional methodology.
- *Enzymes are compatible with each other.* Since enzymes generally function under the same or similar conditions, several biocatalytic reactions can be carried out in a reaction cascade in one flask. Thus sequential reactions are feasible by using multi-enzyme systems in order to simplify reaction processes.

In nature, there exists an enzyme-catalyzed process equivalent to almost every type of organic reaction. On the other hand, the reactions catalyzed by enzymes can often be achieved by classical chemical methods with a large expenditure of time and effort; moreover, some biocatalysts can accomplish reactions impossible to emulate in organic chemistry.

Enzymes display three major types of selectivities:

1. *Chemoselectivity*: Since the purpose of an enzyme is to act on a single type of functional group, other sensitive functionalities, which would normally react to a certain extent under chemical catalysis, do survive. As a result, reactions generally tend to be 'cleaner' and laborious purification of product(s) from impurities can largely be omitted.
2. *Regioselectivity and Diastereoselectivity*: Due to their complex threedimensional structure, enzymes may distinguish between functional groups, which are chemically situated in different regions of the same substrate molecule.
3. *Enantioselectivity:* Almost all enzymes are made from L-amino acids and thus are chiral catalysts. As a consequence, any type of chirality present in the substrate molecule is 'recognized' upon the formation of the enzyme-substrate complex. Thus a prochiral substrate may be transformed into an optically active product through enzymes.

The importance of enantiomerically pure compounds is based on the fact that, all the major biochemical events taking place in an organism are governed by enzymes; and since the majority of enzymes are highly selective with respect to the chirality of a substrate, it is obvious that the enantiomers of a given bioactive compound, such as a pharmaceutical, cause different biological effects. For example, R-enantiomer of the penicillamine molecule has an antiarthritic effect and it is used in drugs with this property; where S-penicillamine is toxic. Of the numerous methods which have been developed in the past decades for the preparation of optically active compounds, the biocatalytic processes are becoming more and more popular since substances can be produced in high enantiomeric excess using enzymes.

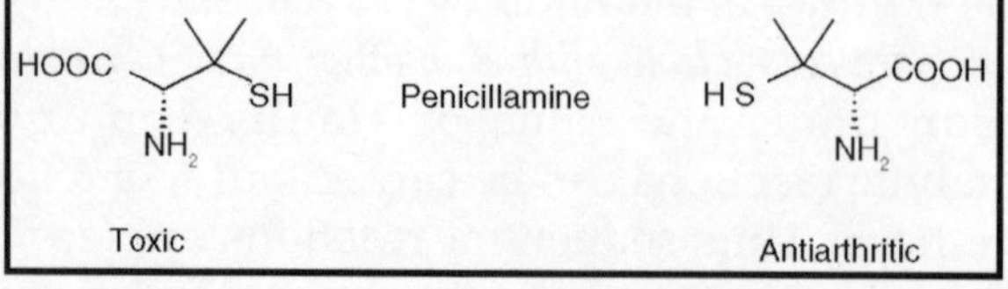

Fig. R- and S- Enantiomers of Penicillamine Molecule.

Benzaldehyde lyase (BAL, EC 4.1.2.38), is used for the synthesis of enantiomerically pure α-hydroxy ketones, an important class of compounds in natural product and drug synthesis. Benzaldehyde lyase is a thiamine pyrophosphate (TPP) dependent enzyme, and it catalyzes the cleavage of the acyloin linkage of benzoin to form two molecules of benzaldehyde.

ENZYMATIC ACTION

Like all catalysts, enzymes accelerate the rates of reactions while experiencing no permanent chemical modification as a result of their participation. Enzymes can accelerate, often by several orders of magnitude, reactions that under the mild conditions of cellular concentrations, temperature, pH, and pressure would proceed imperceptibly (or not at all) in the absence of the enzyme. The efficiency of an enzyme's activity is often measured by the turnover rate, which measures the number of molecules of compound upon which the enzyme works per molecule of enzyme per second. Carbonic anhydrase, which removes carbon dioxide from the blood by binding it to water, has a turnover rate of 10^6. That means that one molecule of the enzyme can cause a million molecules of carbon dioxide to react in one second.

Most enzymatic reactions occur within a relatively narrow temperature range (usually from about 30 and deg C; to 40 and deg C;), a feature that reflects their complexity as biological molecules. Each enzyme has an optimal range of pH for activity; for example, pepsin in the stomach has maximal reactivity under the extremely acid conditions of pH 1–3. Effective catalysis also depends crucially upon maintenance of the molecule's elaborate three-dimensional structure. Loss of structural integrity, which may result from such factors as changes in pH or high temperatures, almost always leads to a loss of enzymatic activity. An enzyme that has been so altered is said to be denatured.

Consonant with their role as biological catalysts, enzymes show considerable selectivity for the molecules upon which they act (called substrates). Most enzymes will react with only a small group of closely related chemical compounds; many demonstrate absolute specificity, having only one substrate molecule which is appropriate for reaction.

Numerous enzymes require for efficient catalytic function the presence of additional atoms of small nonprotein molecules. These include coenzyme, any one of a group of relatively small organic molecules required for the catalytic function of certain enzymes. A coenzyme may either be attached by covalent bonds to a particular enzyme or exist freely in solution, but in either case it participates molecules, many of which only transiently associate with the enzyme. Nonprotein components tightly bound to the protein are called prosthetic group, non-amino acid portions of certain protein molecules. The key part of the prosthetic group may be either organic (such as a vitamin) or inorganic (such as a metal) and is usually required for biological activity, especially when the prosthetic group. The region on the enzyme molecule in close proximity to where the catalytic event takes place is known as the active site. Prosthetic groups necessary for catalysis are usually located there, and it is the place where the substrate (and coenzymes, if any) bind just before reaction takes place.

The side-chain groups of amino acid residues making up the enzyme molecule at or near the active site participate in the catalytic event. For example, in the enzyme trysin, its complex tertiary structure brings together a histidine, organic compound, one of the 22 á-amino acids commonly found in animal proteins. Only the L-stereoisomer appears in mammalian protein. residue from one section of the molecule with glycine, organic compound, one of the 20 amino acids commonly found in animal proteins. Glycine is the only one of these amino acids that is not optically active, *i.e.* serine, organic compound, one of the 20 amino acids commonly found in animal proteins. Only the L-stereoisomer appears in mammalian protein. The side chains of the residues in this particular geometry produce the active site that accounts for the enzyme's reactivity.

GENERAL CHARACTERISTICS OF ENZYMES

Enzymes are usually proteins of high molecular weight that acts as catalysts in biological systems. They accelerate hundreds of reactions taking place simultaneously in the cell, without undergoing any permanent chemical change and without affecting the reaction equilibrium; by lowering the activation energy of the reaction via binding the substrate and forming an enzymesubstrate complex. Enzymes are specific, versatile, and very effective biological catalysts, resulting in much higher reaction rates as compared to chemically catalyzed reactions under ambient conditions. Enzyme specificity is thought to be a consequence of its elaborate three-dimensional conformation which allows formation of the active site responsible for the catalytic ability of the enzyme. Enzymes catalyze only the reactions of very narrow ranges of reactants, which may consist of a small number of closely related classes of compounds, a single class of compounds, or a single compound. This means that the chosen reaction can be catalyzed to the exclusion of side-reactions, eliminating undesirable by-products. Thus, higher productivities may be achieved, reducing material costs.

Another distinguishing characteristic of enzymes is their frequent need for cofactors. A cofactor is a nonprotein compound which combines with an otherwise inactive protein to give a catalytically active complex. The simplest cofactors are metal ions like Ca^{2+}, Zn^{2+}, Co^{2+}, etc. Each enzyme has a specific temperature and pH range where it functions to its optimal capacity, the optima for these proteins usually lie between 37-47°C, and pH optima range from acidic, *i.e.*, 1.0, to alkaline, *i.e.*, 10.5. Biologically active enzymes may be extracted from any living organism.

A very wide range of sources is used for commercial enzyme production- from *Actinoplanes* to *Zymomonas*, from spinach to snake venome. Of the hundred or so enzymes being used industrially, over half are from fungi and yeast and over a third are from bacteria, with the remainder divided between animal (%8) and plant (%4) sources. It is a remarkable fact that enzymes with the same name but obtained from different organisms often have different amino acid sequences and hence different properties and catalytic activities.

Enzymes can be isolated and are active outside the living cell. They are such efficient catalysts that they accelerate chemical reactions measurably, even at concentrations so low that they cannot be detected by most chemical tests for protein. Like other chemical reactions, enzyme-catalyzed reactions proceed only when accompanied by a decrease in free energy; at equilibrium the concentrations of reactants and products are the same in the presence of an enzyme as in its absence. An enzyme can catalyze an indefinite amount of chemical change without itself being diminished or altered by the reaction. However, because most isolated enzymes are relatively unstable, they often gradually lose activity under the conditions employed for their study.

Table. Some Representative Enzymes, their Sources, and Reaction Specificities

Enzyme	Some sources	Reaction catalyzed
Pepsin	Gastric juice	Hydrolysis of proteins to peptides and amino acids
Urease	Jackbean, bacteria	Hydrolysis of urea to ammonia and carbon dioxide
Amylase	Saliva, pancreatic juice	Hydrolysis of starch to maltose

Phosphorylase	Muscle, liver, plants	Reversible phosphorolysis of starch or glycogen to glucose- 1-phosphate
Transaminases	Many animal and plant tissues	Transfer of an amino group from an amino acid to a keto acid
Phosphohexose isomerase	Muscle, yeast	Interconversion of glucose-6- phosphate and fructose-6- phosphate
Pyruvic carboxylase	Yeast, bacteria, plants	Decarboxylation of pyruvate to acetaldehyde and carbon dioxide
Catalase	Erythrocytes, liver	Decomposition of hydrogen peroxide to oxygen and water
Alcohol dehydrogenase	Liver	Oxidation of ethanol to acetaldehyde
Xanthine oxidase	Milk, liver	Oxidation of xanthine and hypoxanthine to uric acid

Chemical Nature

All enzymes are proteins. Their molecular weights range from about 10,000 to more than 1,000,000. Like other proteins, enzymes consist of chains of amino acids linked together by peptide bonds. An enzyme molecule may contain one or more of these polypeptide chains. The sequence of amino acids within the polypeptide chains is characteristic for each enzyme and is believed to determine the unique three-dimensional conformation in which the chains are folded. This conformation, which is necessary for the activity of the enzyme, is stabilized by interactions of amino acids in different parts of the peptide chains with each other and with the surrounding medium.

These interactions are relatively weak and may be disrupted readily by high temperatures, acid or alkaline conditions, or changes in the polarity of the medium. Such changes lead to an unfolding of the peptide chains (denaturation) and a concomitant loss of enzymatic activity, solubility, and other properties characteristic of the native enzyme. Enzyme denaturation is sometimes reversible. Many enzymes contain an additional, nonprotein component, termed a coenzyme or prosthetic group.

This may be an organic molecule, often a vitamin derivative, or a metal ion. The coenzyme, in most instances, participates directly in the catalytic reaction. For example, it may serve as an intermediate carrier of a group being transferred from one substrate to another. Some enzymes have coenzymes that are tightly bound to the protein and difficult to remove, while others have coenzymes that dissociate readily. When the protein moiety (the apoenzyme) and the coenzyme are separated from each other, neither possesses the catalytic properties of the original conjugated protein (the holoenzyme). By simply mixing the apoenzyme and the coenzyme together, the fully active holoenzyme can often be reconstituted. The same coenzyme may be associated with many enzymes which catalyze different reactions. It is thus primarily the nature of the apoenzyme rather than that of the coenzyme which determines the specificity of the reaction.

The complete amino acid sequence of several enzymes has been determined by chemical methods. By x-ray crystallographic methods even the exact three-dimensional molecular structure of a few enzymes has been deduced.

Classification and Nomenclature

Enzymes are usually classified and named according to the reaction they catalyze. The principal classes are as follows.

Oxidoreductases catalyze reactions involving electron transfer, and play an important role in cellular respiration and energy production. Some of them participate in the process of oxidative phosphorylation, whereby the energy released by the oxidation of carbohydrates and fats is utilized for the synthesis of adenosine triphosphate (ATP) and thus made directly available for energy-requiring reactions.

Transferases catalyze the transfer of a particular chemical group from one substance to another. Thus, transaminases transfer amino groups, transmethylases transfer methyl groups, and so on. An important subclass of this group are the kinases, which catalyze the phosphorylation of their substrates by transferring a phosphate group, usually from ATP, thereby activating an otherwise metabolically inert compound for further transformations.

Hydrolases catalyze the hydrolysis of proteins (proteinases and peptidases), nucleic acids (nucleases), starch (amylases), fats (lipases), phosphate esters (phosphatases), and other substances. Many hydrolases are secreted by the stomach, pancreas, and intestine and are responsible for the digestion of foods. Others participate in more specialized cellular functions. For example, cholinesterase, which catalyzes the hydrolysis of acetylcholine, plays an important role in the transmission of nervous impulses.

Lyases catalyze the nonhydrolytic cleavage of their substrate with the formation of a double bond. Examples are decarboxylases, which remove carboxyl groups as carbon dioxide, and dehydrases, which remove a molecule of water. The reverse reactions are catalyzed by the same enzymes.

Isomerases catalyze the interconversion of isomeric compounds.

Ligases, or synthetases, catalyze endergonic syntheses coupled with the exergonic hydrolysis of ATP. They allow the chemical energy stored in ATP to be utilized for driving reactions uphill.

Specificity

The majority of enzymes catalyze only one type of reaction and act on only one compound or on a group of closely related compounds. There must exist between an enzyme and its substrate a close fit, or complementarity. In many cases, a small structural change, even in a part of the molecule remote from that altered by the enzymatic reaction, abolishes the ability of a compound to serve as a substrate. An example of an enzyme highly specific for a single substrate is urease, which catalyzes the hydrolysis of urea to carbon dioxide and ammonia. On the other hand, some enzymes exhibit a less restricted specificity and act on a number of different compounds that possess a particular chemical group. This is termed group specificity.

A remarkable property of many enzymes is their high degree of stereospecificity, that is, their ability to discriminate between asymmetric molecules of the right-handed and left-handed configurations. An example of a stereospecific enzyme is L -amino acid oxidase. This enzyme catalyzes the oxidation of a variety of amino acids of the type $R-CH(NH_2)COOH$. The rate of oxidation varies greatly, depending on the nature of the R group, but only amino acids of the L configuration react.

CLASSIFICATION OF ENZYMES

- Enzymes are classified according to the report of the International Union of Biochemistry. The Enzyme Commission (EC) numbers divide enzymes into six main groups according to the type of reaction catalyzed. For identification purposes, every enzyme has a four-digit number EC A.B.C.D; the following properties are encoded: denotes the main type of reaction.
- Stands for the subtype, indicating the substrate type.
- Indicates the nature of the co-substrate.
- Is the individual enzyme number.

For example, the EC number of benzaldehyde lyase is EC 4.1.2.38, catalyzes the cleavage and synthesis of benzoin.

Table. International classification of enzymes

No	Class	Type of reaction catalyzed
1	Oxidoreductases	Transfer of electrons
2	Transferases	Group-transfer reactions
3	Hydrolases	Transfer of functional groups to water
4	Lyases	Addition of groups to double bonds or the reverse
5	Isomerases	Transfer of groups within molecules to yield isomeric forms
6	Ligases	Formation of C-C, C-S, C-O, and C-N bonds by condensation reactions coupled to ATP cleavage

ENZYME ACTIVITY

The amount of enzyme present or used in a process is difficult to determine in absolute terms (*e.g.* grams), as its purity is often low and a proportion may be in an inactive, or partially active, state. The activity is a measure of enzyme content that is clearly of major interest when the enzyme is to be used in a process.

For this reason, enzymes are usually marketed in terms of activity rather than weight.

Enzyme activity is usually determined from a rate assay and expressed in terms of the activity unit; which was defined by the Commission on Enzymes as: One unit (U) of enzyme activity is defined as the amount which will catalyze the transformation of one micromole of substrate per minute under defined conditions.

A change in temperature, pH, and/or substrate concentration affects the reaction rate; therefore, these parametres must be carefully controlled in order to achieve reproducible results.

Also, a comparison of the activity of different enzyme preparations is only possible if the assay procedure is performed exactly in the same way.

ENZYME KINETICS

It is established that enzymes form a bound complex with their substrates during the course of their catalysis and prior to the release of products. This can be simply illustrated, using a mechanism based on that of Michaelis and Menten for a one-substrate reaction, by the reaction sequence:

$$E + S \underset{K-1}{\overset{k_1}{\rightleftarrows}} ES \xrightarrow{K_{-2}} P$$

where, enzyme E and substrate S combine to form a complex ES, which then dissociates into product P. The Michealis-Menten equation is simply derived from the rate of the above reaction by substituting Km for (k-1+k+2)/k+1:

$$r = \frac{r_{max} C_s}{K_m + C_s}$$

where, r is the volumetric rate of reaction (mol m^{-3} s^{-1}), CS is the concentration of the substrate (mM), rmax is the maximum rate of reaction at infinite reactant concentration, and Km is the Michealis constant for the substrate (mM). Km is a measure of the affinity of an enzyme for a particular substrate, a low Km value representing a high affinity and a high Km a low affinity.

The Michaelis-Menten equation in its original form is not well suited for estimation of the kinetic parametres rmax and Km. Rearrangement of equation gives:

$$\frac{1}{r} = \frac{1}{r_{max}} + \frac{K_m}{r_{max}} \cdot \frac{1}{C_s}$$

So that a plot of 1/r versus 1/CS should give a straight line with slope Km/rmax and intercept 1/rmax.This plot is known as the Lineweaver-Burk plot.

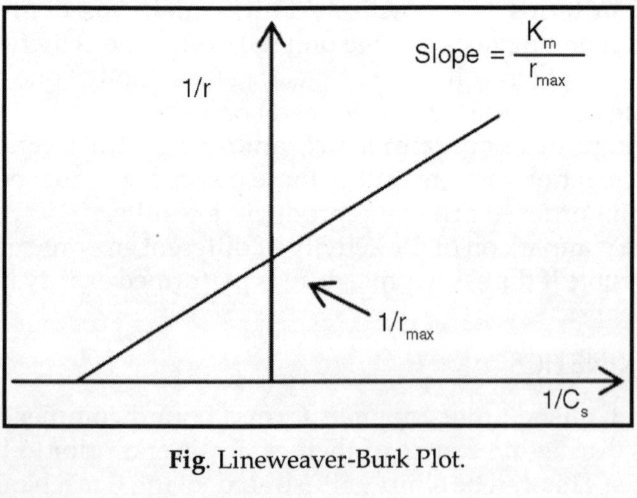

Fig. Lineweaver-Burk Plot.

BENZALDEHYDE LYASE

Benzaldehyde lyase (BAL, EC 4.1.2.38), is a thiamine -rophosphate (TPP) dependent enzyme as a cofactor; and it reversibly -lyzes the cleavage and formation of the acyloin linkage of R- -n. The R- and Senantiomers of benzoin molecule.

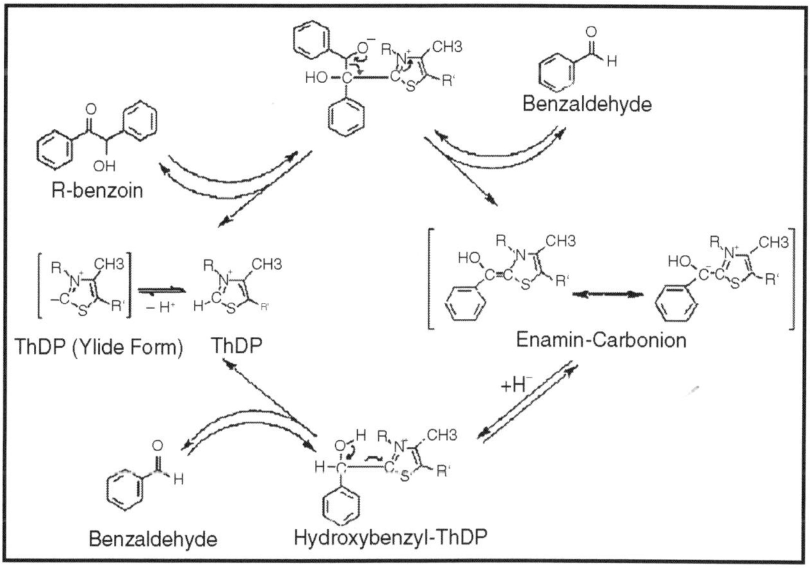

Fig. Benzaldehyde Lyase Catalyzed Cleavage and Synthesis of Benzoin.

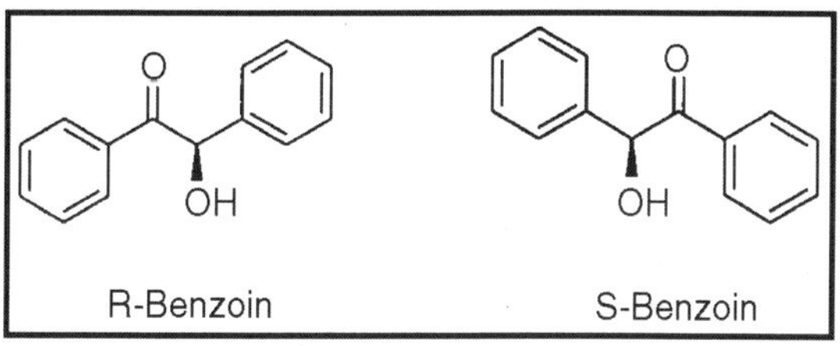

Fig. R- and S- Enantiomers of Benzoin Molecule.

The most important property of this reaction catalyzed by benzaldehyde lyase was recognized by Demir et. al.; which indicates that only R-benzoin is converted into benzaldehyde through benzaldehyde lyase catalysis; where S-benzoin gave no reaction at all. With this property benzaldehyde lyase is used for the synthesis of enantiopure α-hydroxy ketones, an important class of compounds in natural product and drug synthesis. The synthesis of enantiomerically pure compounds are becoming increasingly more important in the production of pharmaceuticals, agrochemicals (*e.g.*, pesticides, fungicides, herbicides), and flavours.

These compounds frequently act by interacting with receptors, enzymes, carrier molecules, and the like. All such interactions are highly stereospecific; which means, one stereoisomer of a drug may interact tightly with a particular receptor, whereas the other streoisomer may have a different target. For example; the R-isomer of the antitubercular drug ethambutol, has potent antitubercular activity, whereas the S-isomer causes degeneration of the optic nerve, leading to blindness. Therefore, the development of efficient and environmentally acceptable processes for the preparation of enantiomerically pure compounds is essential, but still presents a major challenge in organic chemistry. However, enantiomerically pure substances can be produced in high enantiomeric excess by utilizing biocatalytic processes since enzymes are chiral materials. Considering the above mentioned problems, it is obvious that enzymatic methods represent a valuable addition to the existing toolbox available for the enantioselective synthesis of fine chemicals.

Benzaldehyde lyase, which is used for the synthesis of enantiopure α- hydroxy ketones, was firstly reported by Gonzales and Vicuna from *Pseudomonas fluorescens* Biovar I. This strain can grow on benzoin as a sole carbon and energy source due to the benzaldehyde lyase production potential of the microorganism. In this first study, it was found that the molecular weight of the enzyme is 80,000 and and was suggested that the native enzyme may be either a dimer or a single polypeptide with a nonspherical shape. Maximum activity of benzaldehyde lyase was reached at 0.01 mM TPP, while concentrations higher than 0.5 mM were inhibitory.

BIOPROCESS PARAMETRES IN ENZYME PRODUCTION

Any operation involving the transformation of some raw material (biological or non-biological) into some product by means of microorganisms, animal or plant cell cultures, or by materials derived from them (*e.g.* enzymes, organelles), may be termed as a "bioprocess". Until about 1950, the predominant method of producing industrial enzymes was by extraction from animal or plant sources; by 1993, this accounts for less than 10%. Industrial enzymes are now produced by microorganisms grown in aqueous suspension in large vessels, *i.e.*, by fermentation.

Microorganisms are preferred to plants and animals as sources of enzymes because:

- They are generally cheaper to produce;
- Their enzyme contents are more predictable and controllable;
- Reliable supplies of raw material of constant composition are more easily arranged;
- Plant and animal tissues contain more potentially harmful materials, including phenolic compounds (from plants) and endogenous enzyme inhibitors.

Enzymes are usually sensitive to harsh physical and chemical conditions, which demands careful selection of production processes and conditions for each individual enzyme. In aerobic bioprocesses, the important criteria that must be taken into account, in order to have high product yield, are:

- Microorganism
- Medium composition
- Bioreactor operation parametres
  - Temperature
  - pH
  - Oxygen transfer rate
    a. Air inlet rate (Q0/VR)
    b. Agitation rate (N)

MICROORGANISM

In bioprocesses, the selection of host microorganism for production of industrial enzymes is often critical for the commercial success of the product. Potential hosts should give sufficient yields, be able to secrete large amounts of protein, be suitable for industrial fermentations, produce a large cell mass per volume quickly and on cheap media, be considered safe based on historical experience or evaluation by regulatory authorities, and should not produce harmful substances or any other undesirable products. Few microorganisms fulfill all the above criteria and are used for production of industrial enzymes. *Escherichia coli* is one of the most important host microorganisms. Benzaldehyde lyase is naturally produced by wild-type *Pseudomonas fluerescens*. Hinrichsen cloned the gene encoding benzaldehyde lyase to an *Escherichia coli* strain; the studies until then has been performed using this host microorganism.

Escherichia coli

If posttranslational modifications are unnecessary, *E. coli* is most often chosen as the initial host. The main reason for the popularity of *E. coli* is the broad knowledge base for it. *E. coli* physiology and its genetics are probably far better understood than for any other living organism. This large knowledge base greatly facilitates sophisticated genetic manipulations. Also, an important engineering contribution was the development of strategies to grow cultures of *E. coli* to high cell densities.

The average *E. coli* cell is a rod shaped body with diametre and length of approximately 1 and 2 micrometres, respectively. It has a wet mass of 9.5×10^{-13} g, or 2.9×10^{-13} g dry mass. Elemental essay of the dry mass of *Escherichia coli* cells reveals a fairly typical composition of the protoplasm: 50% carbon, 20% oxygen, 14% nitrogen, 8% hydrogen,

3% potassium, 2% sodium, 1% sulfur, 0.05% calcium, 0.05% magnesium, 0.05% chlorine, 0.2% iron and a total of 0.3% trace elements including manganese, cobalt, copper, zinc and molybdenium. *E. coli* is a Gram-negative bacterium. The murein layer of the Gramnegative bacteria is much thinner than that of Gram-positive bacteria, and they make a completely different structure- an outer membrane; which has the ability to resist damaging chemicals. *E. coli* is a facultative aerobe bacterium which can grow in the presence or absence of oxygen. Under anaerobic conditions, *E. coli* can utilize NO_3 as a source of nitrogen, but when growing aerobically it requires a reduced nitrogen source.

The maximum temperature of growth for *E. coli* is approximately 48ºC, its optimal temperature is 37ºC, its minimum temperature is 8ºC, and its normal temperature range extends from about 21ºC to 39ºC. In the absence of methionine, growth stops at 45ºC. *E. coli* grows well near neutrality at pH values from 6.0 to 8.0. Regardless of the value of the external pH, within the cell, pH is maintained at a value quite close to 7.6.

Cell Growth, Kinetics and Yield Factors

For microorganisms, growth is their most essential response to their physiochemical environment. Growth is a result of both replication and change in cell size. In a suitable nutrient medium, organisms extract nutrients from the medium and convert them into biological compounds. Parts of these nutrients are used for energy production and parts are used for biosynthesis and product formation. Microbial growth is a good example of an autocatalytic reaction. The rate of growth is directly related to cell concentration, and cellular reproduction is the normal outcome of this reaction. The rate of microbial growth is characterized by the specific growth rate, ì, which is defined as,

$$\mu = \frac{1}{C_x} \cdot \frac{dC_x}{dt}$$

where CX is the cell mass concentration (kg m^{-3}), t is time (h), and ì is the specific growth rate (h-1). When a liquid nutrient medium is inoculated with a seed culture, the organisms selectively take up dissolved nutrients from the medium and convert them into biomass.

A typical batch growth curve includes the following phases:

- Lag phase; occurs immediately after inoculation and is a period of adaptation of cells to a new environment ($\mu \approx 0$).
- Exponential growth phase; the cells have adjusted to their new environment. Growth achieves its maximum rate ($\mu \approx \mu_{max}$).

- Deceleration phase; growth decelerates due to either depletion of one or more essential nutrients or the accumulation of toxic by-products of growth ($\mu < \mu_{max}$).
- Stationary phase; the net growth rate is zero (no cell division) or the growth rate is equal to the death rate ($\mu = 0$).
- Death phase; cells lose viability and lyse ($\mu < 0$).

Despite the complexity occurring in cell growth, yield principles can be applied to cell metabolism to relate flow of substrate in metabolic pathways to formation of biomass and other products. Yield coefficients allow to quantify the nutrient requirements and production characteristics of an organism. Yield coefficients are defined based on the amount of consumption of another material.

For example, the growth yield in a fermentation is:

$$Y_{x/s} = \frac{\Delta X}{\Delta S}$$

where, $Y_{X/S}$ is the yield coefficient, X and S are mass of cell and substrate, respectively, involved in metabolism. This definition gives an overall yield representing some sort of average value for the entire culture period.

However, in batch processes, the yield coefficients may show variations throughout the process for a given microorganism in a given medium, due to the growth rate and metabolic functions of the microorganism.

Therefore, it is sometimes necessary to evaluate the instantaneous yield at a particular point in time. Instantaneous yield can be calculated as follows:

$$Y_{x/s} = \frac{dx}{ds} = \frac{dX/dt}{dS/dt} = \frac{r_x}{r_s}$$

Yield coefficients based on other substrates or product formation may be defined; for example,

$$Y_{x/o} = \frac{\Delta x}{\Delta O}$$

$$Y_{P/S} = \frac{\Delta P}{\Delta S}$$

For organisms growing aerobically on glucose, YX/S is typically 0.4 to 0.6 g/g for most yeast and bacteria, while YX/O is 0.9 to 1.4 g/g. Anaerobic growth is less efficient, and the yield coefficient reduced substantially.

Table. Definition of Yield Coefficients

Symbol	Definition	Unit
$Y_{x/s}$	Mass of cells produced per unit mass of substrate consumed	Kg cell kg^{-1} substrate
$Y_{x/o}$	Mass of cells produced per unit mass of oxygen consumed	Kg cell kg^{-1} oxygen
$Y_{s/o}$	Mass of substrate produces per unit mass of oxygen consumed	Kgsubstrate kg^{-1} oxygen
$Y_{p/x}$	Mass of product formed per unit mass of substrate consumed	Kg produt kg^{-1} cell
$Y_{p/s}$	Mass of product formed per unit mass of substrate consumed	Kg produt kg^{-1} substrate
$Y_{p/o}$	Mass of product formed per unit mass of oxygen consumed	Kg produt kg^{-1} oxygen

MEDIUM COMPOSITION

Bacteria grow only if their environment is suitable; if it is not optimal, growth may occur at a lower rate or not at all- or the bacteria may die, depending on species and conditions. Essential requirements for growth include a supply of suitable nutrients, a source of energy, water, an appropriate temperature, an appropriate pH, appropriate levels (or absence) of oxygen. Cells need nutrients as raw materials for growth, maintenance and division. The qualitative and quantitative nutritional requirements of cells need to be determined to optimize growth and product formation.

Typical components of a fermentation medium are:

- *Water:* Is the major component of all fermentation media.
- Macronutrients*:* Are needed in concentrations larger than 10^{-4} M. Carbon, nitrogen, oxygen, hydrogen, sulfur, phosphorus, Mg^{2+}, and K+ are major macronutrients, from which living matter is made. Carbon compounds are major sources of cellular carbon and energy. Nitrogen constitutes about 10-14% of cell dry weight. Oxygen is present in all organic cell components and cellular water and constitutes about 20% of the dry weight of cells. Also, molecular oxygen is required as a terminal electron acceptor in the aerobic metabolism.
- Micronutrients*:* Are needed in concentrations of less than 10^{-4} M. Trace elements such as Mo^{2+}, Zn^{2+}, Cu^{2+}, Mn^{2+}, Fe^{2+}, Ca^{2+}, Na^{2+}, vitamins, growth hormones, and metabolic precursors are micronutrients. Trace elements are essential to microbial nutrition. Lack of essential trace elements

increases the lag phase, and may decrease the specific growth rate and the yield.

- *Buffers:* May be necessary to control the pH of the fermentation medium.
- *Inducers:* Must be present in the medium if the product of interest is an induced enzyme which is synthesized only in response to the presence of an inducer in the environment.
- *Antifoams:* Are surface active agents, reducing the surface tension in the foams. They usually have no metabolic effect but may substantially reduce the oxygen transfer rate.

Two major types of growth media are defined and complex media. Defined media contain specific amounts of pure chemical compounds with known chemical compositions. Complex media contain natural compounds whose chemical composition is not exactly known. A complex medium usually can provide the necessary growth factors, vitamins, hormones and trace elements; often resulting in higher cell yields, compared to the defined medium. Often, complex media are less expensive than defined media.

The primary advantage of defined media is that the results are more reproducible and the operator has better control of the fermentation. Further, recovery and purification of a product is often easier and cheaper in defined media. In literature, there is no work on the production of benzaldehyde lyase; hence, there is no reported medium for benzaldehyde lyase production. But related with other biomolecules using the host *E. coli,* Ryan *et al.* and Bhattacharya and Dukey used Luria-Bertani (LB) medium; Castan *et al.,* Leon *et al.* Akesson et. al. and Yang et. al. used glucose based defined medium.

BIOREACTOR OPERATION PARAMETRES

Oxygen transfer, pH, and temperature, which are the major bioreactor operation parametres, show diverse effects on product formation in aerobic fermentation processes by influencing metabolic pathways and changing metabolic fluxes.

Temperature

Generally, for a given type of bacterium, growth proceeds most rapidly at a particular temperature: the optimum growth temperature; the rate of growth tails off as temperatures increase or decrease from the optimum. Rates of all reactions, including those catalyzed by enzymes, rise with increase in temperature in accordance with the Arrhenius equation:

$$k = Ae^{-\Delta G/RT}$$

Typical standard free energies of activation give rise to increases in rate by factors between 1.2 and 2.5 for every 10°C rise in temperature. So, it would be preferable to use enzymes at high temperatures in order to make use of this increased rate of reaction plus the protection it affords against microbial contamination.

For enzymes, however, denaturation begins to occur at 45 to 50°C and is severe at 55°C. One physical mechanism for this phenomenon is obvious: as the temperature increases, the atoms in the enzyme molecule have greater energies and a greater tendency to move. Eventually, they acquire sufficient energy to overcome the weak interactions holding the globular protein structure together, and deactivation follows.

Temperature also affects product formation. However, the optimum temperature for growth and product formation may be different. When temperature is increased above the optimum temperature, the maintenance requirements of cells increase. The yield coefficient is also affected by temperature. Temperature also may affect the rate-limiting step in a fermentation process. At high temperatures, the rate of bioreaction might become higher than the diffusion rate, and diffusion would then become the rate-limiting step. In literature, the production of *E. coli* is performed at 37°C, which is the bacteria's optimal temperature of growth.

Ph

Enzymes are amphoteric molecules containing a large number of acidic and basic groups, situated mainly on the surface. The charges on these groups will vary, according to their acid dissociation constants, with the pH of their environment. This will affect the total net charge of the enzyme and the distribution of charge on its exterior surface, in addition to the reactivity of the catalytically active groups. Thus, the catalytically active enzyme may be a large or small fraction of the total enzyme present, depending upon the pH.

In a similar manner, the charge and the charge distribution on the substrate(s), product(s) and coenzymes (where applicable) will also be affected by pH changes. These charge variations, plus any consequent structural alterations, may be reflected in changes in the binding of the substrate, the catalytic efficiency and the amount of active enzyme. For these reasons, enzymes have a characteristic pH at which their activity is maximal; above or below this pH the activity declines. However, the pHactivity profiles of enzymes are not always bell-shaped. The optimum pH of an enzyme is not necessarily identical with the pH of its normal intracellular surroundings, which may be on the ascending or descending slope of its pHactivity profile. This fact suggests that the pH-activity relationship of an enzyme may be a factor in intracellular control of enzymatic activity. For many bacteria, pH optima ranges from 3 to 8; and many organisms have mechanisms to maintain intracellular pH at a relatively constant level in the presence of fluctuations in environmental pH. When pH differs from the optimal value, the maintenance-energy requirements increase.

In most fermentations, pH can vary substantially. Often the nature of the nitrogen source can be important. Furthermore, pH can change due to the production of organic acids, utilization of acids (particularly amino acids), or the production of bases. Thus, pH control by means of a buffer or an active pH control system is important. Nevertheless, some bioprocesses require controlled pH conditions, while others might require uncontrolled pH operations, in order to increase the product yield and selectivity. The studies in the literature using *E. coli* as a host microorganism, differs in the pH value they utilize.

The effect of pH on the growth of recombinant *E. coli* at three different controlled pH values, *i.e.*, 7.0, 7.4, 8.0; and reported that cell concentration increases when the culture pH decreases, however, enzyme activity attain its highest value at pH values in the vicinity of pH=7.4. The pH constant at 7.0 with aqueous NH_3. pH was kept constant throughout the bioprocess at 7.4 with 10 N NaOH. Also, in some studies of Akesson *et al.* (pH=6.9 by 25% ammonia) and (pH=7.0 by 2 M NaOH) pH value was kept constant throughout the bioprocesses. Hence, there is no systematic investigation in the literature, on the effects of pH on any biomolecule production and by-product distributions by the host *E. coli*.

Oxygen Transfer

The Importance and Mechanism of Oxygen Transfer

As microorganisms require oxygen for their metabolic activities like respiration, growth and product formation in aerobic fermentations, the concentration of the dissolved oxygen in the medium and oxygen uptake rate are very important parametres. Oxygen is present in all organic cell components and cellular water and constitutes about 20% of the dry weight of the cells. Molecular oxygen is required as a terminal electron acceptor in the aerobic metabolism. Gaseous oxygen is introduced into growth media by sparging air; and oxygen transfer rate can be adjusted by either changing the air inlet rate or agitation rate. Cells in aerobic culture take up oxygen from the liquid. The rate of oxygen transfer from gas to liquid is therefore of prime importance, especially at high cell densities when cell growth is likely to be limited by availability of oxygen in the medium. According to the two-film theory, the oxygen must pass through a series of transport resistances, the relative magnitudes of which depend on bubble hydrodynamics, temperature, cellular activity and density, solution composition, interfacial phenomena, and other factors.

- Transfer from the interior of the bubble to the gas-liquid interface,
- Movement across the gas-liquid interface,
- Diffusion through the relatively stagnant liquid film surrounding the bubble,
- Transport through the bulk liquid,
- Diffusion through the relatively stagnant liquid film surrounding the cells,
- Movement across the liquid-cell interface,
- If the cells are in a floc, clump or solid particle, diffusion through the solid to the individual cell,
- Transport through the cytoplasm to the site of reaction.

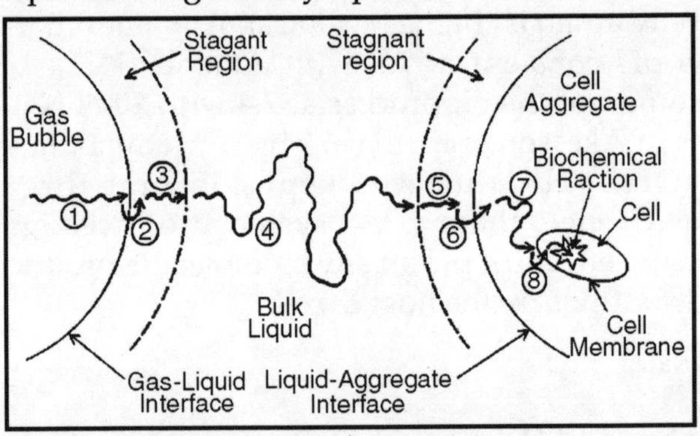

Fig. Schematic Diagram of Steps Involved in Transport of Oxygen from a Gas Bubble to Inside a cell.

When cells are dispersed in the liquid, and the bulk fermentation broth is well mixed, the major resistance to oxygen transfer is the liquid film surrounding the gas bubbles; therefore the rate of oxygen transfer from gas to liquid is of prime importance. An expression for oxygen transfer rate (OTR) from gas to liquid is given by the following equation:

$$OTR = k_L a\left(C_o^* - C_o\right)$$

where, k_L is the oxygen transfer coefficient, a is the gas-liquid interfacial area, kLa is the volumetric oxygen transfer coefficient, C_O* is saturated dissolved oxygen concentration, C_O is the actual dissolved oxygen concentration in the broth.

Since solubility of oxygen in aqueous solutions is very low, the liquid phase mass transfer resistance dominates, and the overall liquid phase mass transfer coefficient, K_La, is approximately equal to liquid phase mass transfer coefficient, k_La.

Oxygen Transfer Characteristics

The transfer of oxygen into the microbial cell in aerobic fermentation processes strongly affects product formation by influencing metabolic pathways and changing metabolic fluxes; therefore, to fine-tune bioreactor performance in relation with the physiology of the microorganism the extent of oxygen transfer requirement needs to be clarified. The rate of oxygen transfer in fermentation broths is influenced by several physical and chemical factors that change either the value of K_La, or the driving force for mass transfer, $(C_O{}^* - C_O)$. Therefore, the oxygen uptake rate of the cells, and the liquid phase mass transfer coefficient, K_La, are important characteristics of oxygen transfer.

The rate at which oxygen is consumed by cells in fermenters determines the rate at which it must be transferred from gas to liquid. Many factors influence oxygen demand; the most important of which are cell species, culture growth phase, and nature of the carbon source in the medium. In batch culture, rate of oxygen uptake varies with time. The reasons for this are twofold. First, the concentration of cells increases during the course of batch culture and the total rate of oxygen uptake is proportional to the number of cells present. In addition, the rate of oxygen consumption per cell, known as the specific oxygen uptake rate (q_O), also varies. Oxygen uptake rate (OUR), $-r_O$, per unit volume of broth is given by:

$$-r_o = q_o C_x$$

The inherent demand of an organism for oxygen (q_O) depends primarily on the biochemical nature of the cell and its nutritional environment. Choice of substrate for fermentation can also significantly affect oxygen demand. Because glucose is generally consumed more rapidly than other sugars or carboncontaining substrates, rates of oxygen demand are higher when glucose is used. The liquid phase mass transfer coefficient, KLa, is an important parametre in bioreactors, which indicates the oxygen transfer rate from gas to the liquid phase. It is important to estimate whether the dissolved oxygen in the medium is successfully transferred to the cells, and it is possible by determining the liquid phase mass transfer coefficient, KLa. Dynamic method is widely used for the determination of the value of KLa experimentally, and it can be applied during the fermentation process.

The Specificity and Mechanism of Enzyme Action

The action of enzymes, unlike that of inorganic catalysts, is strictly specific and depends on the structure of the substrate on which the enzyme acts. A good example of this dependence is the catalyzed reaction of the hydrolytic breakdown of the amino acid arginine into ornithine and urea by arginase:

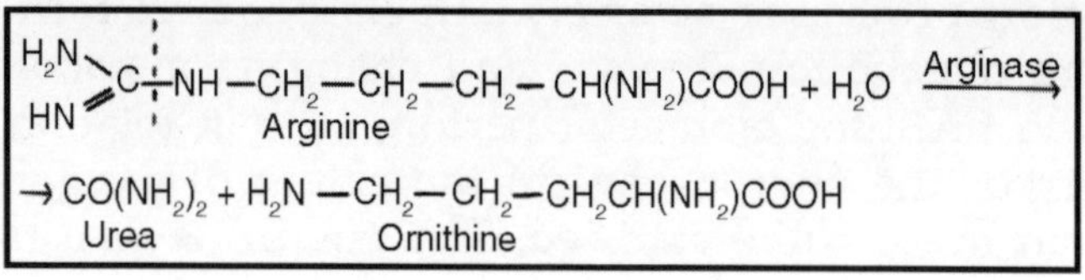

However, arginase does not decompose the methyl ether of arginine:

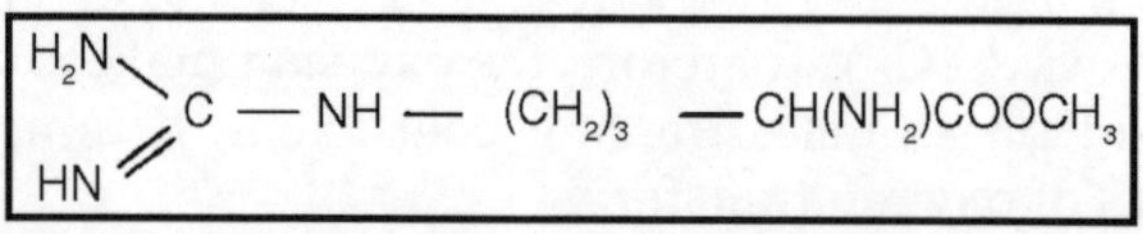

A dipeptide consisting of the radicals of two molecules of arginine yields only half of the theoretical quantity of urea when the dipeptide is acted on by arginase. It is evident that although the decomposition of arginine occurs in an area remote from the carboxyl group COOH (indicated by the broken line), an essential condition for the action of arginase is its combination with the carboxyl group of arginine. For this reason, either displacement of the hydrogen in the carboxyl group to the methyl radical or the bonding of the carboxyl group with the second molecule of arginine strongly influences the action of arginase.

Examples of the specificity of enzyme action may be observed by examining the stereochemical specificity of enzymes, that is, the action of enzymes on stereoisomers. For example, enzymes that oxidize natural L-amino acids do not affect the D-isomers of the same amino acids, and the enzyme dipeptidase, which hydrolyzes dipeptides consisting of radicals of L-amino acids, does not affect dipeptides composed of D-amino acid residues. Observation of the specificity of enzyme action led the German chemist E. Fischer to compare a given substrate and the enzyme catalyzing its transformation to a lock and its key. The stereochemical specificity of enzymes is closely related to one of the basic features of living organisms: their ability to synthesize optically active organic compounds.

Only those functional groups of an enzyme molecule that constitute the molecule's active centre take part in the union between the enzyme and the substrate, a union termed the enzyme-substrate complex. For example, in a molecule of chymotrypsin, an enzyme that hydrolizes proteins and that consists of 246 amino-acid radicals, the active centre is composed of one of the residues of serine (chymotrypsin is one of the serine proteinases) and two radicals of histidine located in sections of the polypeptide chain that are remote from one another.

These functional groups of the active centre converge because of the specific spatial (tertiary) structure that is typical of the chymotrypsin molecule. The disruption of this structure owing to protein denaturation or to other types of chemical modification leads to the alteration or cessation of catalytic activity. In compound enzymes, both the prosthetic group and the functional groups of an apoenzyme take part in forming the enzyme-substrate complex. For example, in the decomposition of pyruvic acid by pyruvate decarboxylase, the substrate is bonded to part of a thiamine pyrophosphate molecule in the following manner:

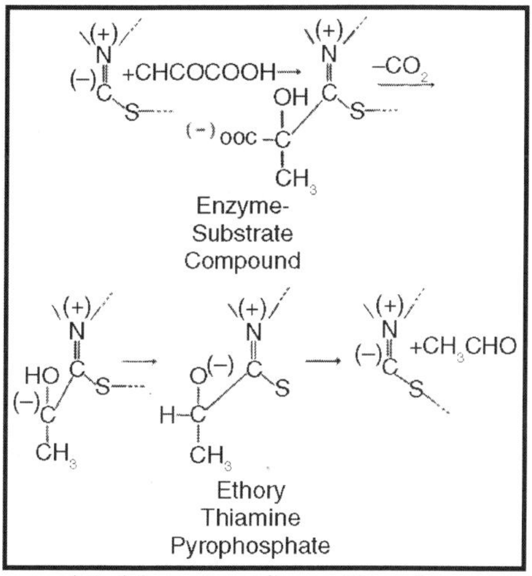

Enzymes have a highly specific action because of their protein nature. For example, pyridoxal enzymes that contain the coenzyme codecarboxylase may belong to different classes and may catalyze diverse reactions. The specificity of these enzymes' action depends on the nature of the given apoenzyme.

Conditions of Enzyme Action

Enzyme action depends on a number of factors, primarily temperature and the reaction of the medium (pH). The temperature conducive to optimum enzymatic activity is generally in the range of 40°–50°C. At lower temperatures the enzymatic reaction is generally slower, and at temperatures approaching 0°C, it virtually ceases.

Temperatures higher than 40°–50°C also slow down the enzymatic reaction and eventually inactivate it. The reduced intensity of enzyme action in the presence of higher temperatures is caused mainly by the incipient destruction (denaturation) of the protein component of the enzyme. Since proteins in a dry state are denatured more slowly than hydrated proteins, which are usually in the form of a gel or solution, the inactivation of an enzyme occurs much more slowly in the dry state than in the presence of moisture. For this reason, dry bacteria spores or dry seeds are able to withstand far higher temperatures than wet spores or seeds.

The Danish physical chemist S. Sarensen was the first to establish that pH is an important influence on enzyme action. Enzymes vary according to the pH value that is optimal for their action. Pepsin, for example, which is contained in gastric juice, is most active in a strongly acidic medium (pH 1–2), and trypsin, a proteolytic enzyme secreted in the pancreas, has an optimum action in a weakly alkaline medium (pH 8–9). The optimum action of papain, a proteolytic enzyme of vegetable origin, occurs in a weakly acidic medium (pH 5–6).

Enzyme action also depends on specific activators and on nonspecific or specific inhibitors. For example, enterokinase, which is secreted in the pancreas, converts inactive trypsinogen into active trypsin. Similar inactive enzymes in the cells and secretions of various glands are called proenzymes. Many enzymes are activated by compounds containing the sulfhydryl group (—SH); among these compounds are the amino acid cysteine and the tri-peptide glutathione, which is found in every living cell. Glutathione has an especially strong activating effect on some proteolytic and oxidizing enzymes.

Nonspecific inhibition of enzymes is caused by substances that form insoluble precipitates with proteins or that block certain groups within proteins, for example, the—SH group. There are also specific inhibitors of enzymes; these inhibitors suppress catalytic functions by bonding with chemical groups in the enzyme's active centre. For example, carbon monoxide (CO) specifically inhibits a number of oxidizing enzymes that contain iron or copper in their active centres. Carbon monoxide forms a chemical compound with these metals, blocks the enzyme's active centre, and consequently halts the enzyme's activity.

The inhibition of enzymes can be reversible or irreversible. In reversible inhibition, for example, in the action of malonic acid on succinic dehydrogenase, the enzyme's activity is restored when the inhibitor is eliminated by dialysis or by another method. In irreversible inhibition the action of the inhibitor, even when the inhibitor is present in very low concentrations, is gradually intensified and eventually halts an enzyme's activity completely.

The inhibition of enzyme action may also be competitive or noncompetitive. In competitive inhibition the inhibitor and the substrate seek to displace each other from the enzyme-substrate complex. High concentrations of a substrate reduce the action of a competitive inhibitor but maintain the action of a noncompetitive inhibitor. The action of activators and inhibitors on enzymes is of great importance in the regulation of enzymatic processes.

Classification and Nomenclature

In 1961 the International Union of Biochemistry recommended that enzymes be divided into six main classes: oxidoreductases, transferases, hydrolases, lyases, isomerases, and ligases. The following numbering system was also recommended: the number, or index, of each enzyme was to contain four digits, separated by periods. The first digit would indicate the class, the second the subclass, the third the sub-subclass, and the fourth the ordinal number in the sub-subclass. According to this system, the enzyme arginase, which decomposes arginine into ornithine and urea, has the number 3.5.3.1. In other words, arginase belongs to the hydrolase class, to the subclass of enzymes that act on the nonpeptide C—N bond, and to the sub-subclass of enzymes that decompose these bonds in linear compounds, rather than in cyclic compounds.

The oxidoreductase class includes enzymes that catalyze oxidation-reduction reactions. This class is divided into 14 subclasses according to the nature of the chemical group in the substrate molecule that undergoes oxidation, for example, the alcohol, aldehyde, or ketone group. The sub-subclasses of the oxidoreductases are numbered according to the type of hydrogen (electron) acceptor taking part in the reaction, for example, a coenzyme, cytochrome, or molecular-oxygen acceptor. The first three digits of the enzyme's number identify the type of enzyme; for example, 1.2.3 designates an oxidoreductase that acts as an electron acceptor on an aldehyde that contains molecular oxygen.

The transferase class, which includes enzymes that catalyze transfer reactions, is divided into eight subclasses according to the nature of the chemical groups transferred. These groups may be monocarbonic or glycosyl radicals, or they may be nitrogen or sulfur-containing groups. In the transferases, the third digit indicates the type of groups transferred; for example, the monocarbonic group may be methyl, carboxyl, or formyl.

The hydrolases are enzymes that catalyze the hydrolytic decomposition of various compounds. The hydrolases are divided into nine subclasses according to the type of bond hydrolyzed: ester, peptide, or glycoside. The third digit of a hydrolase's number defines the type of hydrolyzed bond.

The lyases are enzymes that detach a chemical group from the substrate nonhydrolytically, thus forming double bonds. Lyases also add chemical groups to double bonds. There are five subclasses of lyases. The second digit of a lyase's number refers to the type of bond that undergoes cleavage, for example, carbon-carbon or carbon-oxygen. The third digit refers to the type of chemical group detached.

Isomerases catalyze isomerization reactions and are divided into five subclasses according to the type of reaction catalyzed. Ligases, or synthetases, catalyze the joining together of two molecules, a coupling accompanied by the breakdown of a pyro-phosphate bond in a molecule of adenosine triphosphate (ATP) or of a similar triphosphate. The first digit of a ligase's number refers to the type of bond formed, for example, carbon-nitrogen or carbon-oxygen, and the second digit identifies the nature of the compound formed.

Isomerases catalyze isomerization reactions and are divided into five subclasses according to the type of reaction catalyzed. Ligases, or synthetases, catalyze the joining together of two molecules, a coupling accompanied by the breakdown of a pyro-phosphate bond in a molecule of adenosine triphosphate (ATP) or of a similar triphosphate. The first digit of a ligase's number refers to the type of bond formed, for example, carbon-nitrogen or carbon-oxygen, and the second digit identifies the nature of the compound formed.

Enzymes are also classified according to systematic and trivial (unsystematic) names. For example, the systematic name 2-oxo-acid carboxy-lyase corresponds to the trivial name pyruvate decarboxylase, and the systematic name L-arginine (amidinohydrolase), to the trivial name arginase.

Regulation of Enzymatic Activity

Living organisms have control mechanisms that regulate the synthesis and activity of enzymes. The complex of enzymes within a given organism is determined by the organism's genetic nature. However, this complex may be altered by such internal and external factors as diet, mutations, the effect of ionizing radiation, and the composition of the surrounding air. For example, such molecular diseases as alkaptonuria are caused by mutations. In this hereditary disease, changes in the amino acid tyrosine result in the formation of homogentisic acid. The homogentisic acid accumulates in the body and is excreted with the urine since patients with alkaptonuria lack the ability to synthesize the two enzymes that catalyze the subsequent oxidation of homogentisic acid: para-hydroxyphenyl-pyruvate oxidase and homogentisic acid oxidase.

The influence of nutriments on the organism's enzyme system is particularly easy to trace in microorganisms. For example, a colon bacillus (Escherichia coli) synthesizes only traces of P-galactosidase during its growth on a nutritive medium containing glucose. However, when various β-galactosides are present, large amounts of this enzyme are formed, amounting to 6-7 per cent of all the proteins contained in a cell. Enzymes whose formation or intensified synthesis is caused by the influence of a compound are called induced enzymes.

Other compounds may repress the synthesis of enzymes. In animals, the induction and repression of enzyme synthesis is influenced by hormones as well as by substrates and metabolites. For example, the synthesis of the enzyme glucose-6-phosphatase, which helps synthesize glucose in the liver, is induced by the hormones thyroxine and cortisone but is repressed by insulin. A general theory of the induction and repression of biosynthesis on the genetic level was formulated by the French biologists F. Jacob and J. Monod.

An enzyme may be present in an organism in different molecular forms. The distinct forms of an enzyme that catalyze the same reaction but that differ in physical, chemical, and immunological properties are called isoenzymes. The synthesis of isoenzymes is determined by genetic factors, but it may be altered by environmental factors. Thus, the concentration and activity of an enzyme depend on many diverse environmental factors. These include the water, air, temperature, acidity, and light, the concentration of substrates and of cofactors necessary for enzyme action, the presence of activators and inhibitors, the concentration of metabolites, and, in higher multicellular organisms, the neural and hormonal regulation of enzyme activity.

The Pasteur effect, or the inhibition of fermentation by the action of oxygen, is an example of the effect of the environment on enzyme activity. The activity of many enzymes is regulated according to the allosteric principle. Such enzymes have an allosteric site that is joined to a specific metabolite. The allosteric effector causes changes in the structure of the active centre that reduce or increase the activity of the enzyme.

Some enzymes exist in a cell as components of complexes containing a number of enzymes. In such complexes the activity of each enzyme is controlled by the other enzymes in the complex. An example of such a complex is pyruvate dehydrogenase, which is composed of 16 molecules of pyruvate decarboxylase, eight molecules of dihydrolipoil dehydrogenase, and four aggregates of lipoate acetyltransferase, each of which consists of 16 subunits. Such subcellular structures as the mitochondria, the microsomes, and the lysosomes, as well as the proteolipide membranes that separate these structures from the cytoplasm, help regulate enzyme activity in the cell. Many enzymes are embedded in these membranes as part of complexes of enzymes.

Practical Importance of Enzymes

The fermentation process is fundamental to many industries, including the bread-baking, wine-making, beer-brewing, cheese-making, and distilling industries and is also fundamental to the production of tea and vinegar. Beginning in the early 20th century, the distilling industry was one of a number of industries to use the fermentation methods developed by the Japanese chemist J. Takamine. These methods made use of enzymatic agents obtained from molds or bacteria. Fermentation methods are widely used in a number of countries to saccharify starch by means of amylase into crystalline glucose or to convert starch into alcohol.

When added to bread dough, concentrated amylolytic solutions of enzymes derived from molds improve the quality of the bread and reduce the time needed for rising. Solutions containing proteolytic enzymes derived from microorganisms are used in the leather industry to remove hair and soften hides, and to replace the expensive proteolytic enzyme rennin in the cheese-making industry. Solutions containing microbial pectolytic enzymes are widely used in the preparation of fruit juices; they increase the quantity of juice produced by 10-20 per cent.

Purified enzyme preparations are increasingly used in medicine. In both scientific research and clinical practice, highly purified enzyme preparations are used in biochemical analysis. The use of immobilized enzymes, that is, insoluble complexes of enzymes with some carriers, will undoubtedly increase. The selection of an appropriate carrier results in the formation of a highly active immobilized enzyme that is resistant to denaturants. A column filled with the immobilized enzyme may be used repeatedly to produce a corresponding reaction. Immobilized enzymes are increasingly used in biochemical analysis and in biochemical technology

Chapter 3

Enzyme Immobilization

IMMOBILIZED ENZYMES IN BIOPROCESS

Immobilization of biocatalysts helps in their economic reuse and in the development of continuous bioprocesses. Biocatalysts can be immobilized either using the isolated enzymes or the whole cells. Immobilization often stabilizes structure of the enzymes, thereby allowing their applications even under harsh environmental conditions of pH, temperature and organic solvents, and thus enable their uses at high temperatures in nonaqueous enzymology, and in the fabrication of biosensor probes. In the future, development of techniques for the immobilization of multienzymes along with cofactor regeneration and retention system can be gainfully exploited in developing biochemical processes involving complex chemical conversions. The present review outlines some of the above aspects, and delineates the present status and future potentials of immobilized enzymes and nonviable cells in the emerging biotech industries.

Biotechnology is currently considered as a useful alternative to conventional process technology in industrial and analytical fields. This is mainly because, unlike the chemical catalysts, the biological systems have the advantages of accomplishing complex chemical conversions under mild environmental conditions with high specificity and efficiency. Biological systems help in ingredient substitution, processing aid substitution, more efficient processing, less undesirable products, increased plant capacity, increased product yields, and improved or unique products. The variety of chemical transformations catalysed by enzymes has made these catalysts a prime target of exploitation by the emerging biotech industries.

Despite these advantages, the use of enzymes in industrial applications has been limited by several factors, mainly the high cost of the enzymes, their instability, and availability in small amounts. Also the enzymes are soluble in aqueous media and it is difficult and expensive to recover them from reactor effluents at the end of the catalytic process. This restricts the use of soluble enzymes to batch operations, followed by disposal of the spent enzyme-containing solvent. Over the last few decades, intense research in the area of enzyme technology has provided many approaches that facilitate their practical applications. Among them, the newer technological developments in the field of immobilized biocatalysts can offer the possibility of a wider and more economical exploitation of biocatalysts in industry, waste treatment, medicine, and in the development of bioprocess monitoring devices like the biosensor.

Immobilization means associating the biocatalysts with an insoluble matrix, so that it can be retained in a proper reactor geometry for its economic reuse under stabilized conditions. Immobilization thus allows, by essence, to decouple the enzyme location from the flow of the liquid carrying the reagents and products. Immobilization helps in the development of continuous processes allowing more economic organization of the operations, automation, decrease of labour, and investment/capacity ratio. Immobilized biocatalysts offer several other advantages, notable among them is the availability of the product in greater purity. Purity of the product is very crucial in food processing and pharmaceutical industry since contamination could cause serious toxicological, sensory, or immunological problems. The other major advantages include greater control over enzymatic reaction as well as high volumetric productivity with lower residence time, which are of great significance in the food industry, specially in the treatment of perishable commodities as well as in other applications involving labile substrates, intermediates or products·

IMMOBILIZED NONVIABLE CELLS AS AN ECONOMICAL SOURCE OF ENZYMES

Biocatalysts can be immobilized using either the isolated enzymes or the whole cells or cellular organelles. Immobilization of whole cells has been shown to be a better alternative to immobilization of isolated enzymes· Doing so avoids the lengthy and expensive operations of enzyme purification, preserves the enzyme in its natural environment thus protecting it from inactivation either during immobilization or its subsequent use in continuous system. It may also provide a multipurpose catalyst, specially when the process requires the participation of number of enzymes in sequence. The major limitations which may need to be addressed while using such cells are the diffusion of substrate and products through the cell wall, and unwanted side reactions due to the presence of other enzymes. The cells can be immobilized either in a viable or a nonviable form. Immobilized nonviable cell preparations, which are normally obtained by permeabilizing the intact cells, for the expression of intracellular activity are useful for simple processes that require single-enzyme with no requirement for cofactor regeneration, like hydrolysis of sucrose or lactose· On the other hand immobilized viable cells, which serve as 'controlled catalytic biomass', have opened new avenues for continuous fermentation on heterogeneous catalysis basis by serving as self-proliferating biocatalysts· The present review will focus mainly on immobilized isolated enzyme systems, and the use of immobilized nonviable cells as a source of enzymes. For the aspects on the immobilized viable cell systems, the readers can refer to another article in this issue. Some of the aspects on the applications of immobilized viable cells in biochemical process development and monitoring have also been reviewed recently·

Most of the enzymes used at industrial scale are normally the extracellular enzymes produced by the microbes. This has been mainly due to their ease of isolation as crude enzymes from the fermentation broth. Moreover, the extracellular enzymes are more stable to external environmental perturbations compared to the intracellular enzymes. However, over 90% of the enzymes produced by a cell are intracellular. The economic exploitation of these, having a variety of biochemical potentials, has been limited in view of the high cost involved in their isolation. Also, compared to extracellular enzymes, the intracellular enzymes are more labile. Delicate and expensive separation methods are required to release the enzymes undamaged from the cell, and to isolate them. This increases the labour and the cost of the enzyme. These problems could now be obviated by the use of permeabilized cells as a source of enzyme.

Permeabilization of the cells removes the barrier for the free diffusion of the substrate/product across the cell membrane, and also empties the cell of most of the small molecular weight cofactors, etc., thus minimizing the unwanted side reactions. Side reactions, which can occur due to the presence of other enzymes in a cell, can also be minimized by inactivating such enzymes prior to or after immobilization· Such permeabilized cells, which are often referred to as nonviable or nongrowing cells, can be exploited in an immobilized form as a very economical source of intracellular enzyme for simple bioconversions like hydrolysis, isomerization and oxidation reactions that do not need a cofactor-regeneration system· The decision to immobilize cells either in a viable or nonviable form is very important and depends on their ultimate application. Thus the permeabilized *K. fragilis* cells, which are nonviable, convert lactose in milk to glucose and galactose, whereas the viable (nonpermeabilized) cells convert the lactose to ethanol and are useful in the complete desugaration of milk.

A variety of physical, chemical, and enzymatic techniques have been developed for the permeabilization of cells. Some techniques like entrapment in polyacrylamide itself have been shown to result in permeabilization of the cells. A few typical examples of immobilized enzyme preparation obtained using permeabilized cells.

TECHNIQUES AND SUPPORTS FOR IMMOBILIZATION

A large number of techniques and supports are now available for the immobilization of enzymes or cells on a variety of natural and synthetic supports. The choice of the support as well as the technique depends on the nature of the enzyme, nature of the substrate and its ultimate application.

Therefore, it will not be possible to suggest any universal means of immobilization. It can only be said that the search must continue for matrices which provide facile, secure immobilization with good interaction with substrates, and which conform in shape, size, density and so on to the use for which they are intended. Care has to be taken to select the support materials as well as the reagents used for immobilization, which have GRAS status, particularly when their ultimate applications are in the food processing and pharmaceutical industries.

Macromolecular, colloidal, viscous, sticky, dense or particulate food constituents or waste streams also limit the choice of reactor and support geometries. Commercial success has been achieved when support materials have been chosen for their flow properties, low cost, nontoxicity, maximum biocatalysts loading while retaining desirable flow characteristics, operational durability, ease of availability, and ease of immobilization. The variety of techniques and supports investigated for the immobilization have been reviewed in a number of articles and books. The objective of the preceding part of the paper is to review briefly, in the light of current developments, some approaches that have been used for the immobilization of biocatalysts with reference to the studies carried out in the author's laboratory as well as few others.

Techniques for immobilization have been broadly classified into four categories, namely entrapment, covalent binding, cross-linking and adsorption. A combination of one or more of these techniques has also been investigated. It must be emphasized that in terms of economy of a process, both the activity and the operational stability of the biocatalysts are important. They determine its productivity, which is the activity integrated over the operational time.

Entrapment

Entrapment has been extensively used for the immobilization of cells, but not for enzymes. The major limitation of this technique for the immobilization of enzymes is the possible slow leakage during continuous use in view of the small molecular size compared to the cells. Biocatalysts have been entrapped in natural polymers like agar, agarose and gelatine through thermoreversal polymerization, but in alginate and carrageenan by ionotropic gelation. A number of synthetic polymers have also been investigated. Notable among them are the photo-crosslinkable resins, polyurethane prepolymers, and acrylic polymers like polyacrylamide.

Among these, the most widespread matrix made from monomeric precursors is the polyacrylamide gel. Polyacrylamide may not be a useful support for use in food industry in view of its toxicity, but can have potentials in the treatment of waste and in the fabrication of analytical devices containing biocatalysts. One of the major limitations of entrapment technique is the diffusional limitation as well as the steric hindrance, especially when the macromolecular substrates like starch and proteins are used. Diffusional problems can be minimized by entrapment in fine fibres of cellulose acetate or other synthetic materials or by using an open pore matrix. Recently, the development of so-called hydrogels and thermoreactive water-soluble polymers, like the albumin-poly (ethylene glycol) hydrogel, have attracted attention in the field of biotechnology. In the area of health care, they offer new avenues for enzyme immobilization. Such gels with a water content of about 96% provide a microenvironment for the immobilized enzyme close to that of the soluble enzyme with minimal diffusional restrictions.

COVALENT BINDING

Covalent binding is an extensively used technique for the immobilization of enzymes, though it is not a good technique for the immobilization of cells. The functional groups extensively investigated are the amino, carboxyl, and the phenolic group of tyrosine. Enzymes are covalently linked to the support through the functional groups in the enzymes, which are not essential for the catalytic activity. It is often advisable to carry out the immobilization in the presence of its substrate or a competitive inhibitor so as to protect the active site. The covalent binding should also be optimized so as not to alter its conformational flexibility. Some of these problems however, can be obviated by covalent bonding through the carbohydrate moiety when a glycoprotein is concerned. A number of industrially useful enzymes are glycoproteins wherein the carbohydrate moiety may not be essential for its activity. In general, functional aldehyde group can be introduced in a glycoprotein by oxidizing the carbohydrate moiety by periodate oxidation without significantly affecting the enzyme activity. The enzyme could then be covalently linked to a support containing an alky amine group through Schiffs base reaction. Enzymes like glucose oxidase, peroxidase, invertase, etc. have been immobilized using this technique·

Covalent binding has been extensively investigated using inorganic supports. Enzymes covalently bound to inorganic supports have been used in the industry· Enzymes have also been bound to synthetic membranes, thus integrating biconversion and downstream processing· Large-scale processes using such an approach have been demonstrated for the preparation of invert sugar using invertase·

Cross-linking

Biocatalysts can also be immobilized through chemical cross-linking using homo- as well as heterobifunctional cross-linking agents· Among these, glutaraldehyde which interacts with the amino groups through a base reaction has been extensively used in view of its GRAS status, low cost, high efficiency, and stability· The enzymes or the cells have been normally cross-linked in the presence of an inert protein like gelatine, albumin, and collagen· Studies from our laboratory have shown the use of raw hen egg white as an economic, easily available novel proteinic support rich in lysozyme for the immobilization of enzyme or nonviable cells either in a powder, bead or highly porous foam form, using glutaraldehyde as the cross-linker.

The unique feature of this support is the large concentration of lysozyme naturally present in hen egg white which gets co-immobilized, thus imparting the bacteriolytic property to the support· Adsorption followed by cross-linking has also been used for the immobilization of enzymes. The technique of cross-linking in the presence of an inert protein can be applied to either enzymes or cells. The technique can also be used for the immobilization of enzymes by cross-linking the cell homogenates· Osmotic stabilization of cellular organelles or halophilic cells prior to immobilization using cross-linkers has also shown promise.

Adsorption

This is perhaps the simplest of all the techniques and one which does not grossly alter the activity of the bound enzyme. In case of enzymes immobilized through ionic interactions, adsorption and desorption of the enzyme depends on the basicity of the ion exchanger. Moreover, a dynamic equilibrium is normally observed between the adsorbed enzyme and the support which is often affected by pH as well as the ionic strength of the surrounding medium.

This property of reversibility of binding has often been used for the economic recovery of the support. This has been successfully adapted in industry for the resolution of racemic mixtures of amino acids, using amino acid acylase· A variety of commercially available ion exchangers have been investigated for this purpose· One of the techniques, which has gained importance more recently, is the use of polyethylenimine for imparting polycationic characteristics to many of the neutral supports based on cellulose or inorganic materials· Enzymes with low pI, like invertase, urease, glucose oxidase, catalase, and other enzymes have been bound through adsorption followed by cross-linking on polyethylenimine-coated supports.

Immobilization of enzymes through hydrophobic interaction has also shown promise· One of the important features of this technique, which is of great significance, is that, unlike ionic binding, hydrophobic interactions are usually stabilized by high ionic concentrations, thus enabling the use of high concentrations of substrates as desired in an industrial process without the fear of desorption. Other types of strong interactive binding techniques have also been reported for the reversible immobilization of enzymes.

A typical example is the immobilization of soybean b -amylase on phenyl-boronate agarose, which can be reversed for the recovery of the support using sorbitol· Varieties of biospecific interactions have also been investigated for the reversible immobilization of enzymes by adsorption. Enzymes like acetyl choline esterase, ascorbic acid oxidase, invertase, peroxidase, glucose oxidase, etc. have been immobilized by biospecific-reversible immobilization on lectin-bound supports and invertase, using polyclonal antiinvertase antibodies·

Traditional enzyme immobilization procedures involve isolation of the enzyme, followed by use of several steps for the immobilization. Costs of enzyme purification, the immobilization procedure, bioreactor operational stability, and bioreactor regeneration are the major factors that determine the cost of a bioreactor process. Development of techniques for the simultaneous isolation and immobilization of enzymes from crude extracts has obviated these problems. Some typical examples include the immobilization of a streptavidin-b -galactosidase fusion protein expressed in *Escherichia coli* and bioselectively adsorbed from a crude cell lysate to biotin which is covalently immobilized on controlled pore glass, and the simultaneous purification and immobilization of D-amino acid oxidase from *Trigonopsis variabilis* cell lysate adsorbed on phenyl sepharose· The technique can have future potentials especially in the downstream processing and immobilization of enzyme/proteins obtained by recombinant DNA technology.

Techniques for the adhesion of whole cells on polymeric surfaces are also currently gaining considerable importance. The major advantage for the cells immobilized through adhesion is reduction or elimination of the mass transfer problems associated with the commonly used gel entrapment method. The technique of immobilization usually being followed is the microbial colonization by recycling of the cell suspension along with nutrients, such that a biofilm is gradually formed. This often results in the immobilization of cells in a viable form for use in heterogeneous fermentations·

Useful techniques have been developed also for the immobilization of nonviable cells to be used as an enzyme source for simple chemical conversions. Notable among them include treating the cells or the support with trivalent metal ions like Al or Fe^{3+} or charged colloidal particles' and use of polycationic polymers like chitosan. Novel techniques have been developed to adhere cells strongly on a variety of polymeric surfaces including glass' cotton cloth' cotton threads' and other synthetic and inorganic surfaces using polyethylenimine. This technique has also been recently used for the simultaneous filtration and immobilization of cells from a flowing suspension, thus integrating downstream processing with bioprocessing. This technique may have future potentials for the immobilization of cells for food applications. Polyethylenimine is nontoxic and the United States of America Food and Drug Administration has permitted its use as a direct food additive under the Food Drug and Cosmetic Act.

IMMOBILIZED ENZYMES IN ORGANIC SOLVENTS

In recent years, much research has centred on the conduct of enzyme reactions in organic solvents. It is now well established that hydrolytic enzymes can catalyse esterification and transesterification reactions in monophasic organic solvents and in water-organic biphasic systems. Conventional immobilized enzymes are principally used to facilitate catalysts recovery. Even though enzymes by themselves are insoluble in organic solvents, in addition to others, the prime importance to the use of enzymes in organic media is the necessity to avoid enzyme deactivation or denaturation. A number of enzymes which are used in organic solvents have been immobilized using a variety of techniques with a view to stabilizing them· In systems containing enzymes immobilized on solid supports and working in organic media, the support has a significant influence on the total enzyme activity, and can also displace the reaction equilibrium (hydrolysis towards synthesis) due to the interaction of the support with the water molecules.

Thus the choice of a suitable support material, proper water content, and the selection of the organic solvents are crucial for the use of immobilized enzymes in organic media. Another approach that has been explored is by chemical cross-linking of enzyme crystals, thereby stabilizing the crystalline lattice and its constituent enzyme molecules, which result in forming highly concentrated immobilized enzyme particles that can be lyophilized and stored indefinitely at room temperature. Cross-linked enzyme crystals retain catalytic activity in harsh conditions, including temperature and pH extremes, exogenous proteases, and exposure to organic or aqueous solvents· Lyophilized cross-linked enzyme crystals can be reconstituted easily in these solvents as active monodisperse suspensions. Cross-linked enzyme crystals have shown promise in a variety of applications like the synthesis of aspartame, using thermolysin, and for the resolution of chiral esters· The techniques of immobilization can also be extended to obtain organic solvent-soluble enzymes by treating them with hydrophobic molecules like the lipids·

Immobilization of Enzymes for the Fabrication of Biosensors

The choice of the support and the technique for the preparation of membranes has been dictated by the low diffusional resistance of the membrane coupled with its ability to incorporate optimal amount of enzyme per unit area. In this respect, stable membranes have been prepared by binding glucose oxidase to cheese cloth in the fabrication of a glucose biosensor. Enzymes entrapped inside the reversed micelle have also shown promise in the fabrication of biosensors. Cross-linked enzyme crystals (CLCs) described above provide their own supports and so achieve enzyme concentration close to the theoretical packing limit in excess of even highly concentrated enzyme solutions. In view of this, CLCs are particularly attractive in biosensor applications where the largest possible signal per unit volume is often critical.

Sensors based on small transducer or thinner enzyme immobilized membranes (miniature biosensors) are also emerging. The development of molecular devices incorporating a sophisticated and highly organized biological information processing function, is a long-term goal of bioelectronics. For this purpose, it is necessary in the future to develop suitable methods for microimmobilizing the proteins/ enzymes into an organized array/pattern, as well as designing molecular structures capable of performing the required function. A typical example is the microimmobilization of proteins into organized patterns on a silicone wafer based on a specific binding reaction between strepatavidin and biotin combined with photolithography techniques.Immobilized enzymes have also been used for various other analytical purposes. A recent development has been in obtaining a stable dry immobilized enzyme, like acetylcholineesterase, on polystyrene microtitration plates for mass screening of its inhibitors in water and biological fluids.

Immobilized Multienzymes and Enzyme-cell Co-immobilizates

A common feature of metabolic pathways in intermediary metabolism is that the product of one enzyme in sequence is the substrate for the next and so forth. One of the advantages of such an arrangement is that a favourable high local concentration of intermediates in the microenvironment of an enzyme system can be created. Such an approach perhaps will be useful in the development of enzymatic processes requiring multiple enzymes. The enzymes in question can be simultaneously bound to the same support or could be simultaneously immobilized through entrapment. Kinetic advantages of this system, such as reduction in the lag and significant enhancement in the final product, over mixed soluble enzymes or immobilized and then mixed systems have been demonstrated.

Often a cell may not contain all the necessary enzyme compliments for carrying out a required process. Thus *Saccharomyces cerevisiae,* a potential ethanol fermentor, cannot utilize lactose or starch since it lacks the enzyme b -galactosidase and amylase, respectively. There is considerable interest in genetically modifying such cells for the expression of the deficient enzymes. The success will depend on the plasmid stability, and quite often on the ability of these cells to secrete these enzymes for their action on the raw materials present in the growth medium.

Recently, genetically engineered cells have been obtained which can not only produce the deficient enzyme but can also transport it to the periplasmic place and naturally immobilize it. Another more practical approach is to immobilize the deficient enzyme directly on the cell wall. This approach which is termed as enzyme-cell co-immobilizate permits the tailoring of a whole cell for specific complex chemical conversion by combining the biochemical potential of both the cell and the additional enzyme from another source. A few techniques developed in our laboratory include the binding of the glycoprotein glucose oxidase onto the microbial cell walls, using Con

A. Enzymes can also be immobilized on the microbial cell surfaces coated with polyethylenimine through adsorption followed by cross-linking. The different techniques for the preparation of enzyme–cell co-immobilizate, and their applications have been reviewed.

Immobilized Enzymes for the Regeneration of Cofactors

Enzyme technology has dealt mainly with simple reactions which do not require coenzymes. One of the major challenges for enzyme technologists is in the application of cofactor-dependent enzymes. Over 25% of the known enzymes require cofactors like NAD/NADH, ATP/ADP, etc. which, unlike the enzymes, undergo stoichiometric changes during a biochemical reaction. The living cells have circumvented this problem either by introducing certain cofactor-recycling enzyme systems in their metabolic pathways or through their recycling via the electron transport system in the case of aerobic metabolism. *In vitro* applications of such enzymes will largely depend on developing efficient *in vitro* cofactor recycling systems.

These can be obtained by co-immobilizing another enzyme so that it can recycle the cofactors, *e.g.* alcohol dehydrogenase in the presence of ethanol can be used for the recycling of NAD to NADH in a bioprocess catalysed by a NADH-dependent enzyme. Choice of the coupling enzyme for such application has been made based on its ability to use an economical substrate like ethanol or formate (formate dehydrogenase) as well as one which results preferably in a volatile byproduct like CO_2 or acetaldehyde so as to minimize the downstream processing problems. In addition to the recycling of the cofactor, other important criteria for development of cofactor-dependent biochemical process is also the retention of cofactors.

Currently methods are available for the covalent binding of the cofactors along with the enzymes on polymeric supports or directly on enzymes or on soluble high molecular weight polymers like dextran or polyethylene glycol. Membrane reactors containing continuous cofactor regenerating systems have been used in the large-scale production of sugar derivatives, and optically active a -hydroxy acids and a -amino acids. Studies from our laboratory have shown that such cofactor regenerating system can be obtained by immobilization of permeabilized yeast cells as a source of alcohol dehydrogenase. Multienzymes and cofactors immobilized within microcapsules have been used for the multi-step conversion of liphophilic and lipophobic substrates. A number of approaches have also been investigated for the regeneration of ATP. One of these viable processes makes use of acetate kinase which can phosphorylate ADP using acetyl phosphate as the phosphate donor.

Applications of Immobilized Enzymes

The first industrial use of an immobilized enzyme is amino acid acylase by Tanabe Seiyaku Company, Japan, for the resolution of recemic mixtures of chemically synthesized amino acids. Amino acid acylase catalyses the deacetylation of the L form of the *N*-acetyl amino acids leaving unaltered the *N*-acetyl-d amino acid, that can be easily separated, racemized and recycled. Some of the immobilized preparations used for this purpose include enzyme immobilized by ionic binding to DEAE-sephadex and the enzyme entrapped as microdroplets of its aqueous solution into fibres of cellulose triacetate by means of fibre wet spinning developed by Snam Progetti. Rohm GmbH have immobilized this enzyme on macroporous beads made of flexiglass-like material.

By far, the most important application of immobilized enzymes in industry is for the conversion of glucose syrups to high fructose syrups by the enzyme glucose isomerase. It is evident that most of the commercial preparations use either the adsorption or the cross-linking technique. Application of glucose isomerase technology has gained considerable importance, especially in nontropical countries that have abundant starch raw material.

Unlike these countries, in tropical countries like India, where sugarcane cultivation is abundant, the high fructose syrups can be obtained by a simpler process of hydrolysis of sucrose using invertase. Compared to sucrose, invert sugar has a higher humectancy, higher solubility and osmotic pressure.

Historically, invertase is perhaps the first reported enzyme in an immobilized form. A large number of immobilized invertase systems have been patented. The possible use of whole cells of yeast as a source of invertase was demonstrated by D'Souza and Nadkarni as early as 1978. A systematic study has been carried out in our laboratory for the preparation of invert sugar using immobilized invertase or the whole cells of yeast. These comprehensive studies carried out on various aspects in our laboratory of utilizing immobilized whole-yeast have resulted in an industrial process for the production of invert sugar.

Table. Commercial Immobilized Lactase Preparetions

Company	Enzyme source procedure	Immobilization
Gist brocades	Saceharomyces lactis	Enzyme cntrapped in cellulose triacetate fibres

Snampogetti	Klayveromyces luctis	Enzyme entrapped in cellulose triacetate fibres
Corning glass works	Aspergillus niger	Enzyme covalently bound to silica beads
Valio laboratory	A. niger	Enzyme absorbed/cross-linked to phenol formaldehyde resin
Sturge	A. niger	Enzyme convalently bound to silanized Mn-Zn ferrite
Rchm GmbH	A. oryzac	Enzyme convalently bound to macroporous plexiglass beads
Sumitomo	A. oryzac	Enzyme covalently bound to macroporous amphoteric ion exchange resin
Amerace Corp.	A. oryzac	Enzyme covalently bound to microporous PVC-silica sheet

L-aspartic acid is widely used in medicines and as a food additive. The enzyme aspartase catalyses a one-step stereospecific addition of ammonia to the double bond of fumaric acid. The enzymes have been immobilized using the whole cells of *Escherichia coli*. This is considered as the first industrial application of an immobilized microbial cell. The initial process made use of polyacrylamide entrapment which was later substituted with the carragenan treated with glutaraldehyde and hexamethylenediamine. Kyowa Hakko Kogyo Co. uses Duolite A7, a phenolformaldehyde resin, for adsorbing aspartase used in their continuous process.

Other firms include Mitsubishi Petrochemical Co. and Purification Engineering Inc. Some of the firms, specially in Japan like Tanabe Seiyaku and Kyowa Hakko, have used the immobilized fumarase for the production of malic acid (for pharmaceutical use). These processes make use of immobilized nonviable cells of *Brevibacterium ammoniagenes* or *B. flavus* as a source of fumarase. Malic acid is becoming of greater market interest as food acidulant in competition with citric acid. Studies from our laboratory have shown the possibility of using immobilized mitochondria as a source of fumarase.

One of the major applications of immobilized biocatalysts in dairy industry is in the preparation of lactose-hydrolysed milk and whey, using b -galactosidase. A large population of lactose intolerants can consume lactose-hydrolysed milk. This is of great significance in a country like India where lactose intolerance is quite prevalent.

Lactose hydrolysis also enhances the sweetness and solubility of the sugars, and can find future potentials in preparation of a variety of dairy products. Lactose-hydrolysed whey may be used as a component of whey-based beverages, leavening agents, feed stuffs, or may be fermented to produce ethanol and yeast, thus converting an inexpensive byproduct into a highly nutritious, good quality food ingredient. The first company to commercially hydrolyse lactose in milk by immobilized lactase was Centrale del Latte of Milan, Italy, utilizing the Snamprogetti technology. The process makes use of a neutral lactase from yeast entrapped in synthetic fibres.

Specialist Dairy Ingredients, a joint venture between the Milk Marketing Board of England and Wales and Corning, had set up an immobilized b -galctosidase plant in North Wales for the production of lactose-hydrolysed whey. Unlike the milk, the acidic b -galactosidase of fungal origin has been used for this purpose. An immobilized preparation obtained by cross-linking b -galactosidase in hen egg white (lyophilized dry powder) has been used in our laboratory for the hydrolysis of lactose. A major problem in the large-scale continuous processing of milk using immobilized enzyme is the microbial contamination which has necessitated the introduction of intermittent sanitation steps.

A co-immobilizate obtained by binding of glucose oxidase on the microbial cell wall using Con A has been used to minimize the bacterial contamination during the continuous hydrolysis of lactose by the initiation of the natural lactoperoxidase system in milk. A novel technique for the removal of lactose by heterogeneous fermentation of the milk using immobilized viable cells of *K. fragilis* has also been developed.

One of the major applications of immobilized enzymes in pharmaceutical industry is the production of 6-aminopenicillanic acid (6-APA) by the deacylation of the side chain in either penicillin G or V, using penicillin acylase (penicillin amidase). More than 50% of 6-APA produced today is enzymatically using the immobilized route. One of the major reasons for its success is in obtaining a purer product, thereby minimizing the purification costs. The first setting up of industrial process for the production of 6-APA was in 1970s simultaneously by Squibb (USA), Astra (Sweden) and Riga Biochemical Plant (USSR). Currently, most of the pharmaceutical giants make use of this technology. A number of immobilized systems have been patented or commercially produced for penicillin acylase which make use of a variety of techniques either using the isolated enzyme or the whole cells. This is also one of the major applications of the immobilized enzyme technology in India. Similar approach has also been used for the production of 7-aminodeacetoxy-cephalosporanic acid, an intermediate in the production of semisynthetic cephalosporins.

Immobilized oxidoreductases are gaining considerable importance in biotechnology to carry out synthetic transformations. Of particular significance in this regard are oxidoreductase-mediated asymmetric synthesis of amino acids, steroids and other pharmaceuticals and a host of speciality chemicals. They play a major role in clinical diagnosis and other analytical applications like the biosensors. Future applications for oxidoreductases can be in areas as diverse as polymer synthesis, pollution control, and oxygenation of hydrocarbons. Immobilized glucose oxidase can find application in the production of gluconic acid, removal of oxygen from beverages, and in the removal of glucose from eggs prior to dehydration in order to prevent Maillard reaction. Studies carried out in this direction in our laboratory have shown that glucose can be removed from egg, using glucose oxidase and catalase which are co-immobilized either on polycationic cotton cloth or in hen egg white foam matrix.

Alternatively, glucose can also be removed by rapid heterogeneous fermentation of egg melange, using immobilized yeast. Immobilized D-amino acid oxidase has been investigated for the production of keto acid analogues of the amino acids, which find application in the management of chronic uremia. Keto acids can be obtained using either L- or D-amino acid oxidases. The use of D-amino acid oxidase has the advantage of simultaneous separation of natural L-isomer from DL-recemates along with the conversion of D-isomer to the corresponding keto acid which can then be transaminated in the body to give the Lamino acid.

Of the several microorganisms screened, the triangular yeast *T. variabilis* was found to be the most potent source of D-amino acid oxidase with the ability to deaminate most of the D-amino acids. The permeabilized cells entrapped either in radiation polymerized acrylamide Ca-alginate or gelatin have shown promise in the preparation of a -keto acids. Another interesting enzyme that can be used profitably in immobilized form is catalase for the destruction of hydrogen peroxide employed in the cold sterilization of milk. A few reports are available on its immobilization using yeast cells.

Lipase catalyses a series of different reactions. Although they were designed by nature to cleave the ester bonds of triacylglycerols (hydrolysis), lipase are also able to catalyse the reverse reaction under microaqueous conditions, *viz.* formation of ester bonds between alcohol and carboxylic acid moieties. These two basic processes can be combined in a sequential fashion to give rise to a set of reactions generally termed as interesterification. Immobilized lipases have been investigated for both these processes.

Lipases possess a variety of industrial potentials starting from use in detergents; leather treatment controlled hydrolysis of milk fat for acceleration of cheese ripening; hydrolysis, glycerolysis and alcoholysis of bulk fats and oils; production of optically pure compounds, flavours, etc. Lipases are spontaneously soluble in aqueous phase but their natural substrates (lipids) are not. Although use of proper organic solvents as an emulsifier helps in overcoming the problem of intimate contact between the substrate and enzyme, the practical use of lipases in such psuedohomogeneous reactions poses technological difficulties. Varieties of approaches to solve these, using immobilized lipases, have recently been reviewed.

Significant research has also been carried out on the immobilization and use of glucoamylase. This is an example of an immobilized enzyme that probably is not competitive with the free enzyme and hence has not found large-scale industrial application. This is mainly because soluble enzyme is cheap and has been used for over two decades in a very optimized process without technical problems. Immobilization has also not found to significantly enhance the thermostability of amylase. Immobilized renin or other proteases might allow for the continuous coagulation of milk for cheese manufacture. One of the major limitations in the use of enzymes which act on macromolecular substrates or particulate or colloidal substrates like starch or cellulose pectin or proteins has been the low retention of their realistic activities with natural substrates due to the steric hindrance.

Efforts have been made to minimize these problems by attaching enzymes through spacer arms. In this direction, application of tris (hydroxymethyl) phosphine as a coupling agentmay have future potentials for the immobilization of enzymes which act on macromolecular substrates. Other problem, when particulate materials are used as the substrates for an enzyme, is difficulty in the separation of the immobilized enzyme from the final mixture. Efforts have been made in this direction to magnetize the bicatalyst either by directly binding the enzyme on magnetic materials (magnetite or stainless steel powder) or by co-entrapping magnetic material so that they can be recovered using an external magnet. Magnetized biocatalysts also help in the fabrication of magnetofluidized bed reactor.

A variety of biologically active peptides are gaining importance in various fields including in pharmaceuti-cal industries and in food industries as sweeteners, flavourings, antioxidants and nutritional supplements. Proteases have emerged over the last two decades as powerful catalysts for the synthesis and modification of peptides. The field of immobilized proteases may have a future role in this area. One of the important large scale applications will be in the synthesis of peptide sweetener using immobilized enzymes like the thermolysin. Proteolytic enzymes, such as subtilisin, *a*-chymotrypsin, papain, ficin or bromelain, which have been immobilized by covalent binding, adsorption or cross-linking to polymeric supports are used (Bayer AG) to resolve A *N*-acyl-DL-phenylglycine ester racemate, yielding *N*-acyl-D-esters or *N*-acyl-D-amides and *N*-acyl-L-acids. Immobilized aminopeptidases have been used to separate DL-phenylgycinamide racemates. SNAM-Progetti SpA-UK have used the immobilized hydropyrimidine hydrolase to prepare D-carmamyl amino acids and the corresponding D-amino acids from various substituted hydantoins.

Enzyme Immobilization in Biodegradable Polymers for Biomedical Applications

During the last three decades, enzymology and enzyme technology have progressed considerably and, as a result, there are many examples of industrial applications where enzymes, in the native or immobilized form, are being used. These include food industry, materials processing, textiles, detergents, biochemical and chemical industries, biotechnology, and pharmaceutical uses. The overall impact of enzymes on industrial applications is, however, still quite limited due to their relative instability under operational conditions, which may involve high temperatures, organic solvents, and exposure to other denaturants. Various approaches, including, among others, addition of additives, chemical modification, protein engineering, and enzyme immobilization, have been assessed for their ability to increase the stability of enzymes towards heat or denaturants.

The use of enzymes in medical applications has been less extensive as those for other types of industrial applications. For example, pancreatic enzymes have been in use since the nineteenth century for the treatment of digestive disorders. At present, the most successful applications of enzymes in medicine are extracellular, such as topical uses, removal of toxic substances, and the treatment of life-threatening disorders within the blood circulation. The production of therapeutic enzymes has progressed, but the costs of enzyme production, isolation, and purification are still too high to make them available for clinical applications. Furthermore, the ability to store unstable enzymes for long periods of time is also a limitation for their more widespread use. Most applications in the biomedical field are still in the state of basic studies rather than definite applications, owing to the absence of the necessary information on toxicology, hemolysis, allergenicity, immunological reactions, and chemical stability of the system *in vivo*.

The importance of using enzymes in medical applications and, in particular, the use of immobilized enzymes. For that, some aspects of enzyme immobilization technology, including the traditional physical and chemical methods and new immobilization methodologies, based on biological and genetic engineering approaches, will be reviewed. Several examples will be given of immobilized enzymes in various support materials and using different immobilization strategies, according to the objective of their application in the biomedical field.

ENZYMES IN MEDICINE

Since the mid-1950s there has been a considerable increase in both measurement of enzyme activities and the use of purified enzymes in clinical practice. In recent years, many enzymes have been isolated and purified, and this made it possible to use enzymes to determine the concentration of substrates and products of clinical importance. A further development, arising from the increased availability of purified enzymes, has been targeted to enzyme therapy.

Clinical Diagnosis

The measurement of enzyme activities in serum is of major importance as an aid in diagnosis, being used as means of monitoring progress after therapy, recovery after surgery, and detection of transplant rejection. Urine can be also analysed for determination of enzyme activities since the detection of certain enzymes in urine may indicate kidney damage or failure. On the other hand, the concentration of certain metabolites in serum or urine may be determined by enzymatic methods. The method consists of using an enzyme to transform a metabolite into its product and then estimate the amount of transformed substrate.

The use of enzymatic methods presents several advantages, such as the high specificity of the enzyme to estimate the concentration of the metabolite in the presence of other substances, avoiding the need of purification steps prior to chemical analysis. In addition, enzymatic reactions are performed at mild conditions, allowing the analysis of labile compounds that would be degraded by harsher chemical methods. The cost of purified enzymes may be, however, too high to support routine analysis, but the use of immobilized enzymes allows for enzyme reuse and the application of immobilized enzymes for diagnostic assays and as biosensors. The determination of serum metabolites in serum by enzymatic methods includes a wide range of substances such as glucose, uric acid, urea, cholesterol, cholesterol esters, triglycerides, and creatine, among others.

Enzyme Therapy

Many inborn metabolic disorders are associated with the absence of activity of one particular enzyme normally found in the body. Of the 1250 autosomal recessive human genetic diseases, over 200 involve errors in metabolism that result from specific known enzyme deficiencies. The initial identification of the disease may be difficult to determine and normally requires a tissue biopsy. For some genetic diseases, DNA probes are now available that can be used on small amounts of blood, cells, or amniotic fluid. Albinism, for example, is often caused by the absence of tyrosinase, an enzyme essential for the production of cellular pigments. Tyrosinase is a coppercontaining enzyme that catalyzes the first two rate-limiting steps in the melanin biosynthetic pathway, the oxidation of tyrosine to dopa and the subsequent dehydrogenation of dopa to dopaquinone.

The human tyrosinase gene, encoding 529 amino acids, consists of five exons spanning more than 50 Kb of DNA in chromosomes. When homozygous mutations of the tyrosine gene result in complete absence of melanogenic activity, such a patient, categorized as tyrosinasenegative oculocutaneous albinism, will never develop any melanin pigment in the skin, hair, and eyes throughout his or her life. Some inborn errors in metabolism are relatively harmless, *e.g.*, albinism or alkaptonuria, but others must be detected early if the defect is to be circumvented. This is the case of phenylketonuria, where the enzyme phenylalanine 4-monooxygenase (enzyme that converts phenylalanine into tyrosine) is missing.

Phenylketonuria results in the accumulation of phenylalanine, which may cause mental retardation. Patients with phenylalanine 4-monooxygenase deficiency must follow a phenylalanine-free diet in order to avoid the accumulation of deleterious effects. Phenylalanine is, however, an amino acid essential to maintain growth and protein turnover, and thus it must be supplied in a minimal amount required to maintain normal metabolism. Such a dietary scheme is normally carried out by lowering the amount of protein consumed, but this may cause a deficiency in other essential amino acids. A possible alternative to this therapy is to replace the missing enzyme. It may be difficult to find, however, an enzyme with the same function from a human source since the direct administration of enzymes from other sources into the body would cause an adverse immunological response. A possible approach to circumvent this problem involves the isolation of the enzyme within a microcapsule, fibre, or gel, which will protect the enzyme from proteolysis and avoid the undesirable immunological response.

The pharmacological properties of enzymes have been employed to replace enzymes that are missing or defective as a consequence of an inherited disease or malfunction of an organ where they are normally synthesized or to accomplish a certain biological effect that is dependent on the catalytic activity of the enzyme.

Therefore, depending on the treatment, the administration of enzymes as therapeutic agents can be subdivided into two categories:

- The topical application of an enzyme as an extracellular agent
- The intracellular applications of enzymes to treat metabolic deficiency and related disease.

The main areas where enzyme therapy has been applied are the degradation of necrotic tissue by the use of proteolytic enzymes, removal of toxic compounds from the blood, treatment of genetic deficiency diseases and cancer, and treatment of pancreatic insufficiency. Enzymes may be administered either intra- or extracorporeally, depending on the objective. If the enzyme is to be used for the removal or transformation of a substance present in the blood (*e.g.*, toxic metabolite or a blood clot), then it is only necessary for the enzyme to be present in the blood and not necessary for the enzyme to enter the intracellular compartments.

This type of application may be either intra- or extracorporeal using a bypass as in kidney dialysis. For intracellular therapy, it is necessary for the enzyme to be taken up by the appropriate target cells. Although the attempts made with enzyme therapy in clinical trials have so far had limited success, it is reasonable to assume that the delivery of enzymes would constitute a feasible approach for the treatment of certain diseases in the near future.

ENZYME IMMOBILIZATION TECHNOLOGY

The term *immobilized enzyme* was adopted in 1971 at the first Enzyme Engineering Conference. It describes enzymes physically confined at or localized in a certain region of space with retention of catalytic activity and which can be used repeatedly and continuously. The immobilization of biocatalysts (not only enzymes but also other bioactive molecules such as growth factors and hormones, cellular organelles, microbial cells, and plant and animal cells) is attracting worldwide attention in biotechnology applications. In general, immobilized biocatalysts are more stable and easier to handle compared with their free counterparts.

At present, applications of immobilized biocatalysts include the production of useful compounds by stereospecific or regiospecific bioconversion, the production of energy by biological processes, the selective treatment of specific pollutants to solve environmental problems, continuous analyses of compounds with a high sensitivity and specificity, and medical uses such as new types of drugs for enzyme therapy or artificial organs. Immobilized enzymes are already being used in medical applications for clinical diagnosis and also for intra- and extracorporeal enzyme therapy. Applications in clinical analysis are mainly related to biosensors, which have been used to detect the presence of various organic compounds for many years. For example, glucose oxidase and catalase have been used to measure blood glucose concentration, and cholesterol oxidase and cholesterol esterase to determine cholesterol levels. In addition, enzymes can be immobilized on different prosthetic devices or used extracorporeally (*e.g.*, artificial heart, artificial lung, artificial kidney, equipment for hemodialysis and specific blood purification) as surface modifiers in order to increase the biocompatibility of these devices and to prevent blood clotting.

Methods for Immobilizing Enzymes in Polymeric Carriers

Various methods have been developed for the immobilization of biocatalysts, which are being used extensively today. A wide range of support materials has also been employed for enzyme immobilization. The support type can be classified according to their chemical composition, such as organic or inorganic supports, and the former can be further classified into natural or synthetic matrices. Immobilization techniques can be divided into different categories: physical, chemical, enzymatic, and genetic engineering methods.

Adsorption

The adsorption of an enzyme onto a support or film material is the simplest method of obtaining an immobilized enzyme. Basically, the enzyme is attached to the support material by noncovalent linkages and does not require any preactivation step of the support. The interactions formed between the enzyme and the support material will be dependent on the existing surface chemistry of the support and on the type of amino acids exposed at the surface of the enzyme molecule. Enzyme immobilization by adsorption involves, normally, weak interactions between the support and the enzyme such as ionic or hydrophobic interactions, hydrogen bonding, and van der Waals forces. Most of the support materials available have sufficient surface-charge properties suitable for immobilization by adsorption. They include inorganic carriers (ceramic, alumina, activated carbon, kaolinite, bentonite, porous glass), organic synthetic carriers (nylon, polystyrene), and natural organic carriers (chitosan, dextran, gelatin, cellulose, starch).

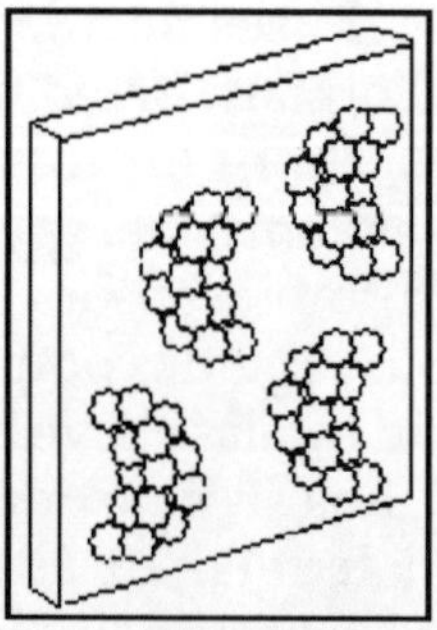

Fig. Biocatalysts Bound to a Carrier by Adsorption.

The method consists of simply mixing an aqueous solution of enzyme with the support material for a period of time, after which the excess enzyme is washed away from the immobilized enzyme on the support. The procedure requires strict control of the pH and ionic strength, because these can alter the charges of the enzyme and the support and therefore affect the level of adsorption. A simple shift in pH can cancel ionic interactions and promote the release of the enzyme from the support.

The main advantages of adsorption are the method simplicity, the little effect on the conformation/activity of the biocatalyst, and the possibility of regenerating inactive enzyme by addition of fresh enzyme. The main disadvantage is the desorption of the biocatalyst from the support due to the weak interactions established. The enzyme desorption can easily occur by changes in the environment medium such as pH, temperature, solvent, and ionic strength or in the case of extended reactions.

Ionic Binding

Immobilization via ionic binding is based, mainly, on ionic binding of enzyme molecules or active molecule to solid supports containing ionic charges. In this method, the amount of enzyme bound to the carrier and the activity after immobilization depends on the nature of the carrier. How the enzyme is bound to the carrier. In some cases, physical adsorption may also take place. The main difference between ionic binding and physical adsorption is the strength of the interaction, which is much stronger for ionic binding, although less strong than covalent binding. The preparation of immobilized enzymes using ionic binding is based on the same procedure as described for physical adsorption.

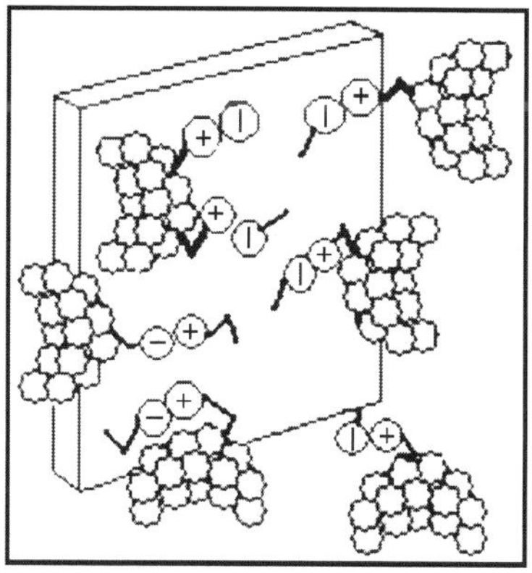

Fig. Biocatalysts Bound to a Carrier by Ionic Binding.

The ionic nature of the binding forces between the enzyme and the support also depends on pH variations, support charge, enzyme concentrations, and temperature. The supports used for ionic binding may be based on polysaccharide derivatives (*e.g.*, diethylamino-ethylcellulose, dextran, chitosan, carboxymethyl-cellulose), synthetic polymers (*e.g.*, polystyrene derivatives, polyethylene vinylalcohol), and inorganic materials (*e.g.*, ambertite, alumina, silicates, bentonite, sepiolite, silica gel, etc.). The immobilization by ionic binding has the advantage that changes in the enzyme conformation only occur in a small extent, resulting in immobilized enzymes with high enzymatic activities. The main disadvantage is the possible interference of other ions, and special attention should be paid in maintaining the correct ionic strength and pH conditions in order to prevent their easy detachment.

Covalent Binding by Chemical Coupling

The covalent binding method is based on the binding of enzymes, or other active molecules, to a support or matrix by means of covalent bonds. The bond is normally formed between a functional group present on the support surface and amino acid residues on the surface of the enzyme. Those which are most often involved in covalent binding are the amino (NH_2) group of lysine or arginine, carboxyl (CO_2H) group of aspartic acid or glutamic acid, hydroxyl (OH) group of serine or threonine, and sulphydryl (SH) group of cysteine. There are many reaction procedures for joining an enzyme to a material with a covalent bond (diazotation, amino bond, Schiff's base formation, amidation reactions, thiol-disulfide, peptide bond, and alkylation reactions). The connection between the support and the biocatalyst can be achieved either by direct linkage between the components or via an intercalated link of different length, the so-called spacer or harm. The advantage of using a spacer molecule is that it gives a greater degree of mobility to the coupled biocatalysts so that its activity can, under certain circumstances, be higher than if it is bound directly to the support. It is important to choose a method that will not involve the reaction with the amino acids present in the active site, since this could inactivate the enzyme. Basically, two steps are involved in the covalent binding of enzymes to a support material.

First, functional groups on the support material are activated by specific reagents (*e.g.*, cyanogen bromide, carbodiimide, aminoalkylethoxysilane, isothiocyanate, and epichlorohydrin, etc.). A large range of support materials is available for covalent binding, and this extensive range reflects the fact that no ideal matrix exists. Therefore, the advantages and disadvantages of a given matrix must be taken into account when considering the appropriate procedure for a given enzyme immobilization. Immobilization of enzymes through covalent attachment has also been demonstrated to induce higher resistance to temperature, denaturants, and organic solvents in several cases.

The extent of these improvements may depend on other conditions of the system, *e.g.*, the nature of the enzyme, type of support, and the method of immobilization. Many factors may influence the selection of the support, and some of the more important are its cost and availability, the binding capacity (amount of enzyme bound per given weight of matrix), hydrophilicity (the ability to incorporate water into the matrix and stability of matrix), structural rigidity, and durability during applications.

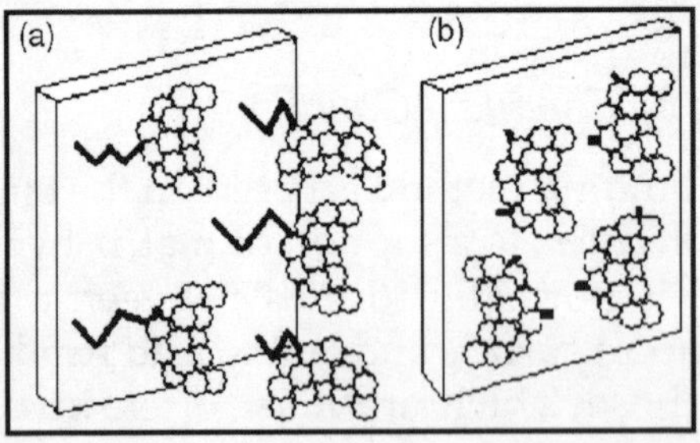

Fig. Covalent Bond Between the Biocatalysts and a Carrier with (a) and without Spacer (b).

Natural polymers, which are very hydrophilic, are popular support materials for enzyme immobilization since the residues in these polymers contain hydroxyl groups, which are ideal functional groups for participating in covalent bonds. A frequently encountered disadvantage of immobilization by covalent binding is that it places great stress on the enzyme. The necessary harshness of the immobilization procedure nearly always leads to considerable changes in conformation and a resultant loss of catalytic activity.

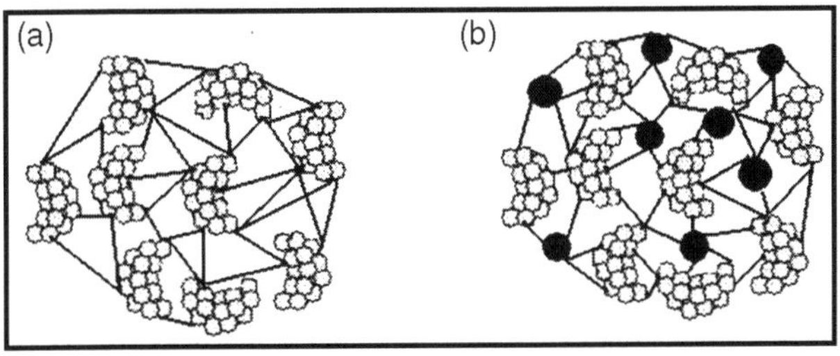

Fig. Biocatalysts Immobilized by Means of Crosslinking (a) and Co-Crosslinking with Inert Molecules Incorporated (b).

Crosslinking

The crosslinking method is based on the formation of covalent bonds between the enzyme or active molecules, by means of bi- or multifunctional reagents. The individual biocatalytic units (enzymes, organelles, whole cells) are joined to one another with the help of bi- or multifunctional reagents (*e.g.*, glutardialdehyde, glutaraldehyde, glyoxal, diisocyanates, hexamethylene diisocyanate, toluene diisocyanate, etc.). Enzyme crosslinking involves normally the amino groups of the lysine but, in occasional cases, the sulfhydryl groups of cysteine, phenolic OH groups of tyrosine, or the imidazol group of histidine can also be used for binding. How the biocatalysts can be linked by a simple crosslinking process and also by co-crosslinking, in which inert molecules are incorporated in the high-polymer network in order to improve the mechanical and enzymatic immobilized preparation. The advantages and disadvantages of a given matrix must be taken into account when considering the appropriate procedure for a given enzyme immobilization. One advantage is the simplicity of the process.

The main disadvantages are the fragility of the particles produced in some cases and diffusion limitations. Since crosslinking and co-crosslinking usually involve covalent bonds, immobilized biocatalysts in this way frequently undergo changes in the conformation with a resultant loss of activity. The isomerization of glucose process is a very important example of the industrial application using biocatalysts crosslinked with glutaraldehyde. Some of the immobilized preparations used in these large-scale processes are produced simply by glutaraldehyde treatment of bacterial cell masses that have formed fine particles.

Entrapment and Encapsulation

The entrapment method for immobilization consists of the physical trapping of the active components into a film, gel, fibre, coating, or microencapsulation. This method can be achieved by mixing an enzyme or active molecule with a polymer and then crosslinking the polymer to form a lattice structure that traps the enzyme. Microencapsulated enzymes are formed by enclosing enzymes solution within spherical semipermeable polymer membranes with controlled porosity.

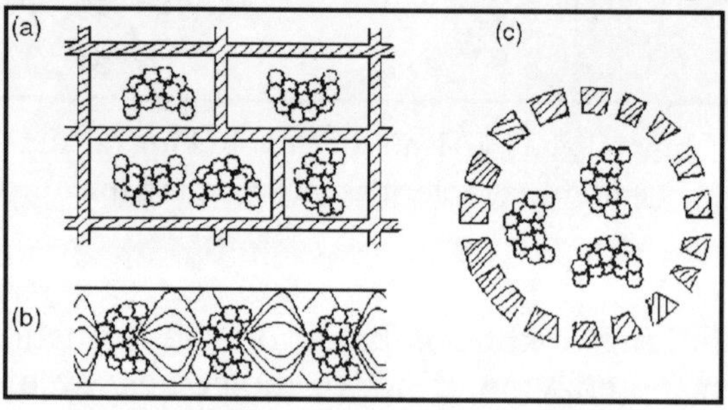

Fig. Enzyme Encapsulation in a Matrix (a), Fibre (b), or Capsule (c).

While the encapsulation of dyes, drugs, and other chemicals has been known for some time, it was not until the mid-1960s that such a method was first applied to enzymes. Since that first report, a number of other enzymes have been successfully immobilized via microencapsulation, using a number of different materials and methods to prepare the microcapsules. The advantages of this immobilization method are the extremely large surface area between the substrate and the enzyme, within a relatively small volume, and the real possibility of simultaneous immobilization. The major disadvantages of this method include the occasional inactivation of enzyme during microencapsulation and the high enzyme concentration required. In addition, to retain the enzyme, the pore size needs to be very low and these systems tend to be very diffusion limited.

Protein Fusion to Affinity Ligands and Enzymatic Conjugation

As described before, there are many methods for protein immobilization, but some of them require chemical modification of the matrix, which may result in material degradation, especially when biodegradable polymers are used. In addition, these modifications, necessary to attach the enzyme to the matrix, often result in the loss of enzyme activity as well as the inclusion of toxic organic compounds, which have to be removed before the system can be used in biomedical applications. In this type of application, the efficacy of immobilized biomolecules for stimulating specific cell responses (*e.g.*, proliferation or differentiation) depends on the mode by which these modulators are presented to the target cells. In these cases, it is important to ensure the correct orientation and full bioactivity of the molecules when they are immobilized. Covalent binding by chemical coupling, again, might hinder ligand–receptor interaction or prevent receptor dimerization and capping on target cells.

In nature, there are certain protein molecules, such as lectins, avidin, immunoglobulin G (IgG) binding domains of protein G and protein A, and carbohydrate binding modules (present in many polysaccharide-degrading enzymes), that bind with high affinity and specifically to certain molecules or solid surfaces. These binding domains may be used as affinity tags for immobilizing proteins to affinity adsorbents.

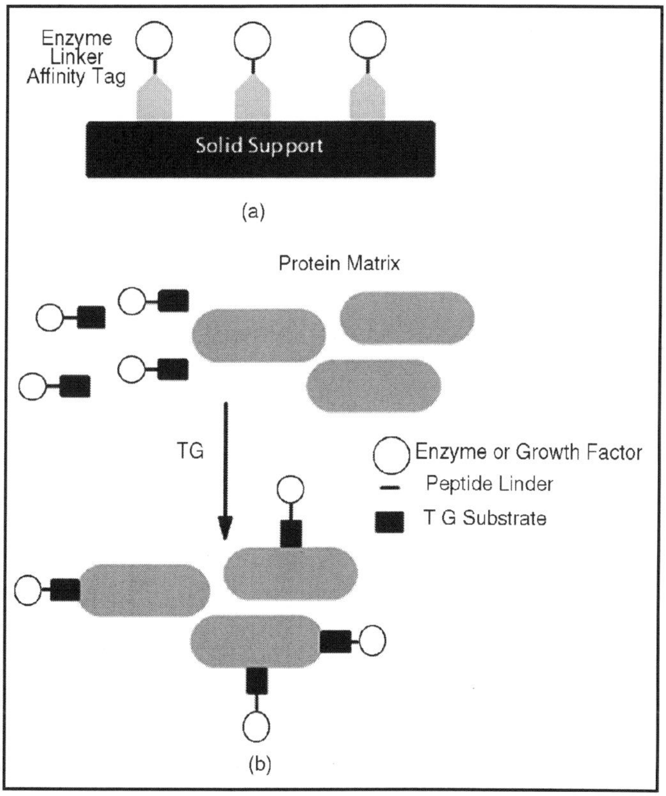

Fig. Enzyme Immobilization to Solid Matrices Via Protein Fusion to an Affinity Ligand (a) and Enzymatic Conjugation Catalyzed by Transglutaminase (b).

In this technique, DNA encoding a polypeptide affinity tag is fused to the gene of interest, and the expression of the gene results in a fusion protein. Such a fusion protein could be immobilized by the specific binding of the affinity tag to an affinity adsorbent. With this method, the conformational changes in the protein upon immobilization are minimal, and the immobilized biomolecule could retain high activity. In addition, since fusion proteins are specifically immobilized on the support materials, these methods normally allow the immobilization of high densities of ligands and can also simplify the immobilization procedure.

The use of genetic engineering techniques to construct chimeric proteins, containing a functional domain displaying bioactivity together with an affinity domain, has proven to be a very useful approach for immobilizing biomolecules on solid materials. Several proteins, including enzymes, antibodies, cytokines, and streptavidin, have been immobilized on the surface of cellulosic matrices by fusing genetically these proteins to cellulose-binding domains. By using protein fusion techniques, bifunctional proteins can be prepared without changing their activity and binding properties.

This approach has found various applications in biotechnology, diagnostics, and medicine for the purification and immobilization of biologically active proteins. It can be used, for instance, to promote the attachment of several mammalian cells to different surfaces by fusing a variety of peptides, growth factors, and cytokines to a specific binding domain, and this might be useful for the activation and growth of progenitor cells in culture. The binding affinity between streptavidin and biotin is among the strongest noncovalent bonds known to exist (*KD*= 10–15*M*). Therefore, the high affinity coupling of the biotin–avidin system has been used to immobilize different biomolecules on the surface of biomaterials and biosensors. It consists of using avidin as a bridge between the biotinylated surface and biotinylated short ligand molecules.

The immobilization of biomolecules on protein matrices may also be achieved via an enzymatic reaction catalyzed by transglutaminase (TG) enzyme. TG catalyzes the acyl transfer reaction between the carboximine group of a peptide-bound glutaminyl residue and a primary amino group of various protein substrates. The result of this reaction is the formation of an irreversible crosslinked, insoluble supramolecular structure. In addition, TG can be used to bind glutamine containing peptides or polypeptides to NH_2 surfaces, and this methodology is schematically represented. This approach was used by Sakiyama *et al.* to incorporate heparinbinding peptides and to design a growth factor delivery system. In the latter work, they developed a growth factor fusion protein, containing a bi-domain peptide consisting of a β-nerve growth factor and an exogenous factor XIIIa (substrate for enzymatic crosslinking). The fusion protein was covalently immobilized within a three-dimensional cell in-growth matrix based on fibrin using the transglutaminase activity of factor XIIIa.

ENZYME IMMOBILIZATION IN BIOMEDICAL APPLICATIONS

Biomaterials can be combined with biomolecules, such as enzymes and growth factors, to yield biologically functional systems. There is a wide and diverse range of materials and methods available for enzyme immobilization on or within the biomaterial. Furthermore, the criteria for selecting the immobilization methods should also take into account that the immobilized enzyme should retain an acceptable level of activity over a certain period of time in terms of economic or clinical aspects.

The methods used for the administration of immobilized enzymes may be divided into two principal groups: immobilized enzymes that are intended for prolonged circulation and enzymes that must be necessarily present in different tissues and organs of the body. In the second case, the immobilized enzyme is intended for local deposition during the treatment of discrete lesions (*e.g.*, thrombi, tumors, atherosclerotic injuries) or of the individual organs.

Biologically Functional Surfaces

Biomaterials, especially when used in tissue engineering applications, must have the capacity to induce tissue regeneration/ repair in order to achieve a more rapid recovery of the defect. At present, the existing scaffolds are not satisfactory in achieving rapid and full recovery of the defect. The attachment of cells to biomaterials, and their subsequent spreading, are mediated by extracellular matrix (ECM) glycoproteins such as fibronectin, collagen, etc. ECM glycoproteins contain short sequences with cell attachment properties, which interact with the integrin family of cell surface receptors. The peptide sequence Arg-Gly-Asp (RGD) has been identified as being capable of interacting with cell surface receptors.

Thus, the immobilization of biologically active molecules on the surface of biomaterials for presenting effectors to target cells or to induce a particular effect is of great interest, since the immobilization of active agents presents the advantage of providing a continuous and localized stimulus for cell proliferation. Unlike nonimmobilized active agents, which are often consumed by cells, immobilized biomolecules remain bound to a substrate that is not consumed by cells and thus remain available to stimulate growth of additional cells.

This is particularly useful in perfusion cultures in which growth medium is continuously added and removed to allow long-term cell proliferation. Biodegradable polymers have been used as scaffolding materials for various tissue engineering applications because they can provide the support on which cells and tissues can adhere, but they can also guide and regulate the proliferation and activities of the adhered cells. However, the intrinsic hydrophobic property of some of these polymers restricts their applications as cell colonizing materials.

Many methods have been used to modify the properties of polymer surface, such as plasma treatment-induced grafting polymerization, ozone oxidation, and immobilization of enzymes, and special biologically active agents have been used to introduce reactive groups onto polymeric surfaces. Various strategies have been developed to incorporate bioactive agents on the surface of biomaterials for controlling cell and tissue responses.

The immobilization process can be involved, enriching surfaces for enhancing the cellular adhesion. Biomolecules such as enzymes, antibodies, antigens, peptides, or drugs have been immobilized on or within polymeric systems. An example of an adhesive protein is fibronectin, which is able to promote cellular adhesion through binding to integrin receptors, and this interaction has also been shown to play a role in cell growth, differentiation, and overall regulation of cell function.

Hern and Hubbell showed that the incorporation of the adhesion peptide RGD into a nonadhesive hydrogel proved to be useful for tissue resurfacing. There are numerous other adhesion peptides for targeting particularly desirable cell types and to modulate biological responses. Urokinase has been widely used for the clinical treatment of thrombogenetic disease and hemorrhoidal disease.

Artificial organ materials, on which urokinase was immobilized for its fibrinolytic activity, have been developed for blood-compatible materials. Immobilized urokinase by encapsulation in poly(2-hydroxyethyl methacrylate) and Introduced urokinase on the surface of the polytetrafluoroethylene using plasma modification technique by covalent bond. Another example of immobilized urokinase application was reported by Kato and coworkers, who had used urokinase immobilized in a Teflon catheter for treatment of thrombosis. Most of the studies found in the literature, regarding the surface functionalization of biomaterials with biological molecules, include the incorporation of adhesion and differentiation factors. The same approach may be used to incorporate specific enzymes able to regulate a number of cell functions. For instance, it is known that mitogen activated protein (MAP) kinase, upon activation by dual phosphorylation at threonine and tyrosine residues, is able to activate downstream targets that have been implicated in controlling gene expression, cell differentiation, and proliferation. This enzyme may be immobilized on the surface of biomaterials to control cell response, but other possibilities using different enzymes remain unexplored.

Enzyme Delivery

Although purified enzymes are now available for some enzyme deficiency diseases there are many problems in delivering the enzyme to the required site under such conditions that it will remain stable and active for a reasonable time. Normally, quite large amounts of enzyme are necessary with high level of purity and in a nonimmunogenic form. In addition, many enzymes when administered are inactivated or degraded fairly rapidly. The delivery of therapeutic molecules requires, therefore, efficient strategies to have a precise control on their release profile according to specific locations.

It might be possible to control the release of such molecules by creating delivery systems sensitive to changes in pH, temperature, or salt concentration or to the feedback provided by cells. The concept of enzyme-activated drugs in therapy is scientifically, as well as clinically, attractive, as it allows the chemist and enzymologist full intellectual rein in designing interlinked systems. As therapeutic drugs, enzymes possess several attributes such as high specificity towards substrate, high solubility for preparing liquid formulations, and optimum activity under physiological conditions. The administration of enzymes, in cases of enzyme deficiency and inborn errors of metabolism and in the treatment of certain types of cancer, appears to offer a successful form of therapy.

Cancer therapy based on the delivery of enzymes to tumor sites has advanced in several directions since antibody-directed enzyme/ prodrug therapy was first described. Nanospheres, nanocapsules, liposomes, micelles, and other nanoparticulates are frequently referred to as carriers for delivery of therapeutic and diagnostics agents. Asparagine is an essential amino acid for certain types of leukemias that lack asparagines synthetase activity. The activity of *L*-asparaginase is to degrade asparagine into aspartate and ammonia. Therefore, asparaginase has been of interest to biochemists and clinicians as a possible cancer therapeutic agent. Some success has been achieved in administering asparaginase in capsules made of nylon and polyurea to mice and rats.

Although asparaginase has been found to be effective in the treatment of some patients, it may have several serious side effects. Relatively high concentrations of the enzyme are needed for it to be clinically effective. These levels cause a wide range of toxic effects on several organs including the liver, pancreas, kidneys, and brain. The enzyme may also be recognized as foreign by the body and potentially severe immunogenic responses will be stimulated, resulting in hypersensitivity reactions.

To overcome these problems, immobilized enzyme derivatives have been prepared on various supports for extracorporeal treatment. Enzymes may be also used in cancer therapy as prodrug activators. This therapy consists basically in using a drug that has been chemically modified so that it remains inactive until specifically activated by an enzyme at the target site. Another example of an anticancer enzyme is hyaluronidase. Hyaluronidase is a globular enzyme of endoglycosidase action, which can depolymerize hyaluronic acid in the organism, decreasing its viscosity and increasing tissue permeability.

Hyaluronidase has been utilized extensively as an adjunct in anticancer chemotherapy regimens, suggesting that hyaluronidase has intrinsic anticancer properties against tumor growth. *In vitro* studies in tissue culture with tumor spheroids and *in vivo* tests using animal models demonstrated the beneficial effect of hyaluronidase for the penetration of drugs into tumor tissue. Later, in a prospective clinical trial, hyaluronidase significantly improved the outcomes of patients with bladder carcinoma if the enzyme was administrated topically together with mitomycin.

Hyaluronidase is also used for local application (subcutaneous injections) during treatment of joint disease, in dermatology, and in ophthalmology. Hyaluronidase had been used in ophthalmology with the aim of formation of a thinner scar and to prevent necrosis after paravataes with zytostatics. Trypsin has also been used to remove dead tissue from wounds, burns, and ulcers to speed the growth of new tissue and skin grafts, as well as to inhibit the growth of some contaminant organisms. The inappropriate activation of trypsinogen within the pancreas leads to development of pancreatitis. Once trypsin is activated, it is capable of activating many other digestive pro-enzymes.

These activated pancreatic enzymes further enhance the auto-digestion of the pancreas. Many materials, such as nylon, polysulfone, glycidyl methacrylate, chitosan, cellulose, and cellulose derivatives, have been used for trypsin immobilization. Enzyme therapy has also been tested for pancreatic insufficiency and cystic fibrosis. Pancreatic insufficiency can be alleviated by administrating orally enteric-coated microspheres containing lipase, amylase, and proteases. A special polymer coating protects the enzymes at low pH, such as in the stomach, and then releases them in the intestine at physiological pH. Lysozyme is a good example of an enzyme that catalyzes chemical reactions in the cell.

Lysozyme acts to kill bacteria by cleaving the covalent bond between the alternating polysaccharides that compose peptidoglycan in bacterial cell walls. The human salivary defence proteins and lysozyme are known to exert a wide antimicrobial activity against a number of bacterial, viral, and fungal pathogens *in vitro.* Therefore, these proteins, alone or in combinations, have been incorporated as preservatives in foods and pharmaceuticals as well as in oral health care products to restore saliva's own antimicrobial capacity in patients with dry mouth. These antimicrobials used in oral health care products, such as dentifrices, mouth rinses, moisturizing gels, and chewing gums, have been purified from bovine colostrum.

Other studies had been reported with lysozyme bound to chitosan, silica gel by means of physical adsorption, crosslinking to a polystyrene divinylbenzene matrix by the formation of ionic bindings, and by covalent attachment to nonporous glass beads. Harada and Kataoka described lysozyme immobilized into poly(ethylene glycol)–poly(aspartic acid) micelle. Lysozyme was selected as a model protein to incorporate into the micelle because it has a high isoelectric point (pI = 11), is positively charged over a wide range of pH, and has practical usage in drug delivery application as a lytic enzyme. Chen and Chen prepared immobilized lysozyme by carbodiimide method to form amide bonds with an enteric coating polymer (hydroxypropyl methyl-cellulose acetate succinate [AS-L]) as the carrier, which shows reversibly soluble-insoluble characteristics with pH changes.

The glucose-6-phosphate dehydrogenase enzyme catalyzes the oxidation of glucose-6-phosphate to 6-phosphogluconate while concomitantly reducing the oxidized form of nicotinamide adenine dinucleotide phosphate (NADP+) to nicotinamide adenine dinucleotide phosphate (NADPH). NADPH, a required cofactor in many biosynthetic reactions, maintains glutathione in its reduced form. Reduced glutathione acts as a scavenger for dangerous oxidative metabolites in the cell. With the help of the enzyme glutathione peroxidase, reduced glutathione also converts harmful hydrogen peroxide to water. Red blood cells rely heavily on glucose-6-phosphate dehydrogenase activity because it is the only source of NADPH that protects the cells against oxidative stresses. People deficient in glucose-6-phosphate dehydrogenase are not prescribed, therefore, with oxidative drugs because their red blood cells undergo rapid hemolysis under this stress.

In Greece, glucose-6-phosphate dehydrogenase deficiency is the main cause of severe neonatal jaundice. The deficiency of this enzyme affects all races; the highest prevalence is among persons of African, Asian, or Mediterranean descent. Study of immobilized glucose-6-phosphate dehydrogenase has been reported. Kotorman *et al.* immobilized glucose-6-phosphate dehydrogenase from yeast on polyacrylamide beads possessing carboxylic functional groups activated by a watersoluble carbodiimide. They verified highest operational stability of immobilized glucose-6-phosphate dehydrogenase. In relation to the use of immobilized enzymes with therapeutic purposes, catalase is one of the most interesting because it is employed to accelerate healing as well as to correct hereditary deficiencies and, in combination with hydrogen peroxide, as an antiseptic against anaerobes. The catalase enzyme has the ability to decompose hydrogen peroxide into oxygen and water, playing a central role in controlling the hydrogen peroxide concentration in human cells. More than 98% of blood catalase is localized in erythtocytes.

These cells, with their high catalase level, provide a general protection against the toxic concentration of this small hydrogen peroxide molecule. The deficiency of catalase could cause acatalasemia. Several methods have been developed for the immobilization of catalase. Immobilization is often accompanied by changes in the enzymatic activity, optimum pH, affinity to the substrate, and stability. The extent of these changes depends on the enzyme, carrier support, and the immobilization conditions. The shift in the optimum pH, from acidic or alkaline to neutral pHs, may be useful for biomedical applications since it will allow the use of some enzymes (more active at low or high pHs) under more physiological conditions. The approach developed by Sakiyama-Elbert *et al.*, consisting in a cell-triggered growth factor delivery system, may also be used for the release of other important therapeutic molecules.

BIOREACTORS FOR EXTRACORPOREAL ENZYME THERAPY

Extracorporeal shunts have been proposed for the treatment of several clinical conditions. The most likely applications for enzymatic treatment are the removal of urea during kidney failure, removal of toxins (*e.g.*, paracetamol) during liver failure, or the reduction of key metabolites from the circulation to treat cancer. Urease is one of the most important enzymes in biomedical applications. Urease is an enzyme that catalyzes the hydrolysis of urea to form ammonia and carbon dioxide. Urea is one of the main metabolic end products, and the removal of its excess has been a major problem for patients suffering from renal failure. Hence, its immobilization by entrapment has been investigated by many workers for applications in biosensors and as artificial kidneys.

The most attention has been given to the development of enzyme reactors, where the urea would be removed and the dialysis fluid prepared for further use. The use of this enzyme is often limited due to its high cost, availability in small amounts, instability, and the limited possibility of feasible recovery of these biocatalysts from a reaction mixture. Numerous synthetic and natural polymeric supports have been used for urease immobilization, and their uses in medical and technical fields are well reported. The covalent bond of urease in different supports has been reported in many studies. Some commonly used supports are chitosan-poly(glycidil methacrylate), carboxymethylcellulose, polyurethane, sepharose-2B, polyacrylamide, ion exchange resins, copolymers of polyglycidylmethacrylate, calcium alginate beads, poly(vinyl alcohol) (PVA), hydroxyapatite, 2-dimethylaminoethylmethacrylate, poly(ethylene glycol dimethacrylate/2-hydroxy ethylene methacrylate) microbeads, poly(caprolactone)/starch, and poly(orthoesters).

L-asparaginase has been used for treating leukemias and disseminating cancers that require asparagines for growth, but this treatment presents several serious side effects. To overcome these problems, immobilized enzyme derivatives have been prepared on various supports for extracorporeal treatment. Blood can be passed over the immobilized enzyme, thus depleting the asparagine supply needed by the cancer cells. The enzyme does not come into direct contact with the organs to which it is toxic, and hypersensitivity reactions do not occur. With this type of treatment, however, the blood plasma must be first separated from the cells to minimize cell damage and then passed through a separate column containing the immobilized enzyme.

This process requires that the blood remain outside the body for relatively long periods of time, resulting in the denaturation and depletion of many plasma proteins. Some techniques have been developed to minimize this problem, such as the use of a porous hollow-fibre plasmapheresis device. With this system, the plasma can be separated from the whole blood and contact with immobilized enzyme in one passage, thus minimizing its time outside the body and then reducing the damage to the plasma proteins. Adsorption techniques have been used to immobilize asparaginase onto hollow fibres after first coating the fibres with albumin and then crosslinking the enzyme with glutaraldehyde.

Maciel and Minim also reported that the use of *L*-asparaginase covalently attached to nylon tubing may constitute a useful system to be used in clinical applications. Bilirubin oxidase is also an example of enzyme used in extracorporeal applications. All human newborns accumulate bilirubin to levels greater that those in adults, and 20% accumulate enough to stain their skin, resulting in jaundice. Bilirubin binds to cellular and mitochondrial membranes, causing cell death in a variety of tissues. Clinically, bilirubin toxicity may lead to mental retardation, cerebral palsy, deafness, seizures, or death. The most common treatments for jaundiced infants are phototherapy and exchange transfusion.

This technique presents serious problems such as hypoglycemia, hypocalcemia, acidosis, transmission of infectious, etc. Lavin *et al.* reported the use of a highly specific enzyme to remove bilirubin from the bloodstream using a small reactor (extracorporeal circuit) containing bilirubin oxidase covalently immobilized in agarose beads. These researchers obtained good results for the removal of the bilirubin in humans and in genetically jaundiced rats.

Heparinase, an enzyme that degrades heparin into small polysaccharides, has also been immobilized into an extracorporeal device (artificial kidney bioreactor) to eliminate the anticoagulant properties of heparin (used to prevent clotting in the device) before the blood returns to the patient.

Diagnostic Assays and Biosensors

Isolated or combined enzymes are being used in medicine as useful tools for clinical analysis. About 50 different enzymes are used in different aspects of clinical diagnoses, and for most of these, much higher levels of purity are required than for most industrial enzymes. Two of the major enzymes used are peroxidase from horseradish and alkaline phosphatase from beef intestinal mucosa, both being required for immunoassays. The enzymes may be used in test strips, ELISA, biosensors, and autoanalysers. The serum uric acid concentration is an important index for clinical diagnosis of gout, leukemia, toxemia of pregnancy, and severe renal impairment.

A number of enzymes are assayed in serum and urine for diagnostic purposes; the more frequently used ones are discussed below. Alkaline phosphatases (ALP) are a group of enzymes found primarily in the liver (isoenzyme ALP-1) and bone (isoenzyme ALP-2). There are also small amounts produced by cells lining the intestines (isoenzyme ALP-3), the placenta, and the kidney (in the proximal convoluted tubules). What is measured in the blood is the total amount of alkaline phosphatase released from these tissues into the blood. As the name implies, this enzyme works best at an alkaline pH (pH 10), and thus the enzyme itself is inactive in the blood. Alkaline phosphatase acts by splitting off phosphorus (an acidic mineral), creating an alkaline pH.

The primary importance of measuring alkaline phosphatase is to check the possibility of bone or liver diseases. Another application of immobilized enzymes is the development of improved sensing devices. Because of their high specificity for given substances, enzymes and monoclonal antibodies are particularly suitable for use as sensors. The membrane-covered electrode described by Clark in 1959 is the dominating sensor for the measurement of dissolved oxygen. Numerous modifications of the original concept have been developed. For instance, biosensors using enzymes have been used to detect the presence of various organic compounds, and recent developments have proven to be both rapid and highly selective. They have been used in important applications such as in clinical laboratories, fermentation processes, and pollution monitoring. Most of them have used a free or immobilized enzyme and an ion-sensitive electrode that measures indirectly (*e.g.*, by temperature or colour changes produced by an enzymatic reaction) the presence of a product whose formation is catalyzed by the enzyme.

The biosensors usually have immobilized biological molecules attached to the surface of a transducer that allows an electronic or optical signal to be converted into an appropriate signal. This type of biosensor could be used to measure glucose, sucrose, lactose, *L*-lactate, galactose, *L*-glutamate, *L*-glutamine, choline, ethanol, methanol, hydrogen peroxide, starch, uric acid, etc., by using specific enzymes. Glucose oxidase is normally used to assay glucose concentration. Glucose sensors are the biosensors that have attracted much interest in both research and applications fields. One particularly important medical application of improved biosensors could be in the treatment of diabetic patients for whom proper levels of insulin and glucose must be maintained. For instance, small implantable devices for sampling blood to determine the levels of glucose and regulate the delivery of insulin could be developed using this enzyme.

A great variety of immobilization methods (*e.g.*, encapsulation, entrapment) and transducers have been developed to construct the glucose sensors with better performance and practicability since the work of Clark and Lyons in 1962. Several materials have been used, such as polyethylene terephthalate (PET), polyacrylamide, *N*-isopolyacrylamide, sol-gel, poly(2-hydroxyethyl methacrylate), alginate, artificial resins, glass, etc. Immobilized alkaline phosphatase on a nanoporous nickel-titanium film for sensor applications. A number of other enzymes have been described with great potential for medical applications, including carboxypeptidases, collagenase, fibrinolysin, pepsin, streptokinase, subtilisin, thrombin, tissue plasminogen activator, amylase, galactosidase, glucoamylase, lactase (-galactosidase), pectinase, pancreatin, phospholipases, cholesterol esterase and other DNases, RNases, phosphatases, esterases, sulfatases, isomerases, glucose isomerase, superoxide dismutase, cholesterol esterease, creatine kinase, and penicillin acylase.

Advantages and Disadvantages of Immobilized Enzymes

The use of immobilized enzymes normally offers several advantages over free enzymes, such as increased stability, localization, and retention of the molecules at the material surface, which enables easier handling, repeated use, and decreased cost. Other important advantages of using therapeutic immobilized enzymes are the prolonged blood circulation lifetime without the loss of specific activity and the lower immunogenicity. This advantage is particularly important for delivering enzymes or other biomolecules and may constitute an alternative and suitable method for the enzyme replacement therapy.

However, some limitations have been attributed to the use of immobilized enzymes in biomedical applications, such as mass transfer resistances (substrate in and product out), adverse biological responses of enzyme support surfaces (*in vivo* or *ex vivo*), fouling by other biomolecules, greater potential for product inhibition, and sterilization difficulties.

Although the preparation of sterile immobilized enzyme systems may be complex, sterilization may be achieved by filtrating all the reagents and protein solutions through 0.2m filters and working under aseptic conditions. A very important issue regarding the use of enzymes or other products derived from biological or biotechnological processes in medical applications is to ensure that these therapeutic products do not contain any pyrogenic material, toxins, or infectious agents able to cause harmful effects.

For that, it is necessary to perform a complete examination of the products to test their safety in terms of local tolerance, toxicity, carcinogenicity, and immunogenicity, among other pharmacological safety tests. Taking into account the diversity in the range of products and the uncertainty about the regulatory status of some of them, it is necessary to design safety evaluation programmes to provide useful information to the responsible of clinical trials and to ensure patient safety.

Chapter 4

RNA Splicing

In molecular biology, splicing is a modification of an RNA after transcription, in which introns are removed and exons are joined. This is needed for the typical eukaryotic messenger RNA before it can be used to produce a correct protein through translation. For many eukaryotic introns, splicing is done in a series of reactions which are catalyzed by the spliceosome, a complex of small nuclear ribonucleoproteins (snRNPs), but there are also self-splicing introns. For most eukaryotic genes (and some prokaryotic ones), the initial RNA that is transcribed from a gene's DNA template must be processed before it becomes a mature messenger RNA (mRNA) that can direct the synthesis of protein. One of the steps in this processing, called RNA splicing, involves the removal or "splicing out" of certain sequences referred to as intervening sequences, or introns. The final mRNA thus consists of the remaining sequences, called exons, which are connected to one another through the splicing process. RNA splicing was initially discovered in the 1970s, overturning years of thought in the field of gene expression.

EARLY STUDIES IN BACTERIA

Gene regulation was first studied most thoroughly in relatively simple bacterial systems. Most bacterial RNA transcripts do not undergo splicing; these transcripts are said to be colinear, with DNA directly encoding them. In other words, there is a one-to-one correspondence of bases between the gene and the mRNA transcribed from the gene (excepting 52 and 32 noncoding regions). However, in 1977, several groups of researchers who were working with adenoviruses that infect and replicate in mammalian cells obtained some surprising results. These scientists identified a series of RNA molecules that they termed "mosaics," each of which contained sequences from noncontiguous sites in the viral genome. These mosaics were found late in viral infection. Studies of early infection revealed long primary RNA transcripts that contained all of the sequences from the late RNAs, as well as what came to be called the intervening sequences (introns). Subsequent to the adenoviral discovery, introns were found in many other viral and eukaryotic genes, including those for hemoglobin and immunoglobulin. Splicing of RNA transcripts was then observed in several in vitro systems derived from eukaryotic cells, including removal of introns from transfer RNA in yeast cell-free extracts. These observations solidified the hypothesis that splicing of large initial transcripts did, in fact, yield the mature mRNA. Other hypotheses proposed that the DNA template in some way looped or assumed a secondary structure that allowed transcription from noncontiguous regions.

SPLICING OCCURS

The biochemical mechanism by which splicing occurs has been studied in a number of systems and is now fairly well characterized. Introns are removed from primary transcripts by cleavage at conserved sequences called splice sites. These sites are found at the 52 and 32 ends of introns. Most commonly, the RNA sequence that is removed begins with the dinucleotide GU at its 52 end, and ends with AG at its 32 end. These consensus sequences are known to be critical, because changing one of the conserved nucleotides results in inhibition of splicing. Another important sequence occurs at what is called the branch point, located anywhere from 18 to 40 nucleotides upstream from the 32 end of an intron. The branch point always contains an adenine, but it is otherwise loosely conserved. A typical sequence is YNYYRAY, where Y indicates a pyrimidine, N denotes any nucleotide, R denotes any purine, and A denotes adenine. Rarely, alternate splice site sequences are found that begin with the dinucleotide AU and end with AC; these are spliced through a similar mechanism.

Splicing occurs in several steps and is catalyzed by small nuclear ribonucleoproteins (snRNPs, commonly pronounced "snurps"). First, the pre-mRNA is cleaved at the 52 end of the intron following the attachment of a snRNP called U1 to its complementary sequence within the intron. The cut end then attaches to the conserved branch point region downstream through pairing of guanine and adenine nucleotides from the 52 end and the branch point, respectively, to form a looped structure known as a lariat. The bonding of the guanine and adenine bases takes place via a chemical reaction known as transesterification, in which a hydroxyl (OH) group on a carbon atom of the adenine "attacks" the bond of the guanine nucleotide at the splice site. The guanine residue is thus cleaved from the RNA strand and forms a new bond with the adenine.

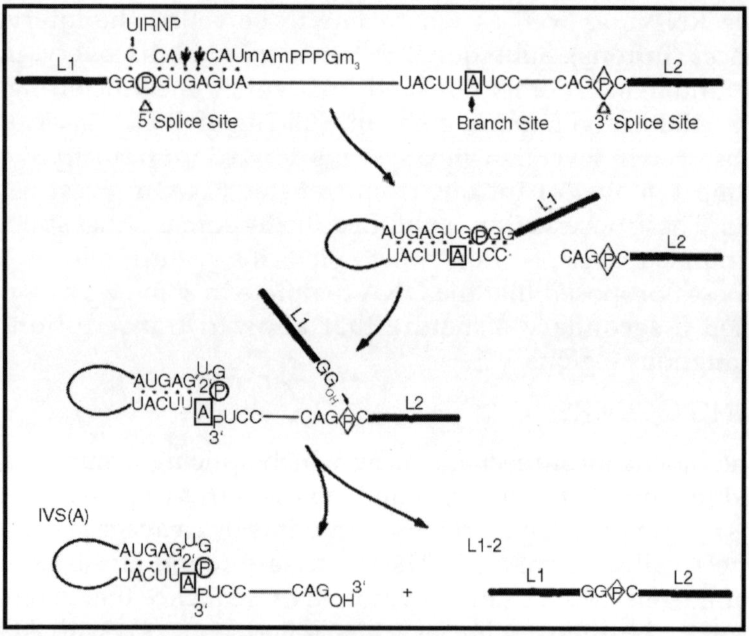

Fig. Pre-mRNA Splicing

Next, the snRNPs U2 and U4/U6 appear to contribute to positioning of the 52 end and the branch point in proximity. Then, with the participation of U5, the 32 end of the intron is brought into proximity, cut, and joined to the 52 end. These steps also take place by transesterification; in this case, an OH group at the 32 end of the exon attacks the phosphodiester bond at the 32 splice site. The adjoining exons are covalently bound, and the resulting lariat is released with U2, U5, and U6 bound to it.

In addition to consensus sequences at their splice sites, eukaryotic genes with long introns also contain exonic splicing enhancers (ESEs). These sequences, which help position the splicing apparatus, are found in the exons of genes and bind proteins that help recruit splicing machinery to the correct site. Most splicing occurs between exons on a single RNA transcript, but occasionally trans-splicing occurs, in which exons on different pre-mRNAs are ligated together.

The splicing process occurs in organelles called spliceosomes, in which the snRNPs are found along with additional proteins. The primary variety of spliceosome is one of the most plentiful organelles in the cell, and recently, a secondary type of spliceosome has been identified that processes a minor category of introns. These introns are referred to as U12-type introns because they depend upon the action of a snRNP called U12 (the common introns described above are called U2-type introns). The role of U12-type introns is not yet defined, but their persistence throughout evolution and conservation between homologous genes of widely divergent species suggests an important functional basis.

SELF-SPLICING AND ALTERNATIVE SPLICING

Some RNA molecules have the capacity to splice themselves; the initial discovery of this self-splicing ability in the protozoan Tetrahymena thermophila was recognized with the Nobel Prize in 1989. The self-splicing introns found in T. thermophila are now referred to as Group I introns; this class also includes other protozoan ribosomal RNA genes, some fungal mitochondrial genes, and some phage genes. Group I introns all fold into a complex secondary structure with nine loops and employ transesterification reactions as described above. On the other hand, Group II self-splicing introns are found in mitochondrial genes and are excised by a mechanism that bears similarities to pre-mRNA splicing, including the production of lariats. For this reason, it has been proposed that perhaps pre-mRNA introns and splicing mechanisms evolved from the Group II introns, but this remains conjecture at this point in time.

Early in the course of splicing research, yet another surprising discovery was made; specifically, researchers noticed that not only was pre-mRNA punctuated by introns that needed to be excised, but also that alternative patterns of splicing within a single pre-mRNA molecule could yield different functional mRNAs. The first example of alternative splicing was defined in the adenovirus in 1977 and demonstrated that one pre-mRNA molecule could be spliced at different junctions to result in a variety of mature mRNA molecules, each containing different combinations of exons.

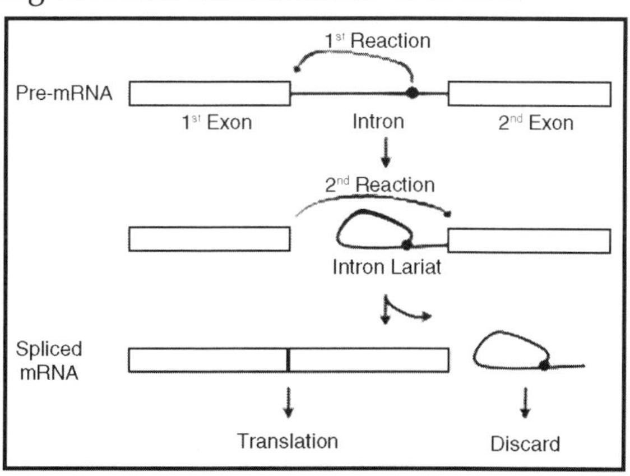

Fig. Alternative Splicing

Shortly afterward, alternative splicing was found to occur in cellular genes as well, with the first example identified in the IgM gene, a member of the immunoglobulin superfamily. Another example of a gene with an impressive number of alternative splicing patterns is the Dscam gene from Drosophila, which is involved in guiding embryonic nerves to their targets during formation of the fly's nervous system. Examination of the Dscam sequence reveals such a large number of introns that differential splicing could, in theory, create a staggering 38,000 different mRNAs. This ability to create so many mRNAs may provide the diversity necessary for forming a complex structure such as the nervous system. In fact, the existence of multiple mRNA transcripts within single genes may account for the complexity of some organisms, such as humans, that have relatively few genes (approximately 20,000).

PRE-MRNA SPLICING MODULATIONS IN SENESCENCE

From the viewpoint of a molecular biologist, aging reflects gradual deterioration of the molecular components, checkpoints and/or events, the concerted functioning of which are vital for cell viability and proliferation. The complexity of alternative splicing of pre-mRNA makes this process particularly vulnerable to senescence, leading to both transient changes and chronic aging-related diseases. With the emergence of microarray technologies, changes in gene expression profiles during aging have recently become the focus of extensive research. However, relatively little is yet known about changes in alternative splicing, a key pre-mRNA processing mechanism during cellular and organismal senescence. The progressive nature of the aging process is often ascribed to passive accumulation of late-onset mutations due to lack of natural selection beyond the age of reproduction. Alternatively, aging is viewed as the outcome of active selection of traits that are beneficial at an early stage of life but have deleterious effects later on.

The idea that aging is a genetically programmed mechanism has also been subject to some debate. Planned or sporadic, passive or active, aging also involves changes in the pathways of gene expression that are not necessarily associated with mutation. This is exemplified in most aging-related diseases, such as Alzheimer's and Parkinson's diseases and tauopathies, which represent the best studied cases of aging, involving aberrations in the alternative splicing of pre-mRNA. Here, we describe this emerging concept and discuss its research implications. We present examples of alternative splicing events during aging, and propose that these examples reflect a generally modified state of the pre-mRNA processing machinery in senescent cells, leading to altered expression levels of premRNA processing factors and their downstream target mRNAs and corresponding proteins during aging.

Aging-related modifications in the alternative splicing of specific gene products were sought, primarily as a means to search for the molecular origin(s) of aging-related deterioration, for example, neural adhesion processes, dopaminergic neurotransmission, insulin responses and gastric cytoprotection. Together, the picture that emerges from these studies is of a general control switch that causes all of these aging-induced impairments in alternative splicing. For example, these may all reflect one or a few changes in regulatory factor(s), such as splicing-related proteins, which modify the delicate regulation of the splicing process. Support for this concept can be found in recent microarray screens for changes in gene expression during aging.

Significant age-related tissuespecific changes were found in the expression levels of several RNA processing genes in aged, compared to young, gastrocnemius muscle, neocortex and cerebellum of C57BL/6 mice. Changes spanned serine, arginine-rich (SR) proteins, heterogeneous nuclear ribonucleoproteins (hnRNPs), as well as 3 ¢ -end processing factors. We present a scheme of the key processes in the pathway of pre-mRNA splicing, highlighting those elements in which age-related changes have been reported.

Pre-mRNA Processing During Aging

SR proteins contain at least one RNA recognition motif (RRM) and an arginine, serine-rich (RS) domain, allowing them to bind both the pre-mRNA and additional proteins. They bind the nascent pre-mRNA at specific sites, which can act as both splicing enhancers and splicing suppressors. These *cis* -acting motifs are usually degenerate and can be found within both exons and introns, exonic splicing enhancers (ESEs) being the more prevalent. Through such interactions with the pre-mRNA, SR proteins can regulate splicing and alternative splicing in a concentration-dependent manner. hnRNPs serve as a crucial checkpoint in pre-mRNA processing. They include a diverse group of proteins containing RNA binding motifs as well as several auxiliary domains.

This allows them to simultaneously bind pre-mRNA and other proteins. Misdirected regulation of the expression levels of SR-proteins and hnRNPs can influence splice site selection or lead to alternative or aberrant splicing of many downstream target sequences. hnRNP A1, for example, was shown to suppress splicing of exon 3 of the HIV-1 tat gene through high-affinity binding to an exonic splicing silencer (ESS) and recruitment of additional hnRNP A1 molecules, while the

SR protein ASF/SF2 prevents this hnRNP A1 accumulation at the ESS and antagonizes the splicing arrest.

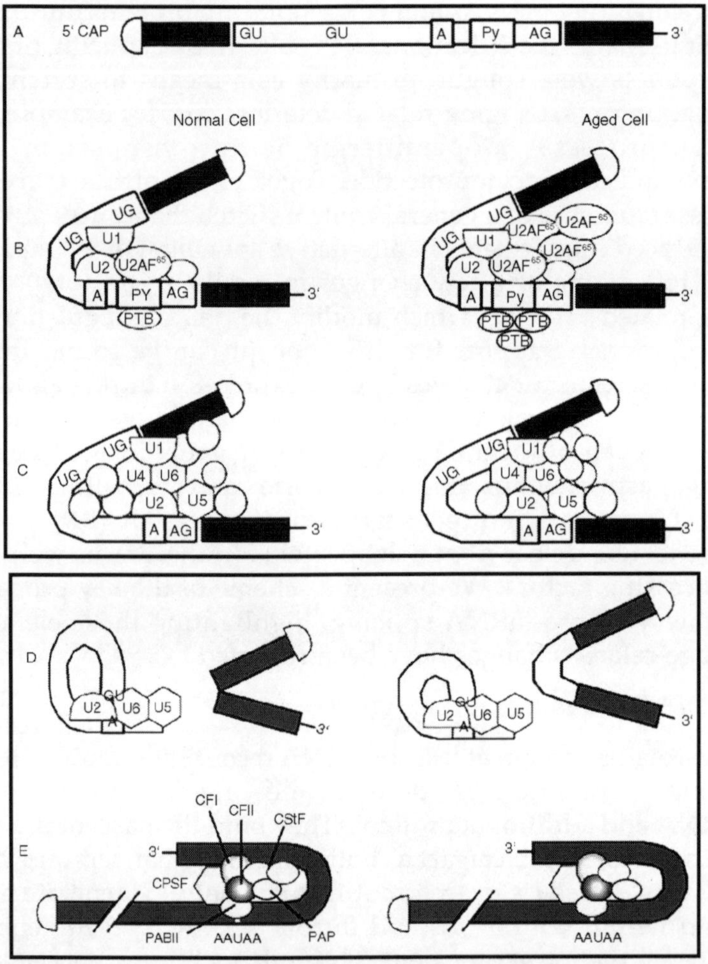

Fig. 1 Misdirected Regulation of Pre-mRNA Processing During Aging.

One of the first essential splicing factors to be identified was the 65-kDa subunit of the U2 auxiliary factor (U2AF). U2AF 65 binding to the conserved poly pyrimidine (Py) tract at the 3 ¢ end of introns is required for the subsequent binding of U2 snRNA to the 5 ¢ splice site (usually a GU dinucleotide). The other subunit, U2AF 35, binds the 3 ¢ splice site (usually an AG). U2AF 65 mRNA levels were found to increase by over three-fold in the gastrocnemius muscle of aged mice. As U2AF 65 contains both the RNA recognition motif and the RS domain of SR-proteins, it can regulate splicing in a concentration-dependent manner, as do the SR proteins. Uncontrolled regulation of the expression level of U2AF 65 may therefore lead to aberrant splicing activity and alternative splicing of numerous target genes.

The Py tract-binding protein (PTB, hnRNP I) also binds intronic Py tracts. PTB was shown to play a direct role in pre-mRNA alternative splicing through antagonistic effects on exon definition, repressing, for example, the inclusion of exon 7 in the β -tropomyosin gene. Its expression levels were reported to increase over two-fold in the cerebellum of aged mice, which can potentially harm finely regulated alternative splicing of many pre-mRNAs by repressing exon inclusion and promoting aberrant splicing. Other factors that may have a direct or indirect role in influencing splicing during aging are hPRP22, hPRP16, SF3A2, DDX18 and hnRNP H3. The expression level of the splicing factor PRP22, for example, was 2.5-fold increased in the neocortex of aged C57BL/6 mice, while the same factor displayed 2.7-fold decrease in the cerebellum of these mice. Although the function of several of these factors is largely obscure, this expanded list points to a central involvement of splicing changes in the aging process.

Initiation and Efficacy of Transcription

Another checkpoint, which may affect aging-related changes in pre-mRNA splicing, occurs prior to the splicing process, at the earlier phases controlling the initiation and efficacy of transcription. This involves the expression levels of a large number of transcription factors, also shown to be modified in aged tissues. Since transcription and splicing are tightly coupled processes, such changes are likely to affect the splicing machinery as well. The transcription factor SOX17 (D49473), for example, with a direct role in pre-mRNA splicing, exhibited 2.4-fold over-expression in the aged gastrocnemius muscle. While changes in expression levels may not correlate directly with changes in activity, altered expression of transcriptionassociated genes during aging adds further support to the concept of pre-mRNA splicing as an age-sensitive process.

Indeed, aging-related diseases often reflect abnormal upstream factors, including impaired regulation or aberrant fine-tuning of mRNA processing in general and pre-mRNA splicing in particular. The 3 ¢ -end processing complex of pre-mRNA is also subject to aging-related changes. This complex consists of cleavage factors 1 and 2 (CF1, 2), cleavage stimulation factor (CStF), cleavage and polyadenylation specificity factor (CPSF), poly(A)-polymerase (PAP), and poly(A)-binding protein II, which acts after the initial, slow polyadenylation phase is complete. CStF was decreased in the neocortex of aged mice, perhaps indicating a depression of the 3 ¢ -terminal processing of nascent mRNAs in conjunction with the aberrant splicing that accompanies aging. The cytoplasmic poly(A)-binding protein (PABI) was decreased to less than half its normal level in the cerebellum of aged mice, while the poly(A)+export protein was similarly increased in the gastrocnemius muscle of these mice. This points to the tissue specificity of age-dependent changes in 3 ¢ -end processing and mRNA export, again, keeping in mind that changes in expression levels may not necessarily directly correlate with changes in activity.

AGE-RELATED DISEASES

Age-related, late-onset diseases are especially vulnerable to splicing variations of different origins. These include mutations affecting the correct splicing of a particular gene. Age-related macular degeneration (AMD) is an example of an aging-related disease that is associated with a late-onset splicing mutation. AMD is the most common cause of acquired visual impairments in the elderly. The occurrence of AMD is often associated with mutations within the Stargardt disease gene (STGD1 or ABCR), coding for a photoreceptor-specific member of the ATP-binding cassette (ABC) superfamily of transporter proteins. One of these mutations, a G for A substitution at position 5196, was found to be a donor splice site mutation.

The late onset of the disease phenotype, in this case, may be explained by age-related modifications in the efficacy and/or composition of pre-mRNA processing factors, since the disease is associated with the consequences of a genomic splice-site mutation that for unknown reasons are apparent only at a later stage in life. Aging-related impairments in pre-mRNA processing need not necessarily involve a protein-modifying mutation as they may be caused by an altered upstream splicing regulator. Amyotrophic lateral sclerosis (ALS) serves as an example. ALS is a late-onset neurodegenerative disease that is characterized by selective degeneration of spinal cord motor neurones. Most ALS patients suffer from a significant loss of the astroglial excitatory amino acid transporter 2 (EAAT2, previously known as GLT-1) in the motor cortex and spinal cord. Affected tissues are highly populated by abnormal EAAT2 mRNA species resulting from aberrant alternative splicing.

Since the EAAT2 gene is not mutated in these patients, it was suggested that a protein involved in pre-mRNA processing is the culprit. Furthermore, alternative splicing of neuronal nitric oxide synthase (nNOS) was demonstrated in the spinal cord of ALS patients: nNOS b and nNOS g, but not nNOS a, were up-regulated. This demonstrates modified alternative splicing of nNOS, in addition to

EAAT2, and strengthens the hypothesis that an as yet unknown aberrant upstream splicing regulator is involved in ALS pathology. Alzheimer's disease (AD) is the most common neurodegenerative disorder of aging, characterized by progressive memory loss and cognitive deterioration. AD involves premature death of selected cholinergic neurones, associated with the formation of amyloid plaques, the appearance of which in the brain of patients is facilitated by mutations in several different genes (*e.g.* APP, PS1, PS2).

Misdirected splicing regulation of relevant gene products either due to specific mutations or to aberrant processing of such products with no known mutation were both demonstrated in AD. Presenilins provide examples of both types of alteration: mutations within the fourth intron of presenilin 1 (PS1) were shown to impair PS1 splicing and cause early onset AD, while an exon 5-deficient splice variant of apparently normal PS2, which accumulates following hypoxia in cultured neuroblastoma cells, was found to be prevalent in the brains of AD patients, indicating that this unique splice variant is inducible, rather than being produced at a constant level due to mutation in the corresponding gene. Tauopathies are a family of late-onset neurodegenerative diseases associated with mutations within the microtubuleassociated protein (MAP) tau gene. Progressive accumulation of filamentous tau inclusions causes neural degeneration in specific brain regions of patients with tauopathies. The late onset of this phenotype suggests that an age-related molecular change is responsible. Tau is alternatively spliced in the adult human brain and transgenic mice over-expressing the shortest human tau variant display agedependent CNS deterioration reminiscent of human tauopathies. Thus, as yet unspecified impairments in pre-mRNA processing and/ or splicing may contribute to these neurodegenerative conditions.

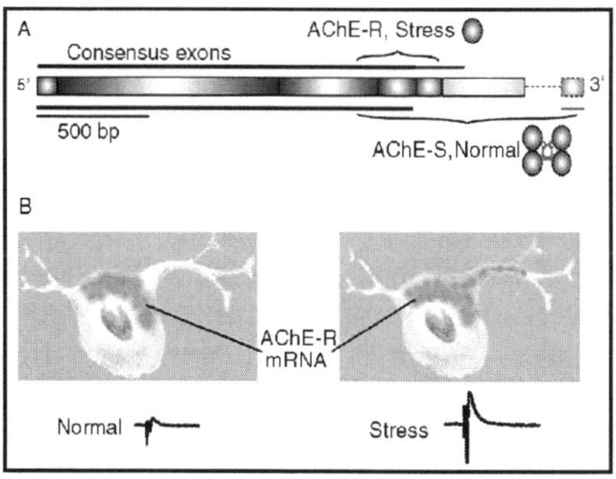

Fig. 2 Stress-induced Alternative Splicing in Neuronal Acetylcholinesterase mRNA.

Frontotemporal dementia with Parkinsonism linked to chromosome 17 (FTDP-17) is an example of a tauopathy associated with changes in alternative splicing. Several mutations within the tau gene were demonstrated for FTDP-17, a large proportion of which affect the splicing pattern of tau exon 10 by influencing exonic/intronic splicing enhancers/suppressors. The aberrant splicing of tau exon 10 alters the ratio of the tau isoforms incorporated into the neuronal tangles that are the neuropathological hallmark of the demented brain. Excessive tau accumulation results in filamentous inclusions within these tangles. In this case as well, the mutated genotype leads to an abnormal phenotype with a delayed onset, hinting at a change in an unidentified age-related factor involved in pre-mRNA processing.

STRESS AND SPLICING

Another consideration with regard to late-onset phenotypes involves the interrelationships between individual genotypes and the environment. Stress-induced changes in the pathway to gene expression, for example, may change the age of onset of an aberrant pattern of pre-mRNA splicing, initiating a cascade of events that expedite the onset of an aging-related disease. Such conditions are evident primarily in the central nervous system because of its role in the initiation of stress responses and because of the range of cell types and complex gene expression profiles in the mammalian brain. Examples include the aberrant splicing of K^+ channels of the Tra1 splicing regulator and of acetylcholinesterase (AChE) mRNA.

The latter example is particularly relevant to the issue at hand, as it directly relates to the behavioural changes and the cognitive impairments which are characteristic of old age and which are implicated in age-related neurodegenerative diseases. The *ACHE* gene gives rise to several mature transcripts. Of these, the primary 'synaptic' AChE-S mRNA constitutes the preponderant AChE mRNA under normal conditions and is translocated into neuronal processes. Under the influence of various stressors, however, a splicing shift occurs which leads to overproduction and neurite translocation of the normally rare 'readthrough' AChE-R mRNA transcript. It is not yet clear why AChE-R mRNA stays in the cell body under normal conditions but travels to neuronal processes under conditions of stress. However, this change alters both the nature of the AChE protein product and its subcellular location.

Thus, neuritic AChES forms synaptic membrane-associated tetramers; in contrast, AChE-R forms soluble, secretory monomers that cannot adhere to the membrane and hence fail to address incoming cholinergic stimuli, leading to prolonged neuronal hypersensitivity. Large excesses of AChE-R, such as those that accumulate in the injured brain, are detrimental: massive overproduction of this variant in the brain of head-injured transgenic mice over-expressing human AChE reduces survival and slows the recovery, whereas antisense intervention with AChE-R production improves survival and recovery of both transgenic mice and their parent strain. As head injury is the largest risk factor known for nonfamilial AD, the splice shift leading to AChE-R overproduction seems to be causally involved in an increased risk for neurodegeneration. That the AD brain displays increases in AChE-R monomers as opposed to decreases in AChE-S tetramers supports this notion.

CELL PROLIFERATION

One final key process, which is now known to be affected by aging and which depends on the regulation of pre-mRNA processing, is cell proliferation. Pre-mRNA processing events, which are subject to aging dependent impairment, participate in control of cell proliferation, and aging has long been known to be associated with aberrant cell proliferation. Therefore, although neuronal stem cells maintain into old age some capacity to proliferate, the efficacy of cell proliferation is likely affected by damaged splicing, with consequences for many physiological functions. The proliferation of neuronal stem cells in the murine hippocampus, for example, emerges as an essential prerequisite for the preservation of newly acquired trace memories. This physiological process is impaired even in non-demented, normal older individuals.

Our current considerations therefore imply that the memory impairments of old age may reflect detrimental changes in pre-mRNA processing and consequent impairment in stem cell proliferation. Most of the examples presented above refer to post-mitotic tissues such as brain and muscle, in which the relevant cells are terminally differentiated. In view of this line of speculation, it would be most interesting to explore the changes during aging that take place in the pre-mRNA processing machinery of proliferating cells of these organs, particularly of the neuronal stem cells.

INTRON

An intron is a DNA region within a gene that is not translated into protein. These non-coding sections are transcribed to precursor mRNA (pre-mRNA) and some other RNAs (such as long noncoding RNAs), and subsequently removed by a process called splicing during the processing to mature RNA. After intron splicing (ie. removal), the mRNA consists only of exon derived sequences, which are translated into a protein.

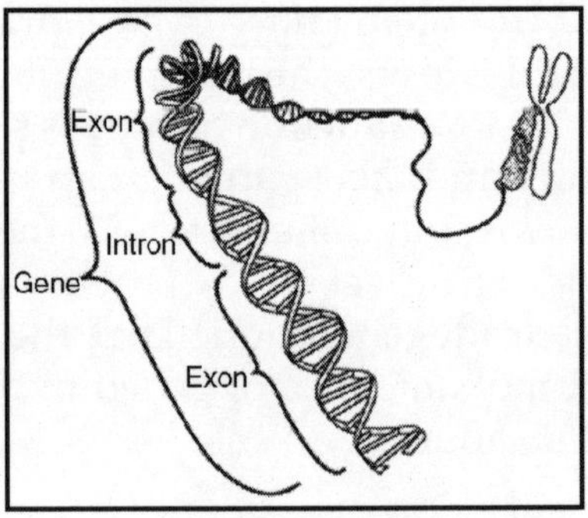

Fig. Intron

Introns are common in eukaryotic pre-mRNA, but in prokaryotes they are only found in tRNA and rRNA; introns have variable length and alternate with exons in intron-containing genes. The number and length of introns varies widely among species, and among genes within the same species. Some eukaryotes, *e.g.* sac fungi, have evolved genomes with few introns, while the genomes of many other eukaryote groups are rich in introns. Alternative splicing of introns within a gene may introduce greater variability of protein sequences translated from a single gene. The control of mRNA splicing is performed by a wide variety of signaling molecules.

THE PROCESS OF REMOVING INTRONS SHOWS EVIDENCE OF DESIGN

When MRNA is transcribed from DNA, both the exon and intron noncoding sequences are transcribed into a macromolecule known as a *heterogeneous nuclear RNA* (hnRNA), or immature RNA. Before the RNA exits the nucleus, the introns must be removed by a precise cleavage-ligation reaction called *splicing*, thereby producing a functional mRNA. Enzymes and small ribonucleoprotein structures exist in all eukaryotic cells to assist in removing introns. In most cases, type I introns are removed in the nucleus by a complex splicing machine composed of *small nuclear ribonuclear proteins* (SNRP). SNRPs consist of 60-300 small nuclear RNA (snRNA) nucleotides in an "intimate alliance with a bouquet of proteins."

The process of intron removal involves a precise looping process controlled by a specific nucleotide sequence that abuts the exons. Almost all known introns are identified by specific *consensus sequences* [GT at the start or *donor* (3') end and AG at the other end, called the *acceptor* (5') site] which help to identify introns for removal. The term "consensus sequence" is employed because, although the sites may consist of more than just GT donor and AG acceptor bases, so far as is known these sequences are common to all eukaryotic organisms. Consensus sequences are DNA segments that use similar base sequences in different genes within a single gene family or across many different species for one specific function.

After the introns are cut out, the exon RNA is spliced back together by *RNA ligase* so that the final mRNA used to code polypeptides normally consists only of exons. In Type II introns, RNA itself functions as an enzyme called a *riboenzyme* or *ribozyme*. These enzymes require a divalent cation (usually magnesium) to function. The process of removing introns involves a complicated reaction pathway that includes RNA intermediates called lariats, branch connecting points, and a large number of accessory proteins.

NUMBER AND LENGTH OF INTRONS

The sections of the human genome sequenced thus far provide an estimate of the genome's composition. It appears that over 90% of the DNA bases consist of repetitive noncoding regions, introns, and regions between genes that have no known function. The number of introns in transcribed RNA ranges from zero in histone genes to an estimated 2.5 million bases in the protein dystrophin. The dystrophin gene has over 75 introns and only 0.5 per cent (11,000 bps) of the gene consists of exons. A defective dystrophin gene is the major cause of Duchene Muscular Dystrophy. Another example is approximately 98% of the gene that causes cystic fibrosis consists of introns. The large number of introns in the genome is revealed by evidence that indicates the *majority* of base pairs in *most* genes consists of introns. The length of one intron varies from 31 nucleotides in an SV-40 gene to over 210,000 in the human dystrophin gene, although 65 to 20,000 nucleotides is more typical.

Role of Introns in the Cell

Indications that introns have a use include the observation that "the cell puts a huge amount of its energy into the creation of these introns, then discards them... Nature would not go to all that trouble without a reason." Why "most genes of higher organisms are broken up into short exons separated by huge stretches of seemingly useless DNA" is particularly puzzling because "the introns are all painfully transcribed into RNA only to be snipped out by snRNPs and thrown away almost immediately." Evolutionist Patrusky says: "Nature, for reasons as yet unknown, created the intron, and evolution has chosen to keep it... and ultimately has found new ways to use it."

An important finding which may help researchers to understand the role of introns is the more primitive and simple the organism generally the fewer the introns. Although the majority of plant and animal genomes have not yet been examined, introns have been found in the vast majority of eukaryotic genes and likely exist in all eukaryotes. Wills speculated that fewer introns exist in prokaryotic organisms because bacteria need streamlined genomes, or may have less need for whatever functions introns serve. He also speculated that the putative early organisms might have lost their introns through evolutionary development.

The fact that all higher organisms possess introns, and the more complex organisms possess a higher proportion of introns, indicates that they serve at least minor, and possibly major, functions. Wills says: "Carrying this enormous extra baggage of unnecessary DNA each generation seems a heavy price to pay for the privilege of the occasional bout of exon shuffling." Many reasons now exist to conclude that introns are not found in cells only because they are "intrinsically capable of replication and therefore can stick around" as once believed. Furthermore, research has found a high level of conservation in some introns, indicating that they have some selective advantage, that is a useful function. How many introns display this conservation is not yet known. This raises the question, "If introns produce a major selection advantage and consequently are characteristic of higher, more developed organisms, what could explain their loss in lower organisms?" Variety is critical for species survival, and producing variety is especially difficult in animals with small genomes, the very creatures which must have possessed the variety required by evolution in order to evolve. Putative ancient bacteria evidently would possess little capacity for variety if the absence of introns (and other structures) in modern bacteria was also the situation in early bacteria.

The former belief that introns were useless was once so prevalent that many felt it would be a major waste of resources for the human genome project to sequence these long stretches of "meaningless DNA." The human genome project involves sequencing, the estimated 35,000 genes and over three billion base pairs at a cost of over $3 billion. The National Science Foundation priority is now on sequencing exons, but eventually the entire code will likely be determined partly because of accumulating evidence that the entire genome has a function. A problem in genetic research is that genes in which introns are left in may be so long and unwieldy that they cannot be carried by a vector virus or plasmid. When the significance of introns is understood better, this problem can be dealt with more effectively.

Alternative Splicing

The problem of identifying introns was compounded by the discovery that alternative readings of the genetic code exist in which "introns" function as exons. Some DNA behave as exons when expressed by one pathway, but as introns when expressed by another pathway. Both pathways can operate simultaneously, resulting in greater protein product variety. Termination codons are also sometimes deliberately bypassed, allowing the coding of a part of an intron in order to produce a specialized protein. Variations in intron removal splicing patterns can be used to create two or more distinct mRNAs which code for different polypeptides. This allows one mRNA code to be used to produce a variety of polypeptides. Splicing variations are controlled by regulator proteins designated as SR proteins. SR proteins determine which splicing pattern predominates in a specific cell at any given time. Another example of an alternate reading system that can make use of introns is frame *shifting*, which causes the normal grouping of nucleotides to be altered so that the DNA is read in a completely different way. A third type of alteration which may involve introns is base hopping in which large DNA segments are skipped when the mRNA is produced.

Exon

An exon is a nucleic acid sequence that is represented in the mature form of an RNA molecule after either portions of a precursor RNA (introns) have been removed by cis-splicing or when two or more precursor RNA molecules have been ligated by trans-splicing. The mature RNA molecule can be a messenger RNA or a functional form of a non-coding RNA such as rRNA or tRNA. Depending on the context, exon can refer to the sequence in the DNA or its RNA transcript.

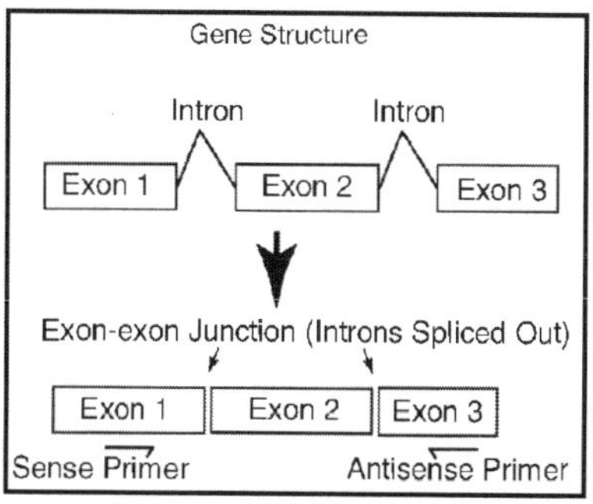

Fig. Exon

Function

In many genes, each exon contains part of the open reading frame (ORF) that codes for a specific portion of the complete protein. However, the term exon is often misused to refer only to coding sequences for thc final protein. This is incorrect, since many noncoding exons are known in human genes. To the right is a diagram of a heterogeneous nuclear RNA (hnRNA), which is an unedited mRNA transcript, or pre-mRNAs. Exons can include both sequences that code for amino acids (red) and untranslated sequences (grey). Stretches of unused sequence called introns (blue) are removed, and the exons are joined together to form the final functional mRNA.

The notation 5' and 3' refer to the direction of the DNA template in the chromosome and is used to distinguish between the two untranslated regions (grey). Some of the exons will be wholly or part of the 5' untranslated region (5' UTR) or the 3' untranslated region (3' UTR) of each transcript.

The untranslated regions are important for efficient translation of the transcript and for controlling the rate of translation and half life of the transcript. Furthermore, transcripts made from the same gene may not have the same exon structure since parts of the mRNA could be removed by the process of alternative splicing. Some mRNA transcripts have exons with no ORF's and thus are sometimes referred to as non-coding RNA.

ALTERNATE SPLICING OF EXONS

A classic example of alternate splicing is the rat muscle protein, troponin T. The gene consists of five exons, each representing a domain of a final protein. These exons are each separated by an intron. The five exons are W, X, Alpha, Beta, and Z. Two types of protein are found. The alpha form consists of exons W, X, alpha and Z. The beta form consists of the W, X, Beta and Z exons.

The two different types of the protein are produced by alternative splicing of the same gene. The two different gene products are produced by selective splicing such that introns three and four and the fourth exon are spliced as one unit. In some manner the 5' GT sequence of intron 3 and the 3' AG sequence of the fourth intron are used during the splcing event.

Alternate Splicing of a Plant mRNA

The plant example of alternate is for the gene encoding the enzyme RUBISCO activate. RUBISCO activase is a soluble, nuclear-encoded chloroplast enzyme. This abundant protein, is about 2% of the soluble leaf protein in spinach. Two molecular forms of the protein are recognized. In spinach, the proteins are 41 kd and 45 kd polypedtides, and the proteins are 44 kd and 47 kd polypeptides in *Arabidopsis*. The monocot corn is different and has only one protein. This enzyme is required to activate the enzymatic activity of RUBISCO, the first step in the conversion of the electromagnetic energy collected during the light reaction of photosynthesis in chemical energy in the forms of sugars.

The RUBISCO activase gene has been cloned. cDNA clones had been isolated previously and revealed two nucleotide binding sites. When the cDNA was expressed in E. coli only one polypeptide was detected. Antibodies from each of the 41 and 45 kd strongly crossed reacted with each other suggesting a high degree of homology between the proteins.

Table. Additional cDNA Clones were Isolated and Sequenced for both Spinach and Arabadopsis.

Species	3' Sequence Comparison	Internal Sequences
Spinach	Identical	22-bp insertion; introduced a new stop codon
Arabadopsis	Identical	11-bp insertion at same location; introduced a new stop

Genomic clones were then isolated for each of the two species and sequenced. The sequence data revealed that each of the genes contained seven exons and six introns. Comparison of the sequences of the genomic and cDNA clones of spinach revealed the following manner in which the two different types of cDNAs could be generated. The full length mRNA that gave rise to the larger protein was generated by splicing at the terminal-most 5' and 3' splicing sites. 22-bp into intron 6 is another 5' splice junction sequence. Alternate splicing occurs at this site to generate the mRNAs that are translated into the shorter proteins.

This 22-bp sequence can be considered an extension of exon 6 and is called an auxon for alternative exon. Alternative splicing events can form different mRNAs from a single primary transcript. Thus an exon can actually be represented as an intron in another transcript.

EXON SHUFFLING

Introns also produce a critically important life trait called phenotypical variety by regulating or facilitating the transposition of exons. Evidence for this role includes the observation that introns commonly are located at or near the boundaries between different domains in the DNA molecule. Called exon shuffling, this mechanism helps to produce genetic and phenotypical variety, in an organized way by producing new exon combinations and thus greater genetic diversification. This would enable the production of a variety of polypeptides from a single gene.

The process of exon shuffling can create a mosaic, or chimeric, protein that is partly built of domains or segments which are similar or identical to domains of other proteins. The mosaic protein is created when an exon from one gene is integrated into another gene's intron. Another kind of exon shuffling is when an exon is duplicated in the same gene. In the context of evolution, exon shuffling is significant due to its ability to quickly create new multidomain proteins, leading to increased variance in species.

tRNA Splicing

Introns are encoded in the genes for tRNA in organisms from all three kingdoms of life. Their removal is an essential step in the maturation of tRNA precursors. In Bacteria, introns are self-splicing and are removed by a group 1 splicing mechanism. In Eukaryotes and Archaea, intron removal is mediated enzymatically by proteins. Recent progress in understanding both eukaryotic and archaeal tRNA splicing has revealed that the two processes, previously thought to be unrelated, are in fact similar. Insight gained from the comparison has provided a clearer understanding of intron recognition, the catalysis of intron removal, and has given new insight into the evolution of the tRNA splicing process.

INTRON-CONTAINING TRNA

TRNA INTRONS IN EUKARYOTES

Intervening sequences (introns) were discovered 20 years ago in the yeast genes for the tyrosine-inserting non-sense suppressor tRNA and for tRNAPhe. With the completion of the yeast genome, it is known that of the 274 yeast tRNA genes 61, or 20%, contain introns. PCR cloning of tRNAs from higher eukaryotes has revealed a similar distribution of intron-containing tRNA. The introns in all of the genes are small (14–60 bases), and they are all located in the same position, one base to the 3_ side of the anticodon. Structure probing revealed that the common "cloverleaf" tertiary structure seen in the crystal structure of tRNAPhe is maintained in intron-containing tRNAs and that the intron, including the splice sites, is the most exposed region of the molecule.

This led to a proposed model for the tertiary structure of pre-tRNA. More recent experiments by Tocchini-Valentini and coworkers have demonstrated that a conserved base pair (the A-I base pair) between a base of the 5_ exon (position 32) immediately following the anticodon stem and a base in the single-stranded loop of the intron (position –3) is required for correct excision at the splice site. Thus, the distinctive feature in eukaryotic intron-containing pre-tRNAs is the position of the intron and the presence of the A-I base pair.

tRNA	Intron Length (Nucleotides)	No. of Genes
$tRNA^{Ser}_{CGA}$	19	1
$tRNA^{Ser}_{GCU}$	19	4
$tRNA^{LYS}_{UUU}$	23	7
$tRNA^{Pro}_{UGG}$	31,30,33	7,2,1
$tRNA^{Trp}_{CCA}$	34	6
$tRNA^{Phe}_{GAA}$	18,19	3,7
$tRNA^{Leu}_{CAA}$	32,33	8,2
$tRNA^{Ile}_{UAU}$	60	2
$tRNA^{Leu}_{UAG}$	19	3
$tRNA^{Tyr}_{GUA}$	14	8
		Total = 61
	Total tRNA Genes = 274	

Fig. Yeast tRNA Precursors Containing Introns

tRNA Introns in Archaea

tRNA introns are found in every archaeal species that has been studied. Although they are often similar, archaeal introns are different from their eukaryotic counterparts. Comparative sequence analysis of a host of archaeal introns has revealed that the splice sites are always located in a 3-nucleotide bulge separated by a 4- bp helix: the bulge-helix-bulge (BHB) motif.

Intron length is variable, from a short intron of 33 nucleotides found in the tRNAMet gene of *Methanococcus jannaschii,* to 105 nucleotides of the tRNATrp gene of *Haloferax volcanii,* but the intron is required to base-pair with the exon to allow formation of the BHB motif. In general, the introns reside one base to the anticodon, but introns have been found elsewhere in the tRNA molecule. One such example is that of the tRNAPro gene of *Methanobacterium thermoautotrophicum,* where there are two introns in the anticodon stem that must be removed in an obligate order to produce the mature tRNA (C.R. Trotta, unpubl.).

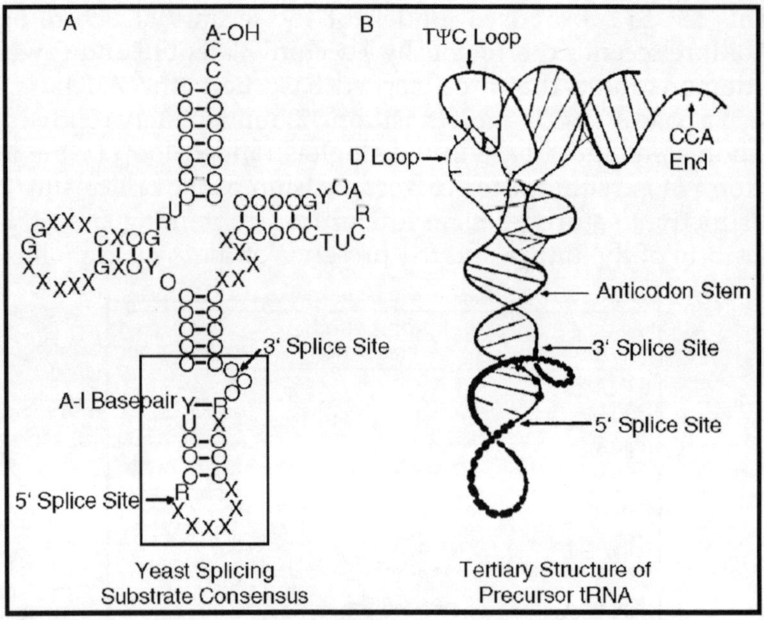

Fig. (A) Consensus of Yeast pre-tRNA
(B) Model of the Tertiary Structure of pre-tRNAs

Another repository for introns in Archaea is the rRNA genes. In fact, one of the earliest discovered archaeal introns, the 622- nucleotide intron of the 23S rRNA of *Desulfurococcus mobilis*, was shown through extensive biochemical characterization to be removed by the archaeal tRNA splicing system with dependence on the BHB motif.

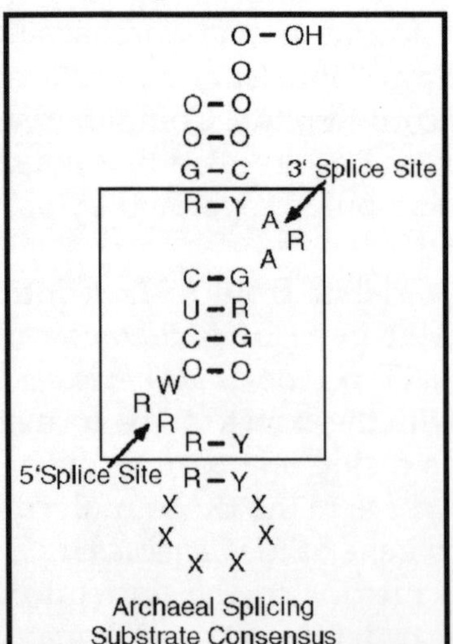

Fig. Consensus Archaeal Substrate.

THE ENZYMATIC GYMNASTICS OF TRNA SPLICING

The Eukaryotic Pathway

The accumulation of pre-tRNAs in the *rna1-1* mutant provided a source of pre-tRNA for assay of tRNA splicing in vitro The system was employed to work out the pathway of tRNA splicing. The first step of the reaction is recognition and cleavage of the pre-tRNA substrate at the 5′ and 3′ splice sites by a tRNA splicing endonuclease. The cleavage by the endonuclease at the splice sites results in 2′,3′ cyclic phosphate and 5_-hydroxyl termini. The tRNA 5′ and 3′ exons are then the substrate for a series of reactions catalyzed by the tRNA splicing ligase (for a detailed review of the ligation mechanism.

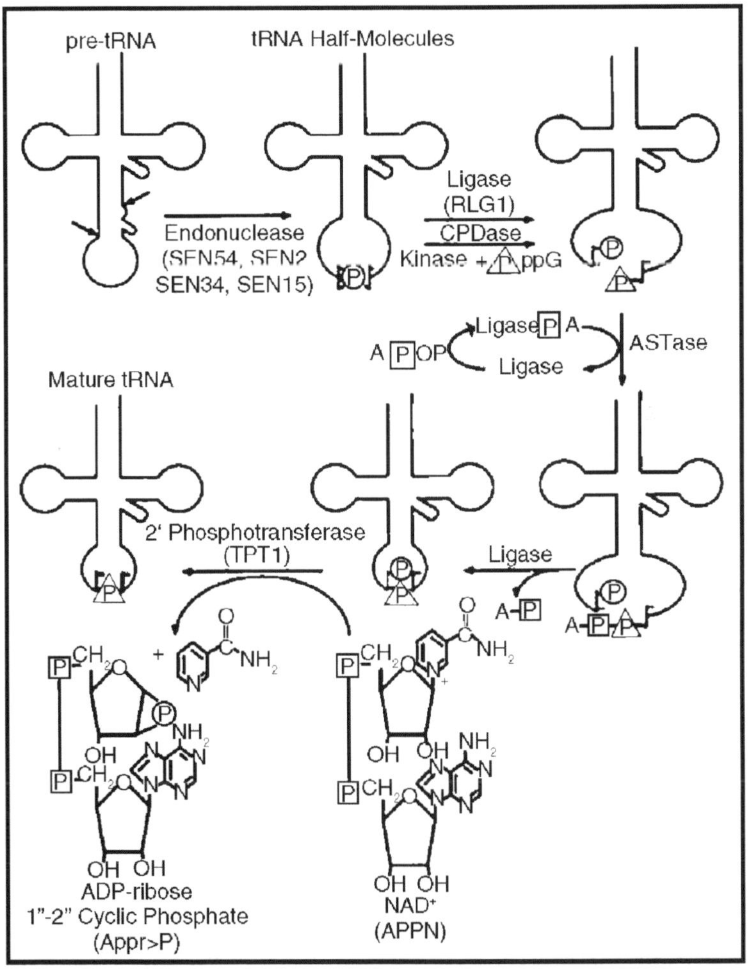

Fig. The tRNA Splicing Pathway of Yeast.

Ligase is a multifunctional enzyme possessing phosphodiesterase, polynucleotide kinase, and RNA ligase activities. In the ligase reaction, phosphodiesterase opens the 2′,3′ cyclic phosphate to give a 2′-phosphate.

Next the 5′-hydroxyl is phosphorylated by transfer of the g-phosphate from an NTP cofactor. Ligase is then adenylylated and the AMP moiety is transferred to the 5′- phosphate on the 3′-exon forming an activated phosphoanhydride.

Finally, ligase joins the two exons by catalyzing an attack of the 3-hydroxyl on the activated donor phosphoanhydride to form a 3′,5′ phosphodiester, 2′-monopho-sphoester bond with the release of AMP. This results in a pretRNA molecule with a 2′-phosphate at the splice junction.

Removal of the 2′-phosphate is accomplished by the activity of a 2′-phosphotransferase that catalyzes the transfer of the phosphate to NAD, releasing nicotinamide and yielding ADP-ribose 1″-2″ cyclic phosphate (Appr > P) and mature tRNA.

Human RNA 52-kinase (hClp1) Can Function as a tRNA Splicing Enzyme

Intron-containing tRNAs are found in all eukaryal and archaeal species. The intron is usually located in the anticodon loop of the pre-tRNA and must be removed for the tRNA to function in protein synthesis.

A tRNA splicing endonuclease recognizes the fold of the pre-tRNA and makes incisions at the exon–intron borders to yield 22,32 cyclic phosphate and 52-OH termini at both break sites. The human tRNA splicing endonuclease is a component of a stable nuclear protein complex that includes human Clp1 (hClp1), a conserved eukaryal polypeptide required for 32 processing of pre-mRNAs. In yeast, the splicing endonuclease resides on the outer mitochondrial surface; consequently, yeast tRNA splicing is a cytoplasmic event. Notwithstanding these differences in localization, the structures and mechanisms of tRNA splicing endonucleases are conserved among lower and higher eukaryal species and archaea.

In contrast, the tRNA sealing reactions of the splicing pathway diverge in archaea and eukarya. The ends of cleaved archaeal tRNAs are rejoined in a single step by attack of the 52-OH on the 22,32 cyclic phosphodiester to form a 32,52 phosphodiester at the splice junction. The responsible archaeal enzyme is not known. Yeast tRNA splicing takes a totally different biochemical route. A single tRNA ligase enzyme, Trl1, heals and seals the broken tRNA ends.

Trl1 performs three reactions, each catalyzed by a discrete protein domain:

- The 22,32 cyclic phosphate of the proximal tRNA half-molecule is hydrolyzed to a 32-OH, 22-PO_4 by a cyclic phosphodiesterase (CPD);
- The 52-OH of the distal half-molecule is phosphorylated by an NTP-dependent polynucleotide kinase;
- The 32-OH, 22-PO_4, and 52-PO_4 ends are sealed by an ATP-dependent RNA ligase to form a spliced tRNA containing an unconventional 22-PO_4, 32-52 phosphodiester at the splice junction. The 22-PO_4 at the splice junction is removed by the 22-phosphotransferase Tpt1.

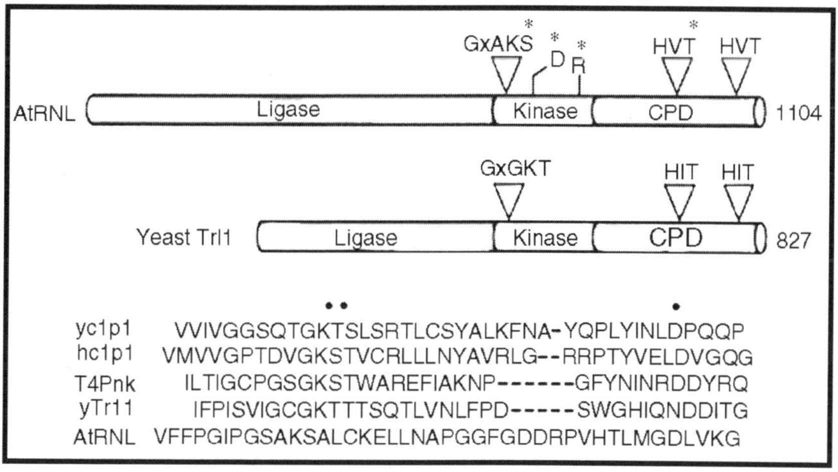

Fig. Domain Organization of Yeast and Plant tRNA Ligases and Conservation of Polynucleotide Kinase Motifs in tRNA Repair Kinases and Yeast and Human Clp1.

Yeast Trl1 consists of an N-terminal ligase domain and a C-terminal end-healing domain. The ligase domain is an ATP-dependent strand-sealing enzyme that belongs to the ligase family exemplified by bacteriophage T4 RNA ligase 1. The end-healing domain is composed of distinct kinase and CPD modules.

The kinase module resembles T4 polynucleotide kinase (Pnk), which belongs to the P-loop phosphotransferase superfamily. The CPD component of yeast tRNA ligase belongs to the 2H phosphotransferase superfamily, which is defined by two HxT motifs, and includes the mammalian 22,32 cyclic nucleotide phosphodiesterase, CNP). An *Arabidopsis* tRNA ligase ortholog (AtRNL) is a 1104-amino acid polypeptide with a similar tripartite domain structure. Plant AtRNL is capable of performing all three essential splicing reactions in vivo in yeast cells that lack Trl1.

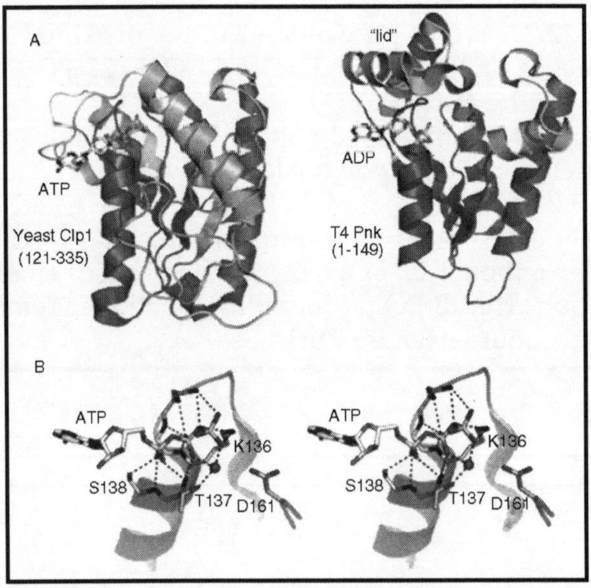

Fig. Structural Homology Between Clp1 and T4 Pnk.

The mechanism of tRNA exon ligation in animal cells is poorly understood. Initial studies demonstrated a direct tRNA end-joining reaction in human cell extracts in which the 22,32 cyclic phosphate is retained as a 32–52 phosphodiester at the splice junction of the mature tRNA. The identity of the responsible enzyme remains a mystery. Later work suggested that human cells also have a yeast-like pathway of tRNA splicing in which the junction 32–52 phosphate in mature tRNA derives from ATP and the starting 22,32 cyclic phosphate is converted to a 22-PO_4 that is ultimately removed. The existence of a yeast-like tRNA repair system in metazoa is consistent with the presence of Tpt1-type phosphotransferase, 22,32 cyclic phosphodiesterase, and RNA-specific 52-OH polynucleotide kinase activities in human cells.

Solving the "division of labour" problem for the two distinct human tRNA exon-joining pathways hinges on identifying the relevant enzymes and probing their function genetically. Recent studies in this vein have cast doubt on the importance of the yeast-type ligation pathway for splicing of animal tRNAs. For example, mammalian CNP and Tpt1 have been shown to complement the essential functions of their yeast homologs, thereby attesting to their capacity to catalyze 32-end-healing and junction dephosphorylation reactions of tRNA splicing in vivo. However, neither CNP nor Tpt1 is essential in mice. The relatively benign effects of completely deleting CNP in mice seem to exclude an essential role for the yeast-type splicing pathway, if one assumes that CNP, which is the source of all measurable 22,32-cyclic phosphodiesterase activity in mouse brain is the only available source of 32-end-healing for the yeast-like splicing pathway.

Similarly, Tpt1 is the sole source of measurable tRNA 22-phosphotransferase activity in mouse cells, yet homozygous *tpt1 "/"* null mice develop normally and are healthy and fertile. These results appear to relegate the yeast-type pathway in mammals to either no role, or a functionally redundant role, in tRNA splicing. Further consolidation of this inference will require genetic assessments of the ligase and kinase components of the yeast-type pathway in mammalian cells. A recent advance on this front was the assignment of RNA 52-OH kinase activity to the human Clp1 protein. An RNA-specific 52-kinase was isolated from HeLa cell nuclei and characterized biochemically nearly 30 years ago. It was speculated then that the human RNA kinase might provide sealable 52-monophosphate ends for an RNA ligase in tRNA splicing, notwithstanding that the 52-OH terminus of the cleaved tRNA ends and requirements for a polynucleotide kinase in yeast tRNA splicing, plant RNA ligation, and bacteriophage T4 tRNA restriction/repair had not yet been established. Interest in metazoan RNA kinase was revived by the realization that exogenous siRNAs acquire a 52-monophosphate end in order to be effective agents of RNAi purified an siRNA kinase from HeLa cells and identified hClp1 (425 amino acids) as the active principle.

A recombinant GST-hClp1 fusion protein sufficed for siRNA kinase activity. Remarkably, the kinase purified from human cells consisted of a complex of hClp1 with all of the subunits of the human tRNA splicing endonuclease, hSEN. The hClp1-hSEN complex could phosphorylate the 52-OH end of cleaved pre-tRNA. Based on siRNA knockdown experiments, Clp1 was suggested to be an agent of mammalian tRNA splicing. The model for hClp1 tRNA splicing function remains to be tested by generating a viable mammalian cell line or organism in which the tRNA kinase activity of Clp1 is eliminated without affecting other aspects of Clp1 function. This is a potentially complicated task, given the essential role of Clp1 in mRNA 32 processing.

Yeast Clp1 (yClp1; 445 amino acids) is essential for the viability of *Saccharomyces cerevisiae*. yClp1 is a subunit of pre-mRNA 32 cleavage factor 1 (CF1); the other subunits are Rna14, Rna14, Hrp1, and Pcf11. The crystal structure of yeast Clp1 bound to a peptide derived from Pcf11 revealed a central P-loop phosphotransferase module to which

ATP and magnesium were bound. Efforts to demonstrate ATP hydrolysis by yClp1 were unsuccessful, which makes sense in light of the subsequent findings that hClp1 catalyzes transfer of the ATP γ-phosphate to a 5' OHRNA terminus, not to water.

The fold of the central yClp1 segment resembles that of bacteriophage T4 polynucleotide kinase. T4 Pnk and yClp1 share a central parallel four-strand β sheet flanked by three homologous á helices. The loop between the first β strand and α helix includes the signature GxGKTS motif that coordinates the ATP phosphates and the divalent cation. The two structures diverge considerably outside the shared core. Yeast Clp1 has no counterpart of the essential T4 Pnk "lid" module that covers the NTP phosphoryl donor. Indeed, yClp1 has a larger central â sheet (seven strands) than does T4 Pnk, and it includes several á helices and loops that have no equivalent in the phage kinase.

To our knowledge, it is not established if the yClp1 ATP-binding site, or an imputed phosphoryl transferase activity of yClp1, is required for yeast viability. A corollary question is whether yClp1, like its human homolog, is an RNA-specific kinase and whether that activity is relevant to RNA metabolism in *S. cerevisiae*. Available evidence would argue against yClp1 being the catalyst of the 52-end-healing step of yeast tRNA splicing, insofar as Trl1 itself has the requisite kinase function and kinase-inactivating mutations in Trl1 are lethal in vivo. However, a different scenario might pertain to human cells.

The findings that tRNA repair enzymes from heterologous sources can complement the catalytic function of one or all of the domains of yeast Trl1 provides a way to interrogate whether candidate proteins can perform tRNA splicing reactions in vivo. Here, we apply this approach to show that hClp1 can indeed execute the 52-end-healing step of tRNA splicing in yeast.

Major functional differences between the yClp1 and hClp1 are revealed by the following findings:

- Complementation of kinase-defective tRNA ligase alleles by hClp1 is abolished by mutations in the hClp1 kinase active site;
- Equivalent mutations of the yClp1 ATP-binding site do not affect yeast viability.
- Overexpression of yClp1 cannot rescue kinase-defective tRNA ligase mutants;
- yClp1 has no detectable RNA kinase activity in vitro;
- hClp1 cannot complement growth of a yeast *clp1*Ä strain.

Chapter 5

Mechanisms of Enzyme Action

Although the catalytic properties of enzymes may seem almost magical, it is simply chemistry—the breaking and making of bonds—that gives enzymes their prowess. Nonetheless, the chemical principles that influence the mechanisms of these few enzymes are universal, and many other cases are understandable in light of the knowledge gained from these examples.

THE BASIC PRINCIPLE

STABILIZATION OF THE TRANSITION STATE

In all chemical reactions, the reacting atoms or molecules pass through a state that is intermediate in structure between the reactant(s) and the product(s). Consider the transfer of a proton from a water molecule to a chloride anion:

$$\underset{\text{Reactants}}{H-O-H + Cl^-} \rightarrow \underset{\text{Transition state}}{H-O^{\delta-}\cdots H\cdots Cl^{\delta-}} \rightarrow \underset{\text{Products}}{HO^- + H-Cl}$$

In the middle structure, the proton undergoing transfer is shared equally by the hydroxyl and chloride anions. This structure represents, as nearly as possible, the transition between the reactants and products, and it is known as the transition state.

Chemical reactions in which a substrate (S) is converted to a product (P) can be pictured as involving a transition state (which we henceforth denote as $X^{\ddagger}$), a species intermediate in structure between S and P.

The catalytic role of an enzyme is to reduce the energy barrier between substrate and transition state. This is accomplished through the formation of an enzyme - substrate complex (ES). This complex is converted to product by passing through a transition state, $EX^{\ddagger}$. As shown, the energy of $EX^{\ddagger}$ is clearly lower than $X^{\ddagger}$. One might be tempted to conclude that this decrease in energy explains the rate acceleration achieved by the enzyme.

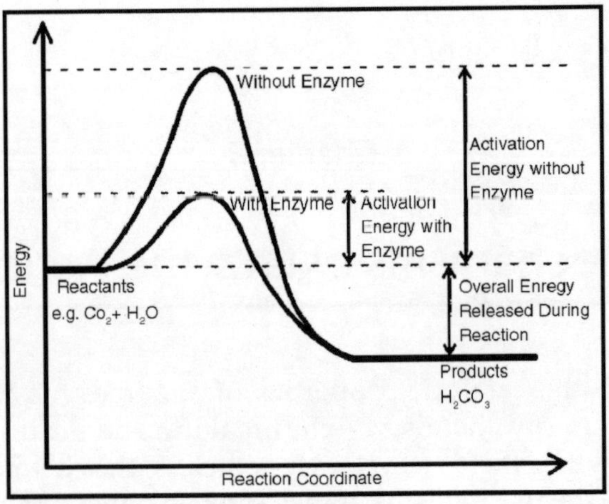

Fig. Enzymes Catalyze Reactions by Lowering the Activation Energy.

The energy barrier for the uncatalyzed reaction is of course the difference in energies of the S and X‡ states. Similarly, the energy barrier to be surmounted in the enzyme-catalyzed reaction, assuming that E is saturated with S, is the energy difference between ES and EX‡. Reaction rate acceleration by an enzyme means simply that the energy barrier between ES and EX‡ is less than the energy barrier between S and X‡. In terms of the free energies of activation, $\Delta Ge^{\ddagger}$, $\Delta Gu^{\ddagger}$.

There are important consequences for this statement. The enzyme must stabilize the transition-state complex, EX‡, more than it stabilizes the substrate complex, ES. Put another way, enzymes are "designed" by nature to bind the transition-state structure more tightly than the substrate (or the product).

The dissociation constant for the enzyme-substrate complex is:

$$K_s = \frac{[E][S]}{[ES]}$$

and the corresponding dissociation constant for the transition-state complex is,

$$K_T = \frac{[E]\left[X^{\ddagger}\right]}{\left[EX^{\ddagger}\right]}$$

Enzyme catalysis requires that $K_T < K_S$. According to transition-state theory the rate constants for the enzyme-catalyzed (k_e) and uncatalyzed (ku) reactions can be related to K_S and K_T by:

$$k_e/k_u > K_S/K_T$$

Thus, the enzymatic rate acceleration is approximately equal to the ratio of the dissociation constants of the enzyme-substrate and enzyme-transition-state complexes, at least when E is saturated with S.

ENZYMES PROVIDE ENORMOUS RATE ACCELERATIONS

Enzymes are powerful catalysts. Enzyme-catalyzed reactions are typically 10^7 to 10^{14} times faster than their uncatalyzed counterparts. These large rate accelerations correspond to substantial changes in the free energy of activation for the urease reaction, for example:

Urea Reactions

Urea Hydrolysis

$$\underset{\text{Urea}}{H_2N\text{-}CO\text{-}NH_2} + H_2O \xrightarrow{\text{Urease}} \underset{\text{Ammonia}}{2\,NH_3} + CO_2$$

Nitrogen Uptake Reaction

$$2\,NH_3 + O_2 + \underset{\substack{\text{Organic}\\\text{Receptor}}}{2\,R\text{-}H} \xrightarrow[\text{Oxidation}]{\text{Plant}} \underset{\substack{\text{Organic}\\\text{Nitrogen}}}{2\,R\text{-}NH_2} + 2\,H_2O$$

Nitrogen Loss Reaction

$$2\,NH_3 + 3/2\,O_2 \xrightarrow[\text{Oridation}]{\text{Microbial}} \underset{\substack{\text{Nitrogen}\\\text{Lost to Air}}}{N_2} + 3\,H_2O$$

An energy of activation some 84 kJ/mol smaller than the corresponding uncatalyzed reaction. To fully understand any enzyme reaction, it is important to account for the rate acceleration in terms of the structure of the enzyme and its mechanism of action. There are a limited number of catalytic mechanisms or factors that contribute to the remarkable performance of enzymes.

These include the following:

- Entropy loss in ES formation
- Destabilization of ES due to strain, desolvation, or electrostatic effects
- Covalent catalysis
- General acid or base catalysis
- Metal ion catalysis
- Proximity and orientation

Any or all of these mechanisms may contribute to the net rate acceleration of an enzyme-catalyzed reaction relative to the uncatalyzed reaction.

A thorough understanding of any enzyme would require that the net acceleration be accounted for in terms of contributions from one or (usually) more of these mechanisms but first it is important to appreciate how the formation of the enzyme-substrate (ES) complex makes all these mechanisms possible.

The Binding Energy of ES Is Crucial to Catalysis

How is it that X‡ is stabilized more than S at the enzyme active site? To understand this, we must dissect and analyse the formation of the enzyme-substrate complex, ES. There are a number of important contributions to the free energy difference between the uncomplexed enzyme and substrate (E + S) and the ES complex. The favourable interactions between the substrate and amino acid residues on the enzyme account for the intrinsic binding energy, DG_b.

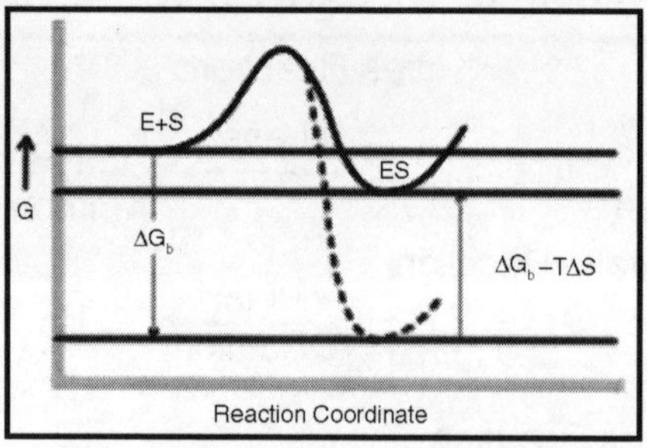

Fig. The Intrinsic Binding Energy of the Enzyme-substrate (ES) Complex.

The intrinsic binding energy ensures the favourable formation of the ES complex, but, if uncompensated, it makes the activation energy for the enzyme-catalyzed reaction unnecessarily large and wastes some of the catalytic power of the enzyme. Because the enzymatic reaction rate is determined by the difference in energies between ES and EX‡, the smaller this difference, the faster the enzyme-catalyzed reaction. Tight binding of the substrate deepens the energy well of the ES complex and actually lowers the rate of the reaction.

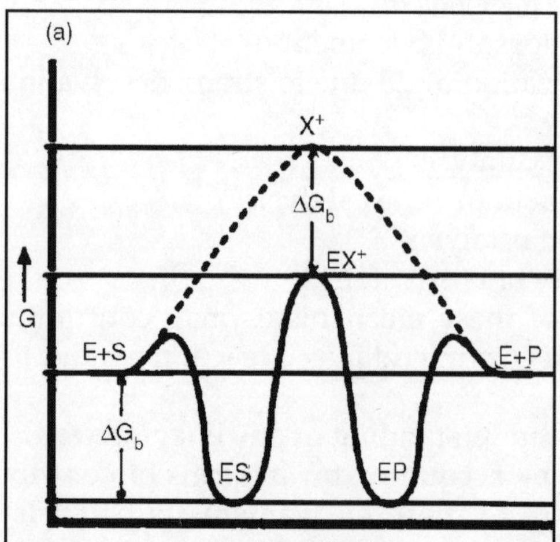

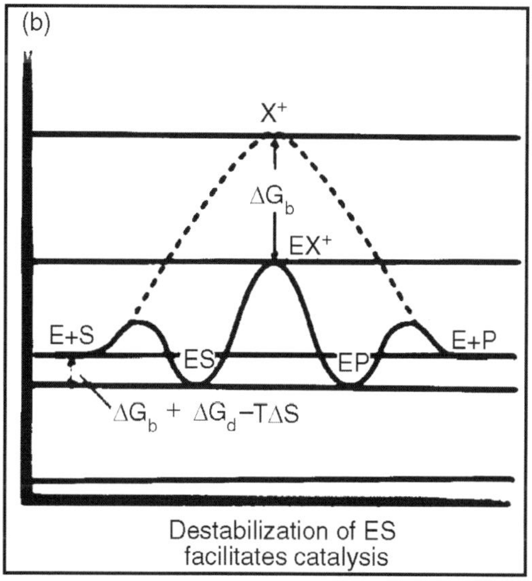

Fig. (a) Catalysis does not Occur if the ES Complex and the Transition State for the Reaction are Stabilized to Equal Extents. (b) Catalysis will Occur if the Transition State is Stabilized to a Greater Extent than the ES Complex.

Entropy Loss and Destabilization of the ES Complex

The message is that raising the energy of ES will increase the enzyme-catalyzed reaction rate.

This is accomplished in two ways:

1. Loss of entropy due to the binding of S to E,
2. Destabilization of ES by strain, distortion, desolvation, or other similar effects.

The entropy loss arises from the fact that the ES complex is a highly organized (low-entropy) entity compared to E + S in solution (a disordered, high-entropy situation). The entry of the substrate into the active site brings all the reacting groups and coordinating residues of the enzyme together with the substrate in just the proper position for reaction, with a net loss of entropy.

The substrate and enzyme both possess translational entropy, the freedom to move in three dimensions, as well as rotational entropy, the freedom to rotate or tumble about any axis through the molecule. Both types of entropy are lost to some extent when two molecules (E and S) interact to form one molecule (the ES complex). Because DS is negative for this process, the term -T D S is a positive quantity, and the intrinsic binding energy of ES is compensated to some extent by the entropy loss that attends the formation of the complex.

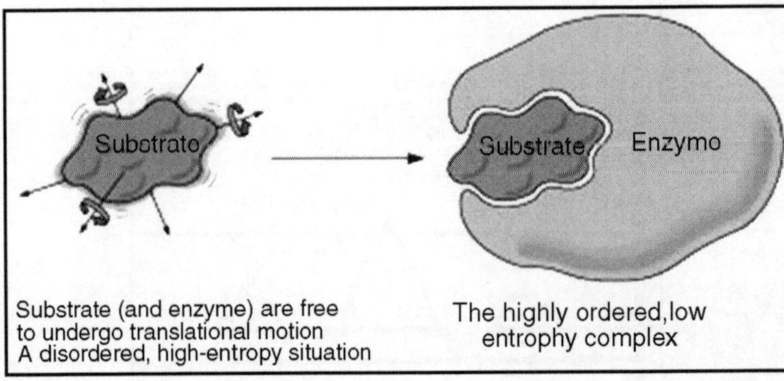

Fig. Formation of the ES Complex Results in a Loss of Entrophy.

Destabilization of the ES complex can involve structural strain, desolvation, or electrostatic effects. Destabilization by strain or distortion is usually just a consequence of the fact (noted previously) that *the enzyme is designed to bind the transition state more strongly than the substrate.* When the substrate binds, the imperfect nature of the "fit" results in distortion or strain in the substrate, the enzyme, or both. This means that the amino acid residues that make up the active site are oriented to coordinate the transition-state structure precisely, but will interact with the substrate or product less effectively. Destabilization may also involve desolvation of charged groups on the substrate upon binding in the active site.

Charged groups are highly stabilized in water. For example, the transfer of Na^+ and Cl^- from the gas phase to aqueous solution is characterized by an enthalpy of solvation, ΔH_{solv}, of -775 kJ/mol. (Energy is given off and the ions become more stable.) When charged groups on a substrate move from water into an enzyme active site, they are often desolvated to some extent, becoming less stable and therefore more reactive.

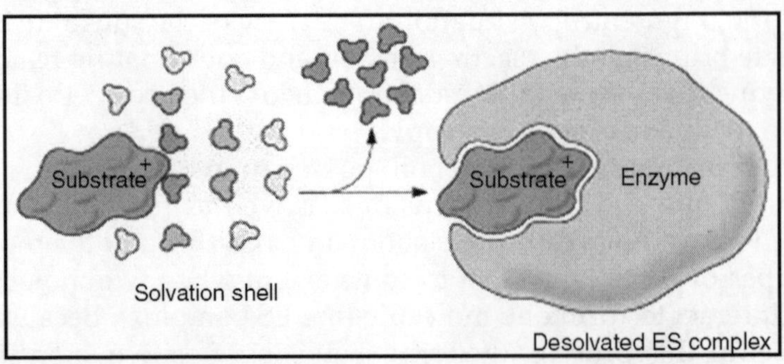

Fig. Substrates Typically lose Waters of Hydration in the

Formation of the ES Complex.

When a substrate enters the active site, charged groups may be forced to interact (unfavourably) with charges of like sign, resulting in electrostatic destabilization. The reaction pathway acts in part to remove this stress. If the charge on the substrate is diminished or lost in the course of reaction, electrostatic destabilization can result in rate acceleration.

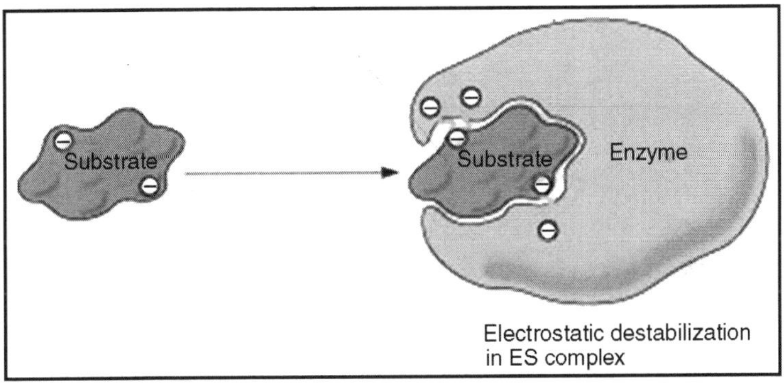

Fig. Electrostatic Destabilization of a Substrate may Arise from Juxtaposition of like Charges in the Active Site.

Whether by strain, desolvation, or electrostatic effects, destabilization raises the energy of the ES complex, and this increase is summed in the term ΔGd, the free energy of destabilization. The net energy difference between E + S and the ES complex is the sum of the intrinsic binding energy, ΔGb; the entropy loss on binding, – T Δ S; and the distortion energy, Gd. ES is destabilized (raised in energy) by the amount ΔGd – TΔS. The transition state is subject to no such destabilization, and the difference between the energies of X‡ and EX‡ is essentially ΔGb, the full intrinsic binding energy.

Transition- State Analogs Bind Very Tightly to the Active Site

It is important to consider the magnitudes of KS and KT. The ratio ke/ku may even exceed 1016, as noted previously. Given a typical ratio of 1012 and a typical KS of 10-3 M, the value of KT should be 10-15 M! This is the dissociation constant for the transition-state complex from the enzyme, and this very low value corresponds to very tight binding of the transition state by the enzyme.

It is unlikely that such tight binding in an enzyme transition state will ever be measured experimentally, however, because the transition state itself is a "moving target." It exists only for about 10-14 to 10-13 sec, less than the time required for a bond vibration. The nature of the elusive transition state can be explored, on the other hand, using transition-state analogs, stable molecules that are chemically and structurally similar to the transition state. Such molecules should bind more strongly than a substrate and more strongly than competitive inhibitors that bear no significant similarity to the transition state. Hundreds of examples of such behaviour have been reported. For example, Robert Abeles studied a series of inhibitors of proline racemase and found that pyrrole-2-carboxylate bound to the enzyme 160 times more tightly than L-proline, the normal substrate.

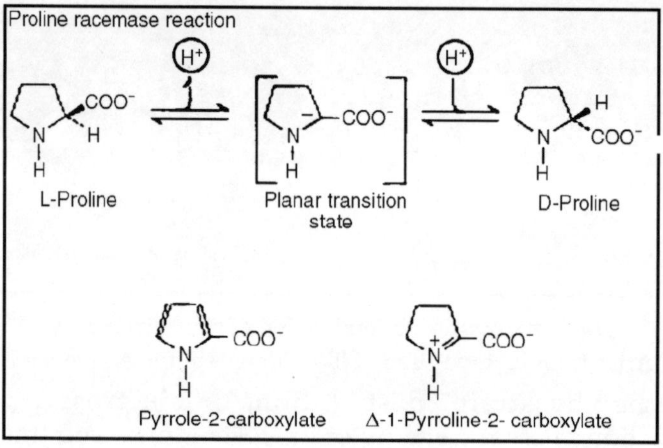

Fig. The Proline Racemase Reaction.

Pyrrole-2-carboxylate and Δ-1-pyrroline-2-carboxylate mimic the planar transition state of the reaction. This analog binds so tightly because it is planar and is similar in structure to the planar transition state for the race-mization of proline. Two other examples of transition-state. Phosphoglycolohydroxamate binds 40,000 times more tightly to yeast aldolase than the substrate dihydroxyacetone phosphate. Even more remarkable, the 1, 6-hydrate of purine ribonucleoside has been estimated to bind to adenosine deaminase with a Ki of 3 × 10-13 M!

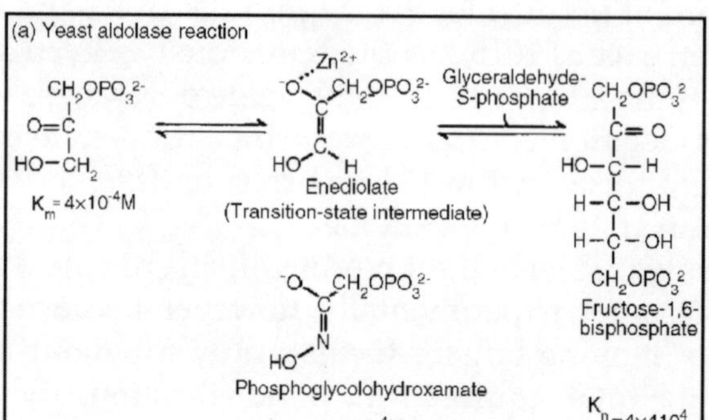

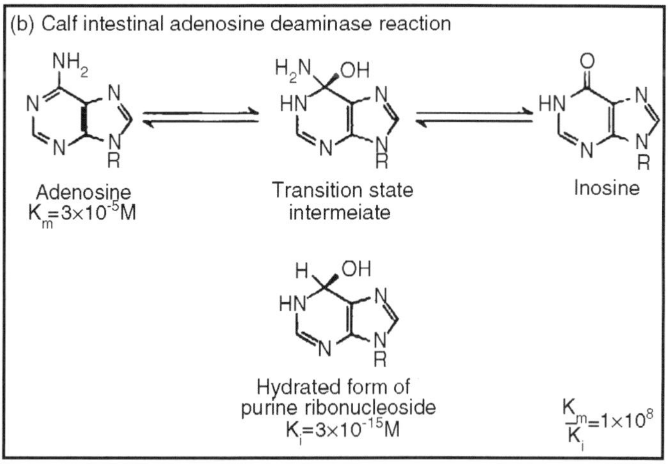

Fig. (a) Phosphoglycolohydroxamate is an Analog of the Enediolate Transition State of the yeast Aldolase Reaction. (b) Purine Riboside, a Potent Inhibitor of the Calf Intestinal Adenosine Deaminase Reaction, Binds to Adenosine Deaminase as the 1, 6-hydrate. The Hydrated form of Purine Riboside is an Analog of the Proposed Transition State for the Reaction.

It should be noted that transition-state analogs are only approximations of the transition state itself and will never bind as tightly as would be expected for the true transition state. These analogs are, after all, stable molecules and cannot be expected to resemble a true transition state too closely.

Covalent Catalysis

Some enzyme reactions derive much of their rate acceleration from the formation of covalent bonds between enzyme and substrate.

Consider the reaction:

$$BX + Y \rightarrow BY + X$$

and an enzymatic version of this reaction involving formation of a covalent intermediate:

$$BX + Enz \rightarrow E{:}\ B + X + Y \rightarrow Enz + BY$$

If the enzyme-catalyzed reaction is to be faster than the uncatalyzed case, the acceptor group on the enzyme must be a better attacking group than Y and a better leaving group than X. Note that most enzymes that carry out covalent catalysis have ping-pong kinetic mechanisms.

The side chains of amino acids in proteins offer a variety of nucleophilic centres for catalysis, including amines, carboxylates, aryl and alkyl hydroxyls, imidazoles, and thiol groups. These groups readily attack electrophilic centres of substrates, forming covalently bonded enzyme-substrate intermediates. Typical electrophilic centres in substrates include phosphoryl groups, acyl groups, and glycosyl groups.

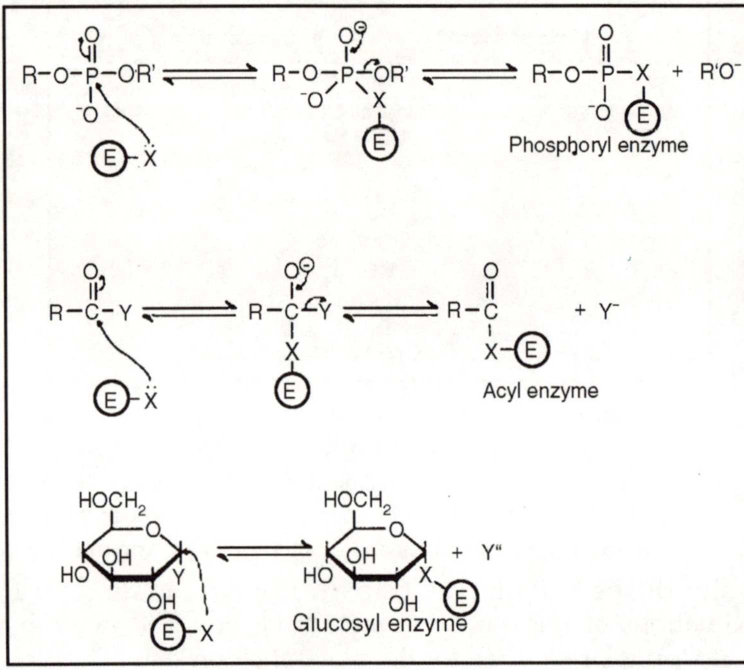

Fig. Examples of Covalent Bond Formation Between Enzyme and Substrate.

The covalent intermediates thus formed can be attacked in a subsequent step by a water molecule or a second substrate, giving the desired product. Covalent electrophilic catalysis is also observed, but usually involves coenzyme adducts that generate electrophilic centres.

General Acid-Base Catalysis

General acid-base catalysis involves a molecule besides water that acts as a proton donor or acceptor during the enzymatic reaction. In order to avoid having an unstable charged intermediates in the reaction, an enzyme has appropriately located functional groups readily to donate a proton or accept a proton to stabilize the transition state. When a functional group accepts a proton, it will release or donate a proton by the end of the catalytic cycle, and vice versa. Functional groups that participate in reaction are His imidazole, alpha-amino group, alpha-carboxyl group, thiol of Cys, R group of carboxyls of Glu and Asp, aromatic OH of Tyr, and guanidino group of ARG. The protonation of these functional groups are dependent on the pH, therefore the enzymatic catalytic activity is sensitive to the pH level.

General acid-base catalysis is involved in a majority of enzymatic reactions, wherein the side chains of various amino acids act as general acids or general basis.

General acid–base catalysis needs to be distinguished from specific acid–base catalysis:

- Specific acid–base catalysis means specifically, –OH or H+ accelerates the reaction. The reaction rate is dependent on pH only (which of course is a function of –OH and H+ concentrations), and not on buffer concentration.
- In General acid–base catalysis, the buffer aids in stabilizing the transition state via donation or removal of a proton. Therefore, the rate of the reaction is dependent on the buffer concentration, as well as the appropriate protonation state.

The mechanism may be a retaining or inverting mechanism. General acid/base catalysis differs from specific acid/base catalysis as in the latter it is the solvent that acts as the acid or base.

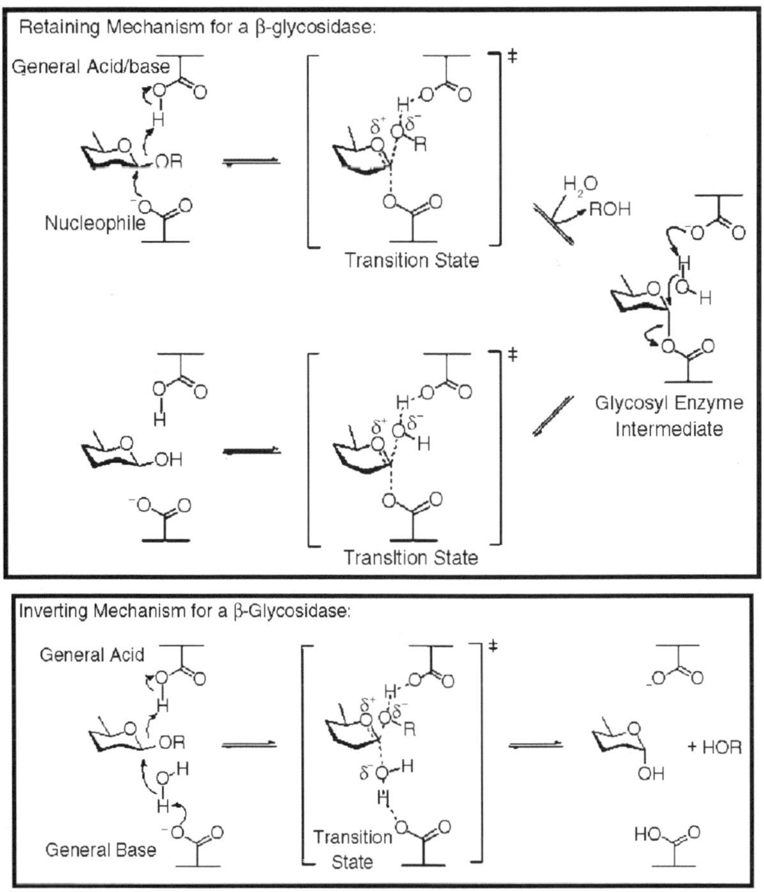

Fig. General Acid-base Catalogies.

METHODS FOR INDENTIFYING THE GENERAL ACID/BASE

Kinetic Analysis of Mutants

Assignment of the acid/base residues may be achieved by the kinetic analysis of enzyme mutants in which the candidate residues have been replaced by a non-nucleophilic residue. Alanine is preferred as a replacement residue owing to its small size. Glycine is non-optimal, as the lack of an alpha-substituent may lead to differences in conformation. Alanine mutants typically have k_{cat} values 1,000-fold lower than the wildtype enzyme when assessed against the natural substrate.This effect is much smaller when the substrate possesses a good leaving group, such as a fluoride or a 2,4-dinitrophenyl group, as in these cases the leaving groups do not require general acid catalysis. In these cases the k_{cat} values are unaffected, but as the deglycosylation step requires general base catalysis the intermediate accumulates, leading to a reduction in the K_M value. A further test for the identity of the acid base residue is the observation of a 'chemical rescue' effect. In the presence of a small anionic nucleophilic group such as azide, fluoride or formate, the glycosyl enzyme will be turned over. In this case the product of 'chemical rescue' should be a glycosyl azide, fluoride or formate with anomeric stereochemistry matching that of the starting glycoside.

Mechanism-based Labelling

Various electrophilic reagents including *N*-bromoacetyl glycosylamines, epoxyalkyl glycosides, and alpha-halo ketones have been used with some success to identify the general acid/base residue. However, such labelling studies are not entirely reliable, and candidate acid/base residues identified through their use should be regarded as preliminary assignments and follow up experiments involving kinetic analysis of mutants should be performed for improving the confidence of the assignment.

GLYCOSIDE HYDROLASES

Glycoside hydrolases are enzymes that catalyze the hydrolysis of the glycosidic linkage of glycosides, leading to the formation of a sugar hemiacetal or hemiketal and the corresponding free aglycon. Glycoside hydrolases are also referred to as glycosidases, and sometimes also as glycosyl hydrolases. Glycoside hydrolases can catalyze the hydrolysis of O-, N- and S-linked glycosides.

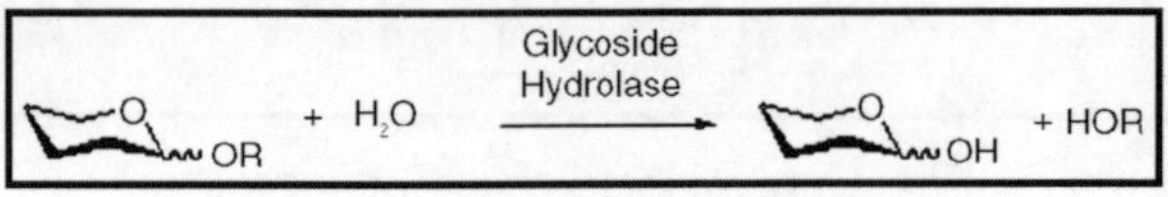

Classification

Glycoside hydrolases can be classified in many different ways. The following paragraphs list several different ways, the utility of which depends on the context in which the classification is made and used.

Endo/exo

exo- and *endo-* refers to the ability of a glycoside hydrolase to cleave a substrate at the end (most frequently, but not always the non-reducing end) or within the middle of a chain. For example, most cellulases are*endo*-acting, whereas LacZ β-galactosidase from *E. coli* is *exo*-acting.

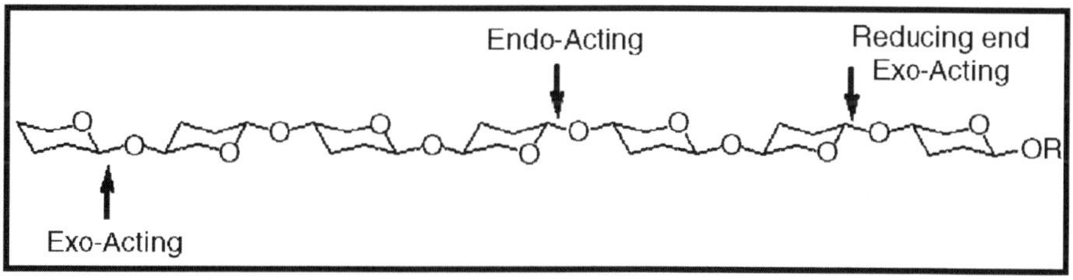

Enzyme Commission (EC) Number

EC numbers are codes representing the Enzyme Commission number. This is a numerical classification scheme for enzymes, based on the chemical reactions they catalyze.

Every EC number is associated with a recommended name for the respective enzyme. EC numbers do not specify enzymes, but enzyme-catalyzed reactions. If different enzymes (for instance from different organisms) catalyze the same reaction, then they receive the same EC number.

A necessary consequence of the EC classification scheme is that codes can be applied only to enzymes for which a function has been biochemically identified. Additionally, certain enzymes can catalyze reactions that fall in more than one class. These enzymes must bear more than one EC number.

Mechanistic Classification

Two reaction mechanisms are most commonly found for the retaining and inverting enzymes, as first outlined by Koshland and as described below.

However several interesting variations on these mechanisms have been found, and one fundamentally different mechanism, catalyzed by an NADH cofactor, has been discovered in recent years, as discussed below.

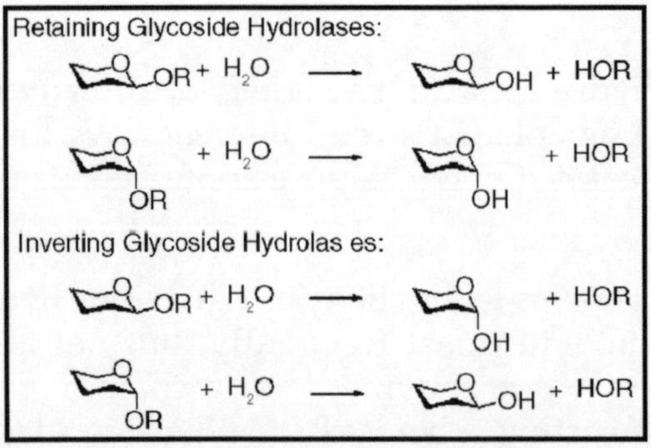

Fig. Mechanistic Classification

Sequence-based Classification

Sequence-based classification uses algorithmic methods to assign sequences to various families. The glycoside hydrolases have been classified into more than 100 families; this is permanently available through the Carbohydrate Active enZyme database. Each family (GH family) contains proteins that are related by sequence, and by corollary, fold. This allows a number of useful predictions to be made since it has long been noted that the catalytic machinery and molecular mechanism is conserved for the vast majority of the glycosidase families as well as the geometry around the glycosidic bond (irrespective of naming conventions). Usually, the mechanism used (ie retaining orinverting) is conserved within a GH family. One notable exception is the glycoside hydrolases of familyGH97, which contains both retaining and inverting enzymes; a glutamate acts as a general base in inverting members, whereas an aspartate likely acts as a catalytic nucleophile in retaining members. Another mechanistic curiosity are the glycoside hydrolases of familes GH4 andGH109 which operate through an NAD-dependent hydrolysis mechanism that proceeds through oxidation-elimination-addition-reduction steps via anionic transition states. This allows a single enzyme to hydrolyze both alpha- and beta-glycosides. Classification of families into larger groups, termed 'clans' has been proposed. A 'clan' is a group of families that possess significant similarity in their tertiary structure, catalytic residues and mechanism. Families within clans are thought to have a common evolutionary ancestry.

MECHANISM

Two reaction mechanisms are most commonly found for the retaining and inverting enzymes, as first outlined by Koshland and as described below. However several interesting variations on these mechanisms have been found, and one fundamentally different mechanism, catalyzed by an NADH cofactor, has been discovered in recent years.

Inverting glycoside Hydrolases

Hydrolysis of a glycoside with net inversion of anomeric configuration is generally achieved via a one step, single-displacement mechanism involving oxocarbenium ion-like transition states, as shown below. The reaction occurs with acid/base assistance from two amino acid side chains, normally glutamic or aspartic acids, that are typically located 6-11 A apart.

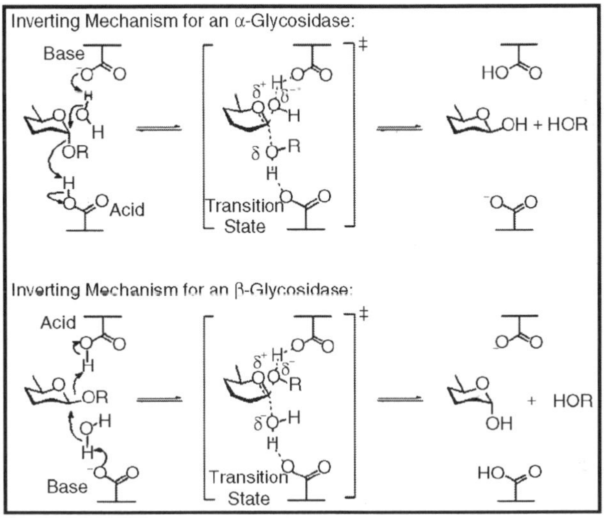

Fig. Inverting glycoside Hydrolases

RETAINING GLYCOSIDE HYDROLASES

Classical Koshland Retaining Mechanism

Hydrolysis with net retention of configuration is most commonly achieved via a two step, double-displacement mechanism involving a covalent glycosyl-enzyme intermediate. Each step passes through an oxocarbenium ion-like transition state. Reaction occurs with acid/base and nucleophilic assistance provided by two amino acid side chains, typically glutamate or aspartate, located 5.5 A apart. In the first step (often called the glycosylation step), one residue plays the role of a nucleophile, attacking the anomeric centre to displace the aglycon and form a glycosyl enzyme intermediate.

At the same time the other residue functions as an acid catalyst and protonates the glycosidic oxygen as the bond cleaves. In the second step (known as the deglycosylation step), the glycosyl enzyme is hydrolyzed by water, with the other residue now acting as a base catalyst deprotonating the water molecule as it attacks. The pKa value of the acid/base group cycles between high and low values during catalysis to optimize it for its role at each step of catalysis. In the case of sialidases, the catalytic nucleophile is a tyrosine residue. This mechanism was originally proposed by Dan Koshland, although at the time the identities of the residues was unclear.

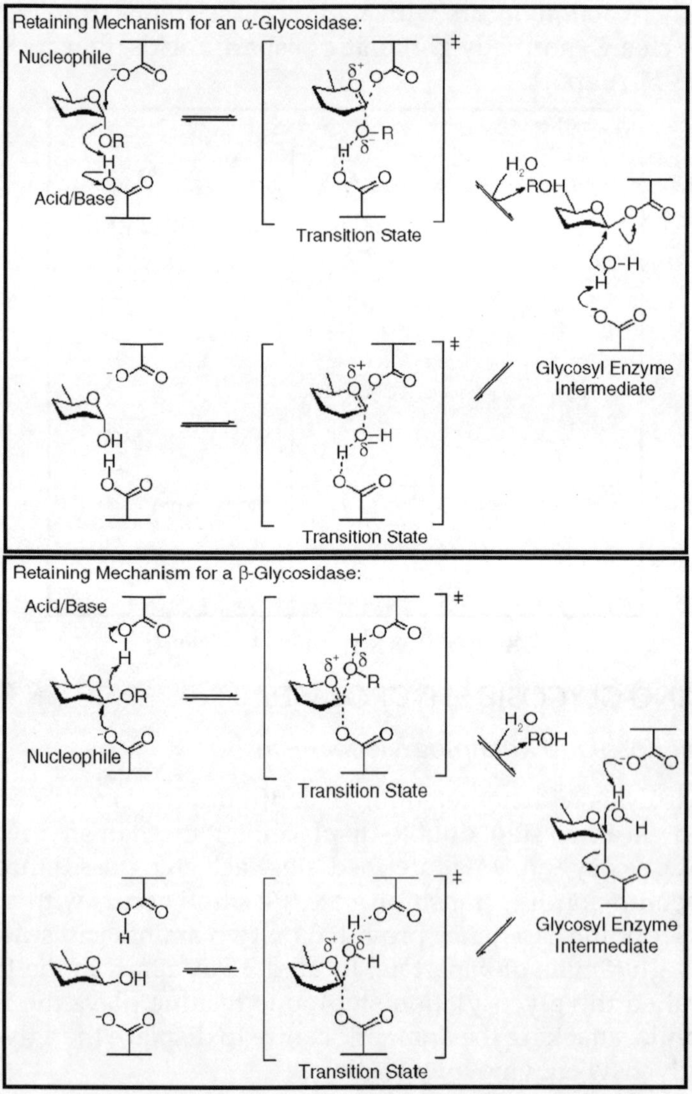

Fig. Classical Koshland Retaining Mechanism

Neighbouring group Participation

Enzymes of glycoside hydrolase families 18, 20,25, 56, 84, and 85 hydrolyse substrates containing an *N*-acetyl (acetamido) or *N*-glycolyl group at the 2-position. These enzymes have no catalytic nucleophile: rather they utilize a mechanism in which the 2-acetamido group acts as an intramolecular nucleophile. Neighbouring group participation by the 2-acetamido group leads to formation of an oxazoline (or more strictly an oxazolinium ion) intermediate. This mechanism was deduced from X-ray structures of complexes of chitinases with natural inhibitors, from the potent inhibition afforded by a stable thiazoline analogue of the oxazoline, and from detailed mechanistic analyses using substrates of modified reactivity. Typically, a stabilizing residue (a carboxylate) stabilizes the charge development in the transition state. Not all enzymes that cleave substrates possessing a 2-acetamido group utilize a neighbouring groups participation mechanism; enzyme of glycoside hydrolase families 3 and 22 utilize a classical retaining mechanism with an enzymic nucleophile. Other hexosaminidases such as those of Glycoside Hydrolase Family 19 utilize an inverting mechanism.

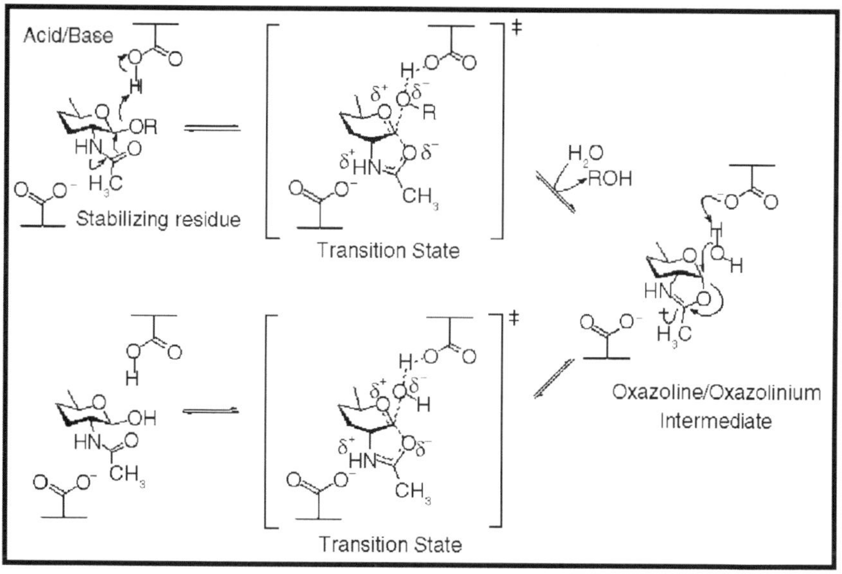

Fig. Neighbouring group Participation

Exogenous Bases: Myrosinases

Glycoside hydrolases termed myrosinases catalyze the hydrolysis of anionic thioglycosides (glucosinolates) found in plants. They are found in Glycoside Hydrolase Family 1. In these enzymes the usual acid/base glutamic acid found in other members of this family is replaced by a glutamine. This likely reduces charge repulsion between the anionic aglycon sulfate. The unusual aglycon is a sufficiently good leaving group to be able to cleave in the first glycosylation step to form the glycosyl enzyme intermediate without the requirement of an acid catalyst. However, since a base catalyst is required for the second step (hydrolysis or deglycosylation) these enzymes require an alternative basic group. This is provided by the co-enzyme L-ascorbate.

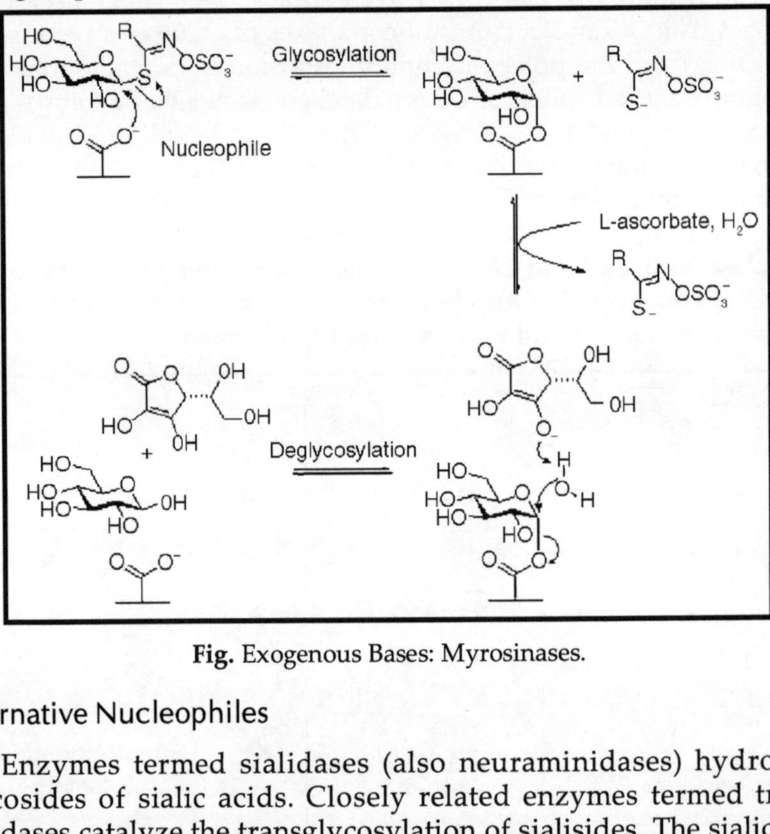

Fig. Exogenous Bases: Myrosinases.

Alternative Nucleophiles

Enzymes termed sialidases (also neuraminidases) hydrolyze glycosides of sialic acids. Closely related enzymes termed trans-sialidases catalyze the transglycosylation of sialisides. The sialidases and trans-sialidases of glycoside hydrolase families 33 and 34 utilize a tyrosine as a catalytic nucleophile, which is activated by an adjacent base residue. A rationale for this unusual difference is that the use of a negatively charged carboxylate as a nucelophile will be disfavoured as the anomeric centre is itself negatively charged, and thus charge repulsion interferes. A tyrosine residue is a neutral nucleophile, but requires a general base to enhance its nucleophilicity. This mechanism was implied from X-ray structures, and was supported by experiments involving trapping of the intermediate with fluorosugars followed by peptide mapping and then crystallography, as well as via mechanistic studies on mutants.

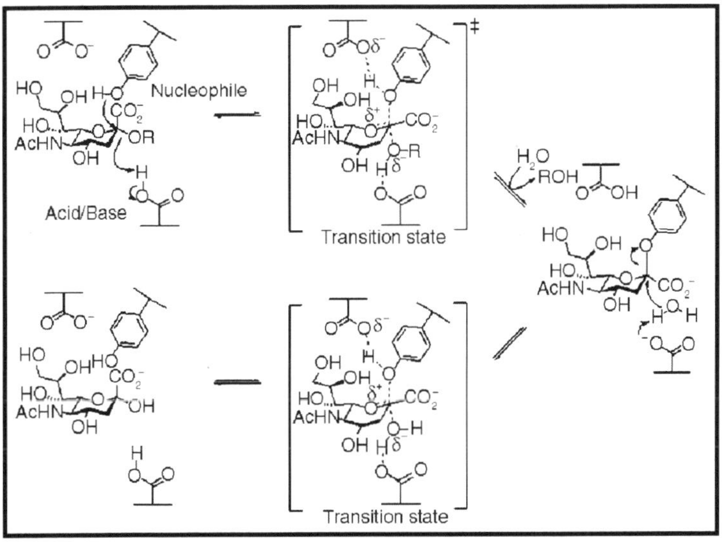

Fig. Alternative Nucleophiles.

NAD-dependent Hydrolysis

The glycoside hydrolases of family 4 and 109 use a mechanism that requires an NAD cofactor, which remains tightly bound throughout catalysis.

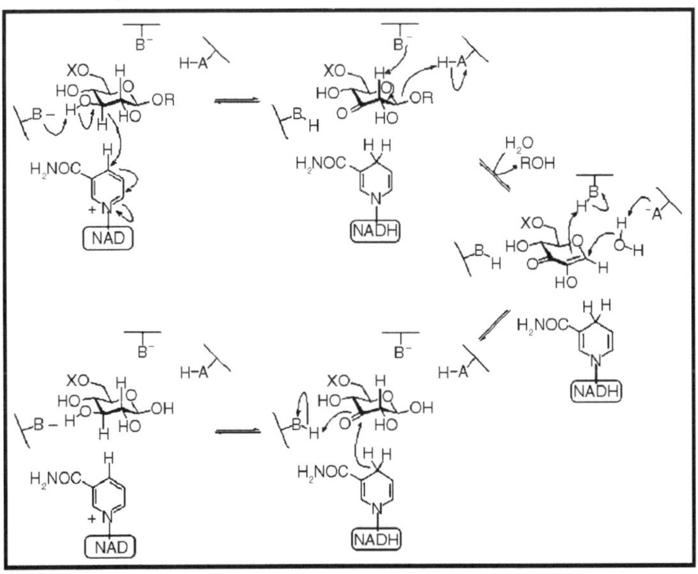

Fig. NAD-dependent Hydrolysis.

The mechanism proceeds via anionic transition states with elimination and redox steps rather than the classical mechanisms proceeding throughoxocarbenium ion-like transition states. As shown below for a 6-phospho-beta-glucosidase, the mechanism involves an initial oxidation of the 3-hydroxyl of the substrate by the enzyme-bound NAD cofactor. This increases the acidity of the C2 proton such that an $E1_{cb}$ elimination can occur with assistance from an enzymatic base. The alpha-beta unsaturated intermediate formed then undergoes addition of water at the anomeric centre and finally the ketone at C3 is reduced to generate the free sugar product. Thus, even though glycosidic bond cleavage occurred via an elimination mechanism, the overall reaction is hydrolysis. This mechanism was elucidated through a combination of stereochemical studies by NMR, kinetic isotope effects, linear free energy relationships, X-ray crystallography and UV/Vis spectrophotometry.

METAL ION CATALYSIS

Many enzymes require metal ions for maximal activity. If the enzyme binds the metal very tightly or requires the metal ion to maintain its stable, native state, it is referred to as a metalloenzyme. Enzymes that bind metal ions more weakly, perhaps only during the catalytic cycle, are referred to as metal activated. One role for metals in metal-activated enzymes and metalloenzymes is to act as electrophilic catalysts, stabilizing the increased electron density or negative charge that can develop during reactions. Among the enzymes that function in this manner is liver alcohol dehydrogenase. Another potential function of metal ions is to provide a powerful nucleophile at neutral pH. Coordination to a metal ion can increase the acidity of a nucleophile with an ionizable proton:

$$M^{2+} + NucH \rightleftharpoons M^{2+}(NucH) \rightleftharpoons M^{2+}(Nuc^-) + H^+$$

A metal such as Cu2+ or Zn2+ can also stabilize the TS. The metal must be able to be bound to the charged intermediate and hence the TS. The tetrahedral oxyanion intermediate of the reaction of an electrophilic carbonyl C can interact with a metal if there is an O on an adjacent atom which can help coordinate the metal ion. This charge stabilization of the developing negative in the TS and the full negative in the intermediate is often called electrostatic catalysis. This method is likely to be found in many enzymes since nearly 1/3 of all enzymes require metal ions. A classic example of an enzyme using metal ion catalysis is carboxypeptidase A.

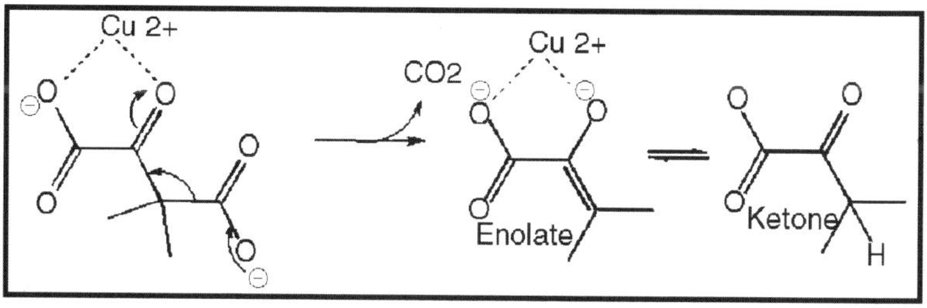

Fig. Metal Ion Or Electrostatic Catalysis - Stabilization Of Ts Charge

Metals can also act in a different way. They may coordinate a water and by further polarizing the H-O bond increase the acidity of the bound water. For instance, the water molecule in the pentammineaquacobalt(III) ion has a pKa of 6.6, compared to pure water, with a pKa of 15.7. (To calculate the latter, write the equilibrium expression for water: H2O + H2O <==> H3O+ + OH-. Then write the Ka expression, as for a generic acid, which is [H3O+][OH-]/[H2O] = 10-14/55.5. The pKa is 15.7). The complexed hydroxide is a better nucleophile than bulk water. An example of an enzyme whose bound metal increases the nucleophilicity of water is carbonic anhydrase.

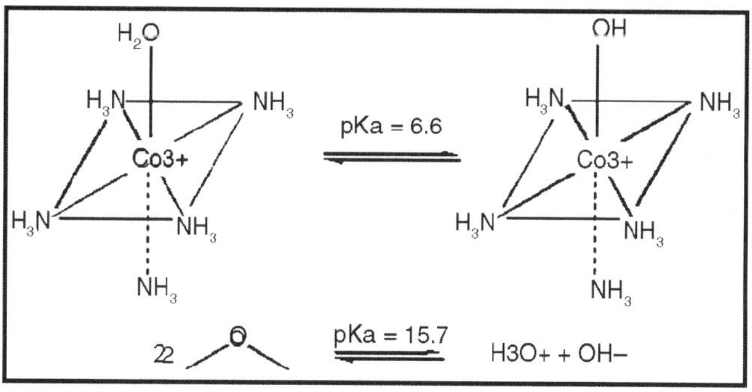

Fig. Metal Ion Catalysis - Decreasing Pka Of Water

COVALENT OR NUCLEOPHILIC CATALYSIS

One way to change the activation energy of the reaction is to change the reaction mechanism in ways which introduces new steps with lower activation energy. A typical way is to add a nucleophilic catalyst which forms a covalent intermediate with the reactant. The original nucleophile can then interact with the intermediate in a nucleophilic substitution reaction. If the nucleophilic catalyst is a better nucleophile than the original nucleophile (usually water) then the reaction is catalyzed. The nucleophilic catalyst and the original nucleophile usually interact with a carbonyl C in a substitution reaction, initially forming the tetrahedral oxyanion intermediate.

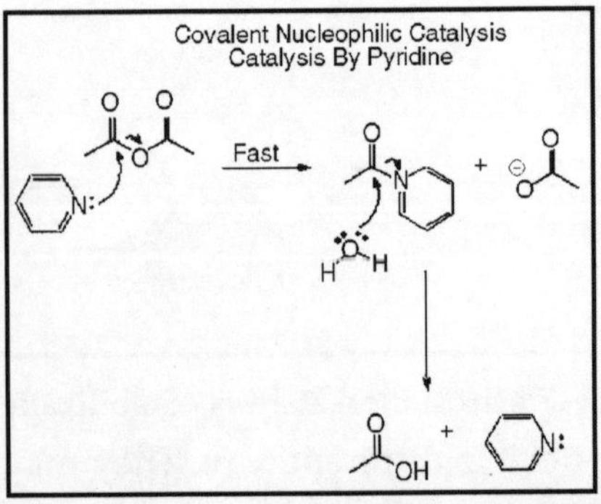

Fig. Nucleophilic Covalent Catalysis By Pyridine

If an amine is used as the nucleophilic catalyst, then the initial addition product (a carbinolamine) can become dehydrated, since the free pair of electrons on the N are more likely to be shared with the carbon to form a double bond than electrons from the original carbonyl O, which is more electronegative than the N). An imine or Schiff Base forms, with a pKa of about 7. This is easily protonated to form a positively charged N at the former carbonyl O centre. This serves as an excellent electron sink for decarboxylation reactions of beta-keto acids and illustrates an important point. Electrons in chemical reactions can be viewed as flowing from a source (such as a carboxyl group) to a sink.

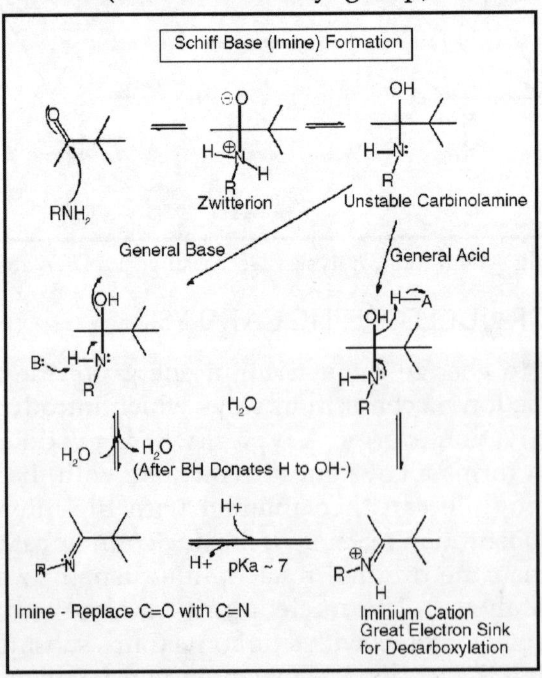

Fig. Mechanism Of Nucleophilic Catalysis by Amines - Schiff Base Formation.

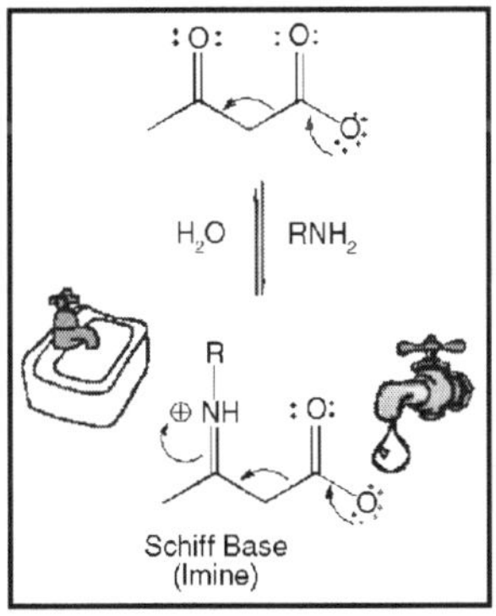

Fig. Electron Flow: Source To Sink

An intriguing question arises: how much of the structure of a large protein is really needed for catalysis? Much work has been directed to the development of small molecule catalysis mimetics of large protein enzymes. Just how small can you go in reducing the size of a protein and still get catalysis. One important feature of enzyme catalysis is that they catalyze reactions in which only one enantiomer is produced. That is, the synthesis is assymertric. This is typically a consequence of the asymmetric enzyme (itself chiral) binding only one enantiomer as a reactant and/or the imposition of steric restrictions on the possible reactions of the bound substrate. Recently, it has been show that L-Pro alone can act as such an assymetric catalyst in an aldol condensation reaction.

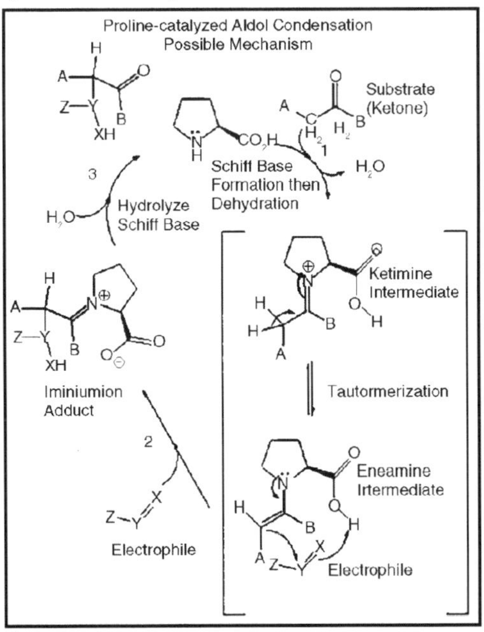

Fig. L-Pro Catalysis Of An Aldol Condensation: Possible Mechanism

INTRAMOLECULAR CATALYSIS

Consider the hydrolysis of phenylacetate. This reaction, a nucleophilic subsitution reaction, could be catalyzed by the addition to solution of the general base acetate, as described above.

Since this reaction would double with the doubling of the solution acetate, the reaction is bimolecular (first order in reactant and catalyst). Now consider the same reaction only when the the general base part of the catalyst, the carboxyl group, is part of the reactant phenylacetate.

Such a case occurs in the acetylated form of salicylic acid - *i.e.* aspirin. When the carboxy group is ortho compared to the acetylated phenolic OH, it is in perfect position to accept a proton from water, decreasing the charge development on the O in the transition state. The general base does not have to diffuse to the appropriate site when it is intramolecular with respect to the carbonyl C of the ester link. The rate of this intramolecular base catalysis is about 100 fold greater than of an intermolecular base catalyst like acetate. It is as if the effective concentration of the intramolecular carboxyl base catalyst is much higher due to its proximity to the reaction site.

Another type of reactions involving a carboxyl group (in addition to simple proton transfer) is when the negatively charged carboxyl O acts as a nucleophile and attacks an electrophilic carbonyl carbon. When the carbonyl is part of an ester, the carboxyl group engages in a nucleophilic substitution reaction, expelling the alcohol part of the ester as a leaving group.

The remaining examples below consider the nucleophilic (carboxyl) substitution on phenylesters, with phenolate as the leaving group. The reactions in effect transfer an acyl group to the carboxyl group to create an anhydride.

First consider acyl transfer with aspirin derivatives. Aspirin, as you know, contains a carboxyl group ortho to an ester substitutent. Hence the carboxyl group can act as a nucleophile and attack the carbonyl carbon of the ester in a nucleophilic substitution reaction. The net effect is to transfer the acetyl group from the phenolic OH to the carboxyl group converting it to an anhydride. This is an intramolecular reaction. Compare this reaction to a a comparable bimolecular reaction shown below.

The first order rate constant of the intramolecular transfer of the acetyl group to the carboxyl group, $k_1 = 0.02\ s^{-1}$. The analogous bimolecular reaction rate constant $k_2 \sim 10^{-10}\ M^{-1}s^{-1}$. Dividing k_1/k_2 gives the relative rate enhancement of the intramolecular over the intermolecular reaction. With units of molarity, this ratio can be interpreted as the relative effective concentration of the intramolecular nucleophile. This makes the effective concentration of the carboxylate in the aspirin derivative 2 × 10^7 M.

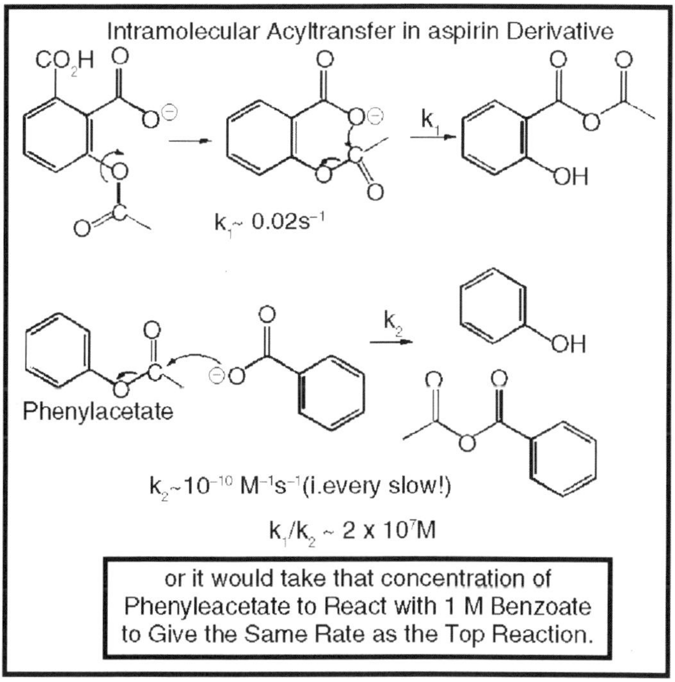

Fig. Acyl Transfer In Aspirin Derivatives.

Now consider the cleavage of phenylacetate using acetate as the nucleophile. The products are acetic anhydride and phenolate. This is a bimolecular reaction (a slow one at that), with a bimolecular rate constant, k_2 which I will arbitrarily set to 1 for comparison to some similar reactions.

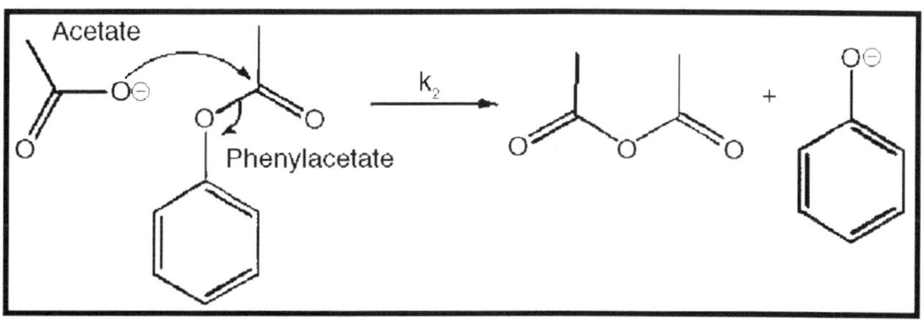

Fig. Reaction of Acetate With Phenylacetate

Now consider a monoester derivatives of succinic acid - phenyl succinate - in which the free carboxyl group of the ester attacks the carbonyl carbon of the ester derivative.

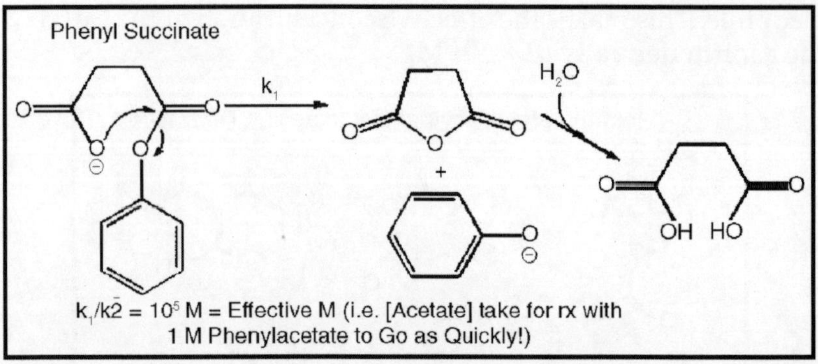

Fig. Intramolecular Reaction of Phenylsucciate.

If you assign a second order rate constant k_2 = 1 $M^{-1}s^{-1}$ to the analogous intermolecular reaction of acetate with phenylacetate (as described above), the first order rate constant for the intramolecular reaction of phenylsuccinate is 10^5 s^{-1}.

The ratio of rate constants, $k_1/k_2 = 10^5$ M. That is it would take 10^5 M concentration of acetate reacting with 1 M phenylacetate in the first bimolecular reaction to get a reaction as fast as the intramolecular reaction of phenylsuccinate. An even more sterically restricated bicyclic phenylcarboxylate shows a $k_1/k_2 = 10^8$ M.

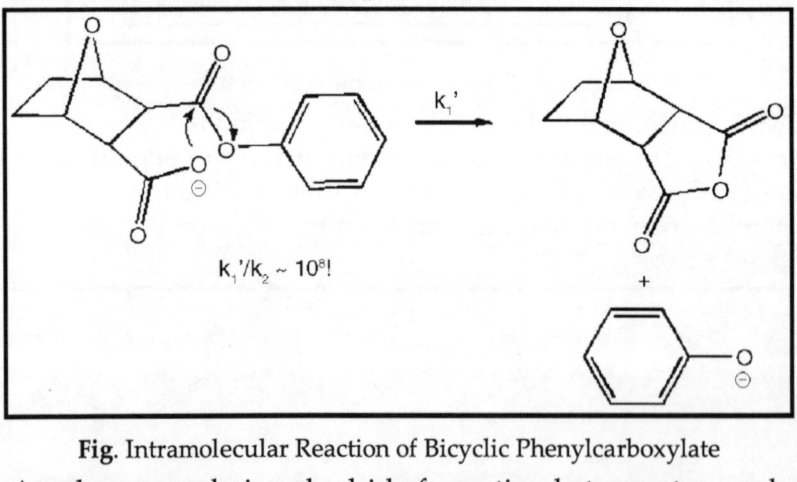

Fig. Intramolecular Reaction of Bicyclic Phenylcarboxylate

Another example is anhydride formation between two carboxyl groups. The DG° for such a reaction is positive, suggesting an unfavourable reaction. Consider two acetic acid molecules condensing to form acetic anhydride.

For this intermolecular reaction, $K_{eq} = 3x10^{-12} M^{-1}$. Now consider the analogous intramolecular reaction of the dicarboxylic acid succinic acid. It condenses in an intramolecular reaction to form succinic anhydride with a $K_{eq} = 8 \times 10^{-7}$ (no units). The ratio $K_{eq\text{-}intra}/K_{eq\ inter} = 3 \times 10^5$ M. It is as if the effective concentration of the reacting groups. because they do not have to diffuse together to react, is 3×10^5 M.

How does this apply to enzyme catalyzed reaction? Enzymes bind substrates in physical steps which are typically fast. The slow step is chemical conversion of the bound substrate, which is effectively intramolecular.

These three kinds of reactions, intermolecular, intramolecular, and enzyme-catalysed can be broken down into two hypothetical steps, a binding followed by catalysis.

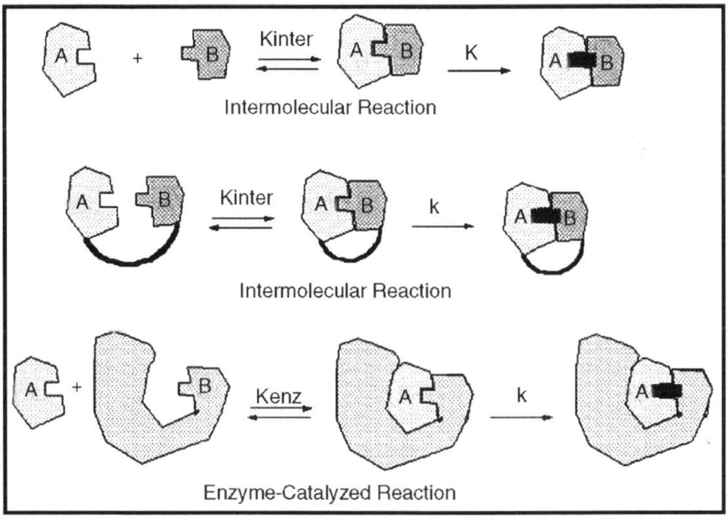

Fig. Three kinds of reactions, Intermolecular, Intramolecular, and Enzyme-Catalysed

If the rate constants for the chemical steps are all identical, the advantage of the intramolecular and enzyme-catalyzed reaction over the intermolecular reaction is K_{INTRA}/K_{INTER} and K_{ENZ}/K_{INTER}, respectively.

Calcium in solution exists as a octahedrally coordinated complex with water occupying all the coordination sites. EDTA, a multidentate ligand, first interacts through one of its potential six electron donors to Ca in a reaction which is entropically disfavoured from the the Ca-EDTA perspective, although one water is released. Once this first intramolecular complex is formed, the rest of the ligands on the EDTA rapidly coordinate with the Ca and release bound water. The former is no longer entropically disfavoured since it is now an intramolecular process while the later is favoured through the release of the remaining five water molecules.

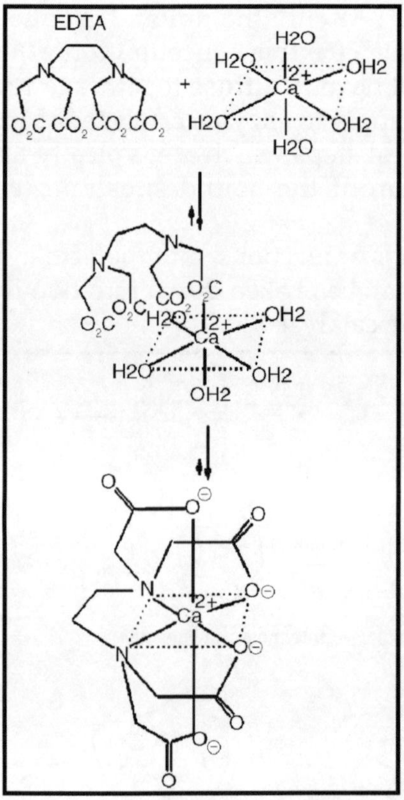

Fig. Binding of Ca2+ And Edta.

We modeled the catalytic advantage offered by intramolecular reaction in terms of a dramatic increase in the effective concentration of reactants, which sometimes reached levels of 10^8 M. Another way is to look at entropy changes associated with dimer formation.

TRANSITION STATE STABILIZATION

Linus Pauling postulated long ago that the only thing that a catalyst must do is bind the transition state more tightly than the substrate. That this must be the case can be seen from the diagram below, which shows how S and S^* (the transition state) can react with E to form a complex which then proceeds to product, or can go to product in the absence of E. From this diagram, it should be evident that c – a = d –b, where a is the DGº for the binding of S to E, and b is the DGº for the binding of S^* to E. For an enzyme to be a catalyst the activation energy for the reaction in the presence of E, d, must be less than in the absence of enzyme, c. Therefore c-d = a-b > 0. Since DGº = -RTln K_{eq}, K_{eq} for binding of S^* to E is greater than for S binding to E.

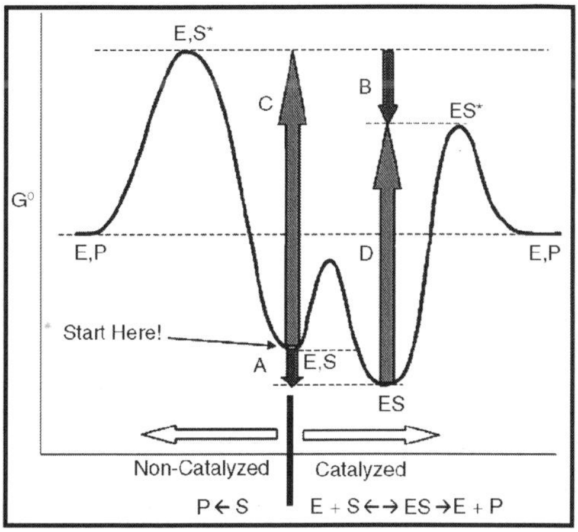

Fig. Enzymes Bind The Transition State More Tightly than The Substrate

The stability of the transition state also affects the reaction kinetics (which makes sense given that the activation energy clearly affects the speed of a reaction). As you probably remember from organic chemistry, S_N2 reactions are slow when the central atom where the substitution will occur is surrounded by bulky substitutents. (Sterics once again.) We discussed this in context to nucleophiliic substitution on a sp^2 hybridized carbonyl carbon in carboxylic acid derivatives versus on a sp^3 hybridized phosphorous in phosphoesters and diesters. The explanation for this phenomena has usually been attributed to hindered access of the central atom caused by bulky substituents (intrinsic effects).

Is this true? Recent studies on S_N2 reactions of methylchloroacetonitrile and t-butylchloroacetonitrile (with the reagent labeled with ^{35}Cl) using $^{37}Cl^-$ as the incoming nucleophile in the gas phase shown that the more hindered t-butyl derivative's activation energy was only 1.6 kcal/mol higher than the methyl derivative, but in aqueous solution, the difference is much greater for comparable reactions. They attributed the differences to solvation effects of the transition state. The bulkier the substituents on the central atom, the more difficult it is to solvate the transition state since water can't reorient around it as well. In effect there is steric hindrance for both reactant and solvent.

Abyzmes - Antibody Catalysis

What does it take for a macromolecule (M) to be a catalyst - an enzyme.

It seems the minimum criterion are:

- M binds a reactant
- M binds the transition state more tightly than the substrate

Anything above these is just "icing on the cake". If different functional group are present in the "active" site of the enzyme that would allow electrostatic, intramolecular, covalent, general acid and/ or base catalysis, the better the catalyst. Linus Pauling recognized the two key factors decades ago. He made the following hypothesis: Antibody molecules (immune system proteins that bind foreign molecules) that can be made to bind to transition state analogs of a substrate, should also presumably catalyze the conversion of substrate, through the transition state, to product. About a decade ago, his prediction was verified. Lerner *et al.* made a transition state analog of an ester. When an ester is hydrolyzed, the sp^2 hybridized carbonyl carbon is converted to an sp^3 hybridized centre in the intermediate, with the carbonyl oxygen becoming an oxyanion. The transition state presumably looks more like this unstable intermediate (sp^3, oxyanion). Lerner synthesized a phosphonate, an ester mimic with a sp^3 hybridized phosphorous replacing the carbonyl C. It also has a negatively charged oxygen as does the intermediate for the ester. This phosphonate ester is very resistant to hydrolysis. When injected into a mouse (after first being covalently attached to a carrier protein so the small molecule becomes "immunogencic"), the mouse makes a protein antibody which binds to the phosphonate. When the corresponding carboxylic acid ester is added to the antibody, it is cleaved with nominal k_{cat} and Km values. Site specific mutagenesis can then be done to make it an even better catalyst! The antibody enzymes have been called abzymes.

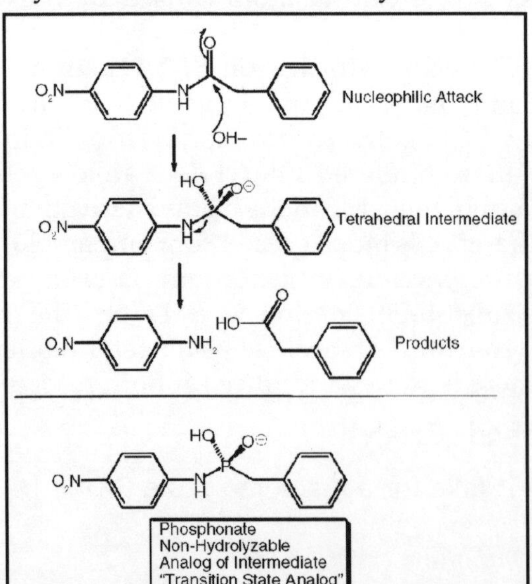

Fig. Phosphonates: Transition State Analogs

PROXIMITY

Chemical reactions go faster when the reactants are in proximity, that is, near each other. In solution or in the gas phase, this means that increasing the concentrations of reacting molecules, which raises the number of collisions, causes higher rates of reaction. Enzymes, which have specific binding sites for particular reacting molecules, essentially take the reactants out of dilute solution and hold them close to each other. This proximity of reactants is said to raise the "effective" concentration over that of the substrates in solution, and leads to an increased reaction rate. In order to measure proximity effects in enzyme reactions, enzymologists have turned to model studies comparing intermolecular reaction rates with corresponding or similar intramolecular reaction rates. A typical case is the imidazole-catalyzed hydrolysis of p-nitrophenylacetate. Under certain conditions the rate constant for this bimolecular reaction is 35 M-1 min-1. By comparison, the first-order rate constant for the analogous but intramolecular reaction is 839 min^{-1}.

The ratio of these two rate constants,

$$(839\ min^{-1})/(35\ M^{-1}\ min^{-1}) = 23.97\ M$$

has the units of concentration and can be thought of as an effective concentration of imidazole in the intramolecular reaction. Put another way, a concentration of imidazole of 23.9 M would be required in the intermolecular reaction to make it proceed as fast as the intramolecular reaction.

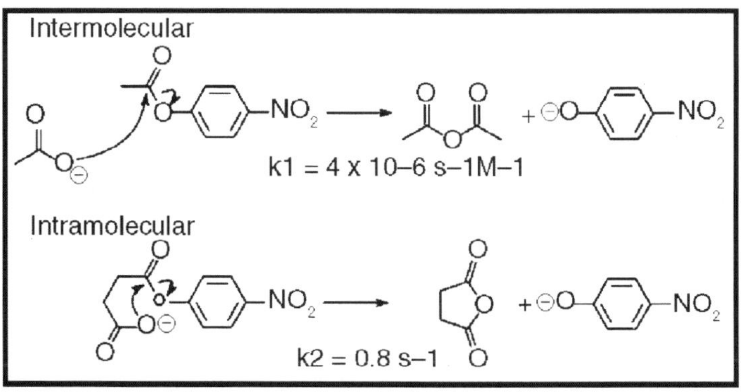

Fig. Proximity Effects in Catalysis

There is more to this story, however. Enzymes not only bring substrates and catalytic groups close together, they orient them in a manner suitable for catalysis as well. Comparison of the rates of reaction of the molecules makes it clear that the bulky methyl groups force an orientation on the alkyl carboxylate and the aromatic hydroxyl groups that makes them approximately 250 billion times more likely to react. Enzymes function similarly by placing catalytically functional groups (from the protein side chains or from another substrate) in the proper position for reaction.

Clearly, proximity and orientation play a role in enzyme catalysis, but there is a problem with each of the above comparisons. In both cases, it is impossible to separate true proximity and orientation effects from the effects of entropy loss when molecules are brought togetherare much smaller than the values. Simple theories based on probability and nearest-neighbour models, for example, predict that proximity effects may actually provide rate increases of only 5- to 10-fold. For any real case of enzymatic catalysis, it is nonetheless important to remember that proximity and orientation effects are significant.

TYPICAL ENZYME MECHANISMS

These particular cases are well understood, because the three-dimensional structures of the enzymes and the bound substrates are known at atomic resolution, and because great efforts have been devoted to kinetic and mechanistic studies. They are important because they represent reaction types that appear again and again in living systems, and because they demonstrate many of the catalytic principles cited above. Enzymes are the catalytic machines that sustain life, and what follows is an intimate look at the inner workings of the machinery.

SERINE PROTEASES

Serine proteases are a class of proteolytic enzymes whose catalytic mechanism is based on an active-site serine residue. Serine proteases are one of the best-characterized families of enzymes. This family includes *trypsin, chymotrypsin, elastase, thrombin, subtilisin, plasmin, tissue plasminogen activator,* and other related enzymes. The first three of these are digestive enzymes and are synthesized in the pancreas and secreted into the digestive tract as inactive proenzymes, or zymogens.

Within the digestive tract, the zymogen is converted into the active enzyme form by cleaving off a portion of the peptide chain. Thrombin is a crucial enzyme in the blood-clotting cascade, subtilisin is a bacterial protease, and plasmin breaks down the fibrin polymers of blood clots. Tissue plasminogen activator (TPA) specifically cleaves the proenzyme *plasminogen,* yielding plasmin. Owing to its ability to stimulate breakdown of blood clots, TPA can minimize the harmful consequences of a heart attack, if administered to a patient within 30 minutes of onset. Finally, although not itself a protease, *acetylcholinesterase is a serine esterase* and is related mechanistically to the serine proteases. It degrades the neurotransmitter acetylcholine in the synaptic cleft between neurons.

The Digestive Serine Proteases

Trypsin, chymotrypsin, and elastase all carry out the same reaction — the cleavage of a peptide chain — and although their structures and mechanisms are quite similar, they display very different specificities. Trypsin cleaves peptides on the carbonyl side of the basic amino acids, arginine or lysine. Chymotrypsin prefers to cleave on the carbonyl side of aromatic residues, such as phenylalanine and tyrosine. Elastase is not as specific as the other two; it mainly cleaves peptides on the carbonyl side of small, neutral residues. These three enzymes all possess molecular weights in the range of 25,000, and all have similar sequences and three-dimensional structures.

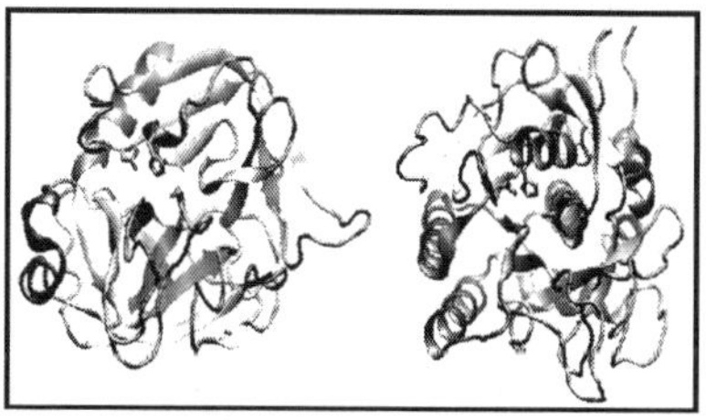

Fig. Structure of Chymotrypsin

The molecule is ellipsoidal in shape and contains an α-helix at the C-terminal end (residues 230 to 245) and several β-sheet domains. Most of the aromatic and hydrophobic residues are buried in the interior of the protein, and most of the charged or hydrophilic residues are on the surface. Three polar residues — His^{57}, Asp^{102}, and Ser^{195} — form what is known as a catalytic triad at the active site.

CATALYTIC TRIAD

A *catalytic triad* is a group of three amino acids that are found in the active sites of some proteases involved in catalysis. Three different proteases that have catalytic triads are: chymotrypsin, trypsin and elastase. In chymotrypsin, the catalytic triad is made from *serine 195, histidine 57, and aspartate 102*. The side chain of serine is bonded to the imidazole ring of the histidine residue which accepts a proton from serine to form a strong alkoxide nucleophile in the presence of a substrate for attack. The aspartate residue orients histidine to make it a better proton acceptor via hydrogen bonding and electrostatic reactions.

The active site of chymotrypsin is marked by serine 195. Serine lies in a small pocket on the surface of the enzyme. Serine is bonded to histidine 57 which is then bound to aspartate 102. All three of these residues are hydrogen bonded at this pocket. These three residues participate in concerted mechanisms that allows chymotrypsin and other proteases to be activated by incoming substrates. This is called the catalytic triad.

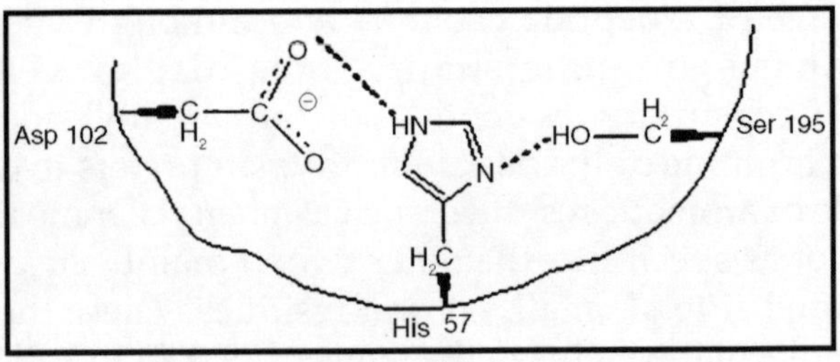

Fig. Catalytic Triad of Chymotrypsin

The histidine residue forces serine into a position that facilitates nucleophilic attack later on through the process of catalysis by approximation. In the presence of a substrate, a chain reaction occurs. First since asparate is acidic, it will be deprotonated first by bases. Aspartate that flanks the histidine residue also provides it with favourable electrostatic effects and makes it a better proton acceptor. So after asparate is deprotonated, proton transfer from histidine goes to aspartate. Now that histidine is deprotonated, it grabs the proton from serine's hydroxyl group. This creates a much more reactive alkoxide group on serine.

Now that the serine is activated we can proceed onto peptide hydrolysis. The alkoxide can attack an incoming substrate to form a tetrahedral intermediate. In this stage we form a resonating oxyanion hole which is a common motif in these kind of reactions. The oxyanion hole stabilizes the tetrahedral intermediate by distributing the negative charge around. The catalytic triad actually reveals a deep hydrophobic pocket where serine is sticking out in the centre. This pocket positions incoming side chains of a substrate. There is a lot of specificty involved as chymotrypsin has a specific pocket with serine while other enzymes such as trypsin and elastase have different composition of pockets. Therefore we can now know that chymotrypsin likes large aromatic or long, nonpolar side chain.

Catalytic triads also exist in trypsin and elastase. Instead of serine, in trypsin, the centre amino acid at the pocket is aspartate. Therefore, its pocket is specific to positively charged species of side chains. Elastase has a pocket that contains two residues of valine, which makes it very hard for big bulky side chains to enter the pocket; therefore, it favours small side chains. Trypsin and elastase are obviously homologs of chymotrypsin. They have 40% similarity in composition and have similar structures.

Proteases: Facilitating a Difficult Reaction

Protein turnover is an important process in living systems. Proteins that have served their purpose must be degraded so that their constituent amino acids can be recycled for the synthesis of new proteins. Proteins ingested in the diet must be broken down into small peptides and amino acids for absorption in the gut.

Proteases cleave proteins by a hydrolysis reaction—the addition of a molecule of water to a peptide bond:

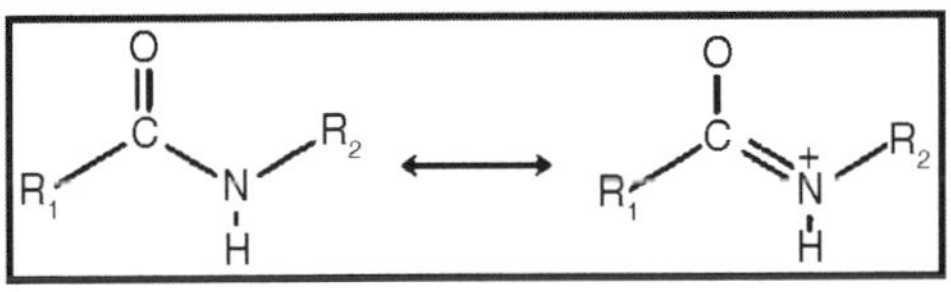

Although the hydrolysis of peptide bonds is thermodynamically favoured, such hydrolysis reactions are extremely slow. In the absence of a catalyst, the half-life for the hydrolysis of a typical peptide at neutral pH is estimated to be between 10 and 1000 years. Yet, peptide bonds must be hydrolyzed within milliseconds in some biochemical processes. The chemical bonding in peptide bonds is responsible for their kinetic stability. Specifically, the resonance structure that accounts for the planarity of a peptide bond also makes such bonds resistant to hydrolysis. This resonance structure endows the peptide bond with partial double-bond character:

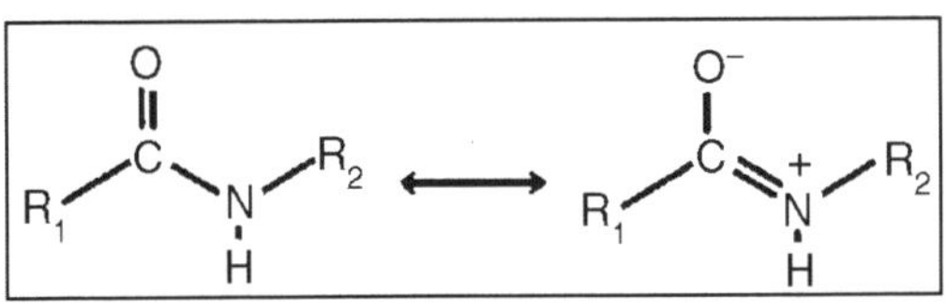

The carbon-nitrogen bond is strengthened by its double-bond character, and the carbonyl carbon atom is less electrophilic and less susceptible to nucleophilic attack than are the carbonyl carbon atoms in compounds such as carboxylate esters. Consequently, to promote peptide-bond cleavage, an enzyme must facilitate nucleophilic attack at a normally unreactive carbonyl group.

Chymotrypsin Possesses a Highly Reactive Serine Residue

A number of proteolytic enzymes participate in the breakdown of proteins in the digestive systems of mammals and other organisms. One such enzyme, chymotrypsin, cleaves peptide bonds selectively on the carboxylterminal side of the large hydrophobic amino acids such as tryptophan, tyrosine, phenylalanine, and methionine. Chymotrypsin is a good example of the use of *covalent modification* as a catalytic strategy. The enzyme employs a powerful nucleophile to attack the unreactive carbonyl group of the substrate. This nucleophile becomes covalently attached to the substrate briefly in the course of catalysis.

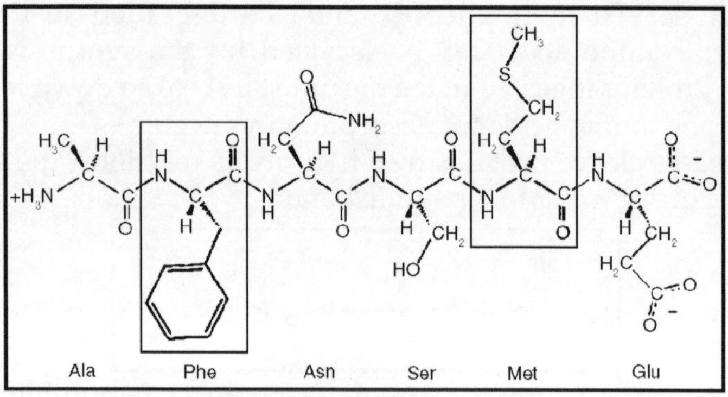

Fig. Specificity of Chymotrypsin

What is the nucleophile that chymotrypsin employs to attack the substrate carbonyl group? A clue came from the fact that chymotrypsin contains an extraordinarily reactive serine residue. Treatment with organofluorophosphates such as diisopropylphosphofluoridate (DIPF) was found to inactivate the enzyme irreversibly. Despite the fact that the enzyme possesses 28 serine residues, only one, serine 195, was modified, resulting in a total loss of enzyme activity. This *chemical modification reaction* suggested that this unusually reactive serine residue plays a central role in the catalytic mechanism of chymotrypsin.

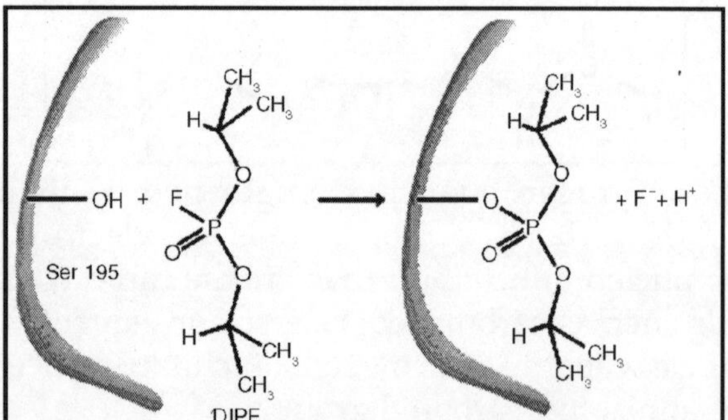

Fig. Mechanism of Chymotrypsin

An Unusually Reactive Serine in Chymotrypsin

Catalytic Triads Are Found in Other Hydrolytic Enzymes

Many other proteins have subsequently been found to contain

catalytic triads similar to that discovered in chymotrypsin. Some, such as trypsin and elastase, are obvious homologs of chymotrypsin. The sequences of these proteins are approximately 40% identical with that of chymotrypsin, and their overall structures are nearly the same.

Fig. Structural Similarity of Trypsin and Chymotrypsin

These proteins operate by mechanisms identical with that of chymotrypsin. However, they have very different substrate specificities. Trypsin cleaves at the peptide bond after residues with long, positively charged side chains—namely, arginine and lysine—whereas elastase cleaves at the peptide bond after amino acids with small side chains—such as alanine and serine. Comparison of the S_1 pockets of these enzymes reveals the basis of the specificity. In trypsin, an aspartate residue (Asp 189) is present at the bottom of the S_1 pocket in place of a serine residue in chymotrypsin. The aspartate residue attracts and stabilizes a positively charged arginine or lysine residue in the substrate. In elastase, two residues at the top of the pocket in chymotrypsin and trypsin are replaced with valine (Val 190 and Val 216). These residues close off the mouth of the pocket so that only small side chains may enter.

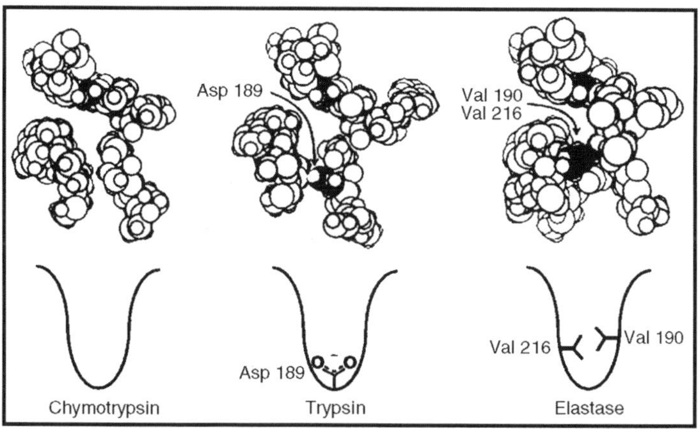

Fig. The S_1 Pockets of Chymotrypsin, Trypsin, and Elastase

Other members of the chymotrypsin family include a collection of proteins that take part in blood clotting. In addition, a wide range of proteases found in bacteria and viruses also belong to this clan. Furthermore, other enzymes that are not homologs of chymotrypsin have been found to contain very similar active sites. The presence of very similar active sites in these different protein families is a consequence of convergent evolution. Subtilisin, a protease in bacteria such as *Bacillus amyloliquefaciens*, is a particularly well characterized example.

The active site of this enzyme includes both the catalytic triad and the oxyanion hole. However, one of the NH groups that forms the oxyanion hole comes from the side chain of an asparagine residue rather than from the peptide backbone. Subtilisin is the founding member of another large family of proteases that includes representatives from Archaea, Eubacteria, and Eukarya.

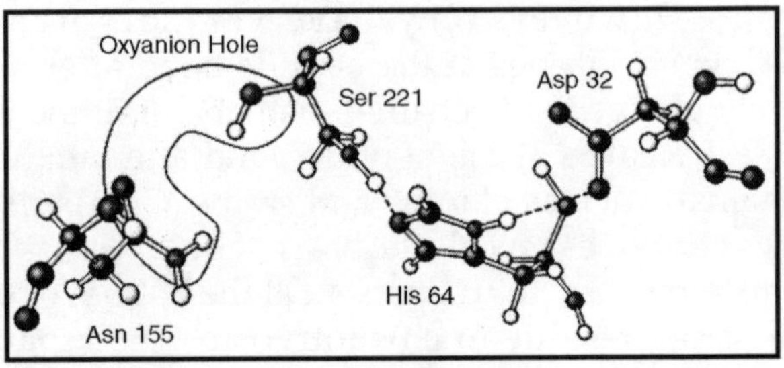

Fig. The Catalytic Triad and Oxyanion Hole of Subtilisin

Yet another example of the catalytic triad has been found in carboxypeptidase II from wheat. The structure of this enzyme is not significantly similar to either chymotrypsin or subtilisin. This protein is a member of an intriguing family of homologous proteins that includes esterases such as acetylcholine esterase and certain lipases. These enzymes all make use of histidine-activated nucleophiles, but the nucleophiles may be cysteine rather than serine.

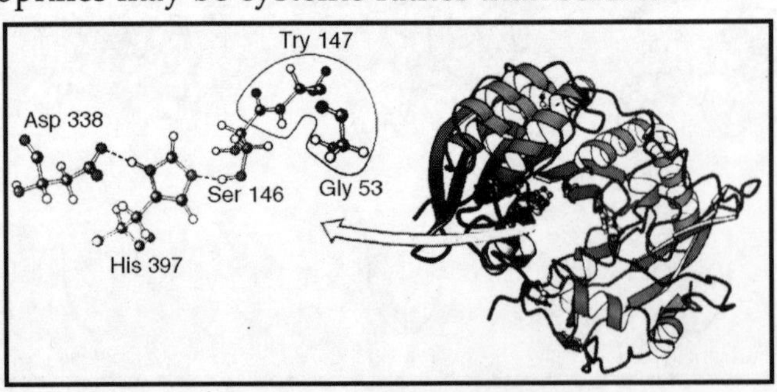

Fig. Carboxypeptidase II

Finally, other proteases have been discovered that contain an active-site serine or threonine residue that is activated not by a histidine-aspartate pair but by a primary amino group from the side chain of lysine or by the N-terminal amino group of the polypeptide chain. Thus, the catalytic triad in proteases has emerged at least three times in the course of evolution. We can conclude that this catalytic strategy must be an especially effective approach to the hydrolysis of peptides and related bonds.

LYSOZYME

Lysozyme is 129 aminoacid residues enzyme hydrolase which catalyzes hydrolysis of 1,4-beta-linkages between N-acetylmuramic acid and N-acetyl-D-glucosamine residues in peptidoglycan and between N-acetyl-D-glucosamine residues in chitodextrins. Molecular weight of Lysozyme is an approximately 14.7 kDa. Alternative name for Lysozyme are 1,4-N-acetylmuramidase, L-7001, N,O-diacetylmuramidase, PR1-Lysozyme, Globulin G1, Globulin G, Lysozyme g, Mucopeptide N-acetylmuramoylhydrolase, Mucopeptide glucohydrolase and Muramidase. This catalytic activity is non-specifically targeted to the bacterial cell membranes and related with general non-specific organism defence.

Lysozyme is present in the mucosal secretion such as saliva and tears. In high concentration, about 3% from all proteins, Lysozyme is present in chicken egg-white. This enzyme is only effective against Gram positive bacterial cells. Gram negative bacteria and yeast are completely resistant to lysing by it. In viruses (or bacteriophages), Lysozyme is used as an agent to break into the host bacterial cell. Lysozyme from the tail of the virus (or bacteriophage) destroys the peptidoglycan bacterial cell wall and then virus can injects its DNA. After multiplication in bacteria, many Lysozyme molecules are created to lyse the bacterial cell wall and release new viruses.

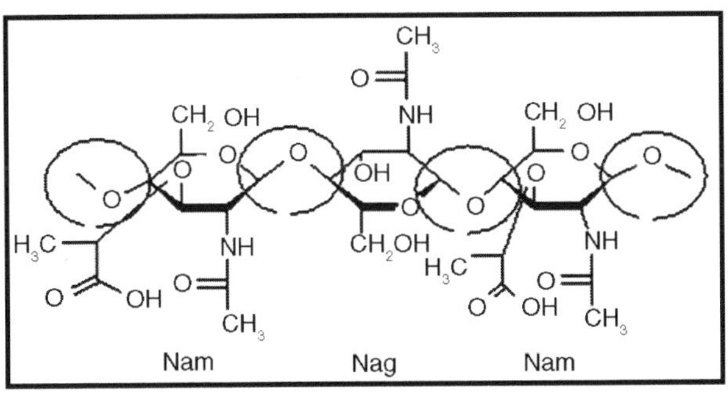

Fig. The Lysozyme Reaction.

Lysozyme is a small globular protein composed of 129 amino acids (14 kD) in a single polypeptide chain. It has eight cysteine residues linked in four disulfide bonds. The structure of this very stable protein was determined by X-ray crystallographic methods in 1965 by David Phillips. Although X-ray structures had previously been reported for proteins (hemoglobin and myoglobin), lysozyme was the first enzyme structure to be solved by crystallographic (or any other) methods. Although the location of the active site was not obvious from the X-ray structure of the protein alone, X-ray studies of lysozyme-inhibitor complexes soon revealed the location and nature of the active site. Since it is an enzyme, lysozyme cannot form stable ES complexes for structural studies, because the substrate is rapidly transformed into products.

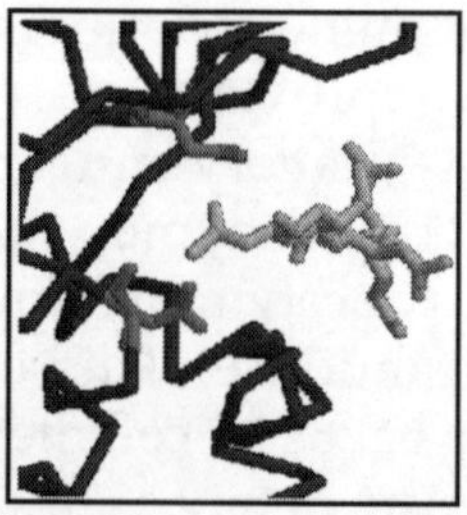

Fig. Structure of Lysozyme. Glu^{35} and Asp^{52}

FUNCTION

The enzyme functions by attacking peptidoglycans (found in the cell walls of bacteria, especially Gram-positive bacteria) and hydrolyzing the glycosidic bond that connects N-acetylmuramic acid with the fourth carbon atom of N-acetylglucosamine. It does this by binding to the peptidoglycan molecule in the binding site within the prominent cleft between its two domains. This causes the substrate molecule to adopt a strained conformation similar to that of the transition state. According to Phillips-Mechanism, the lysozyme binds to a hexasaccharide. The lysozyme then distorts the fourth sugar in hexasaccharide (the D ring) into a half-chair conformation. In this stressed state, the glycosidic bond is easily broken.

The amino acid side-chains glutamic acid 35 (Glu35) and aspartate 52 (Asp52) have been found to be critical to the activity of this enzyme. Glu35 acts as a proton donor to the glycosidic bond, cleaving the C-O bond in the substrate, whereas Asp52 acts as a nucleophile to generate a glycosyl enzyme intermediate. The glycosyl enzyme intermediate then reacts with a water molecule, to give the product of hydrolysis and leaving the enzyme unchanged.

Role in Disease

Lysozyme is part of the innate immune system. Reduced lysozyme levels have been associated with bronchopulmonary dysplasia in newborns. Children fed infant formula lacking lysozyme in their diet have three times the rate of diarrheal disease. Since lysozyme is a natural form of protection from pathogens like Salmonella, E.coli, and Pseudomonas, a deficiency due to infant formula feeding can lead to increased incidence of disease.

Whereas the skin is a protective barrier due to its dryness and acidity, the conjunctiva (membrane covering the eye) is, instead, protected by secreted enzymes, mainly lysozyme and defensin. However, when these protective barriers fail, conjunctivitis results.

GLYCOSIDE HYDROLASE

Glycoside hydrolases (also called glycosidases) catalyze the hydrolysis of the glycosidic linkage to generate two smaller sugars. They are extremely common enzymes with roles in nature including degradation of biomass such as cellulose and hemicellulose, in anti-bacterial defence strategies (eg lysozyme), in pathogenesis mechanisms (eg viral neuraminidases) and in normal cellular function (eg trimming mannosidases involved in N-linked glycoprotein biosynthesis). Together with glycosyltransferases, glycosidases form the major catalytic machinery for the synthesis and breakage of glycosidic bonds.

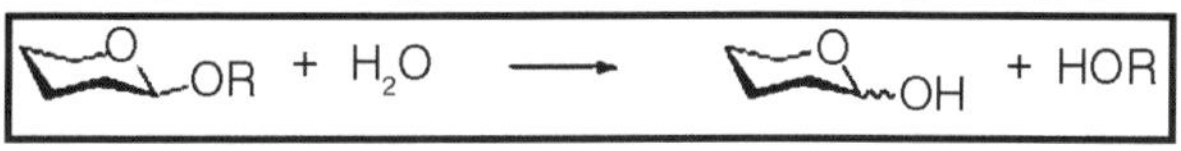

Occurrence and Importance

Glycoside hydrolases are found in essentially all domains of life. In bacteria and prokaryotes, they are found both as intracellular and extracellular enzymes that are largely involved in nutrient acquisition. One of the important occurrences of glycoside hydrolases in bacteria is the enzyme beta-galactosidase (LacZ), which is involved in regulation of expression of the lac operon in E. coli. In higher organisms glycoside hydrolases are found within the endoplasmic reticulum and Golgi apparatus where they are involved in processing of N-linked glycoproteins, and in the lysozome as enzymes involved in the degradation of carbohydrate structures. Deficiency in specific lysozomal glycoside hydrolases can lead to a range of lysosomal storage disorders that result in developmental problems or death.

Glycoside hydrolases are found in the intestinal tract and in saliva where they degrade complex carbohydrates such as lactose, starch, sucrose and trehalose. In the gut they are found as glycosylphosphatidyl anchored enzymes on endothelial cells. The enzyme lactase is required for degradation of the milk sugar lactose and is present at high levels in infants, but in most populations will decrease after weaning or during infancy, potentially leading to lactose intolerance in adulthood. The enzyme O-GlcNAcase is involved in removal of N-acetylglucoamine groups from serine and threonine residues in the cytoplasm and nucleus of the cell. The glycoside hydrolases are involved in the biosynthesis and degradation of glycogen in the body.

Classification

Glycoside hydrolases are classified into EC 3.2.1 as enzymes catalyzing the hydrolysis of O- or S-glycosides. Glycoside hydrolases can also be classified according to the stereochemical outcome of the hydrolysis reaction: thus they can be classified as either retaining or inverting enzymes. Glycoside hydrolases can also be classified as exo or endo acting, dependent upon whether they act at the (usually non-reducing) end or in the middle, respectively, of an oligo/polysaccharide chain. Glycoside hydrolases may also be classified by sequence or structure based methods.

Sequence-based Classification

Sequence-based classifications are among the most powerful predictive method for suggesting function for newly sequenced enzymes for which function has not been biochemically demonstrated. A classification system for glycosyl hydrolases, based on sequence similarity, has led to the definition of more than 100 different families. This classification is available on the CAZy(CArbohydrate-Active EnZymes) web site. The database provides a series of regularly updated sequence based classification that allow reliable prediction of mechanism (retaining/inverting), active site residues and possible substrates. The online database is supported by CAZypedia, an online encyclopedia of carbohydrate active enzymes. Based on three dimensional structural similarities, the sequence-based families have been classified into 'clans' of related structure. Recent progress in glycosidase sequence analysis and 3D structure comparison has allowed the proposal of an extended hierarchical classification of the glycoside hydrolases.

Mechanisms

Inverting Glycoside Hydrolases

Inverting enzymes utilize two enzymic residues, typically

carboxylate residues, that act as acid and base respectively, as shown below for a β-glucosidase:

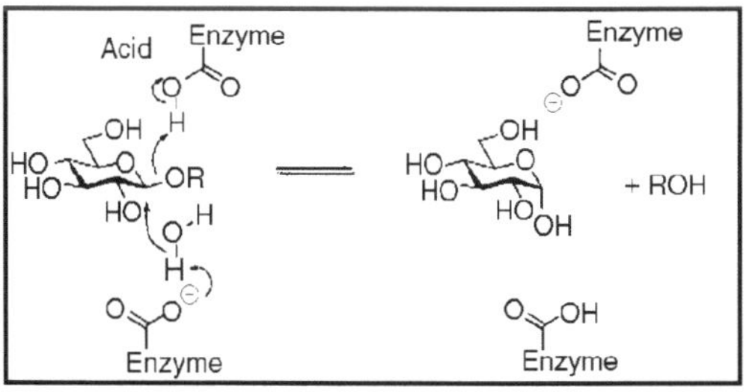

Fig. Inverting Glycoside Hydrolases

Retaining Glycoside Hydrolases

Retaining glycosidases operate through a two-step mechanism, with each step resulting in inversion, for a net retention of stereochemistry. Again, two residues are involved, which are usually enzyme-borne carboxylates. One acts as a nucleophile and the other as an acid/base. In the first step the nucleophile attacks the anomeric centre, resulting in the formation of a glycosyl enzyme intermediate, with acidic assistance provided by the acidic carboxylate. In the second step the now deprotoned acidic carboxylate acts as a base and assists a nucleophilic water to hydrolyze the glycosyl enzyme intermediate, giving the hydrolyzed product. The mechanism is illustrated below for hen egg white lysozyme.

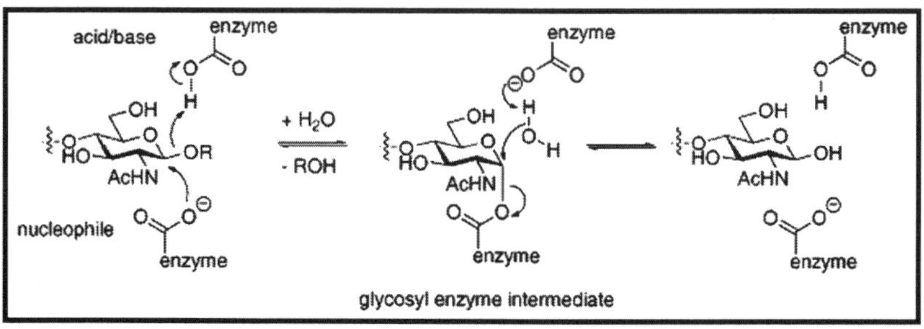

Fig. Retaining Glycoside Hydrolases

An alternative mechanism for hydrolysis with retention of stereochemistry can occur that proceeds through a nucleophilic residue that is bound to the substrate, rather than being attached to the enzyme. Such mechanisms are common for certain N-acetylhexosaminidases, which have an acetamido group capable of neighbouring group participation to form an intermediate oxazoline or oxazolinium ion. Again, the mechanism proceeds in two steps through individual inversions to lead to a net retention of configuration.

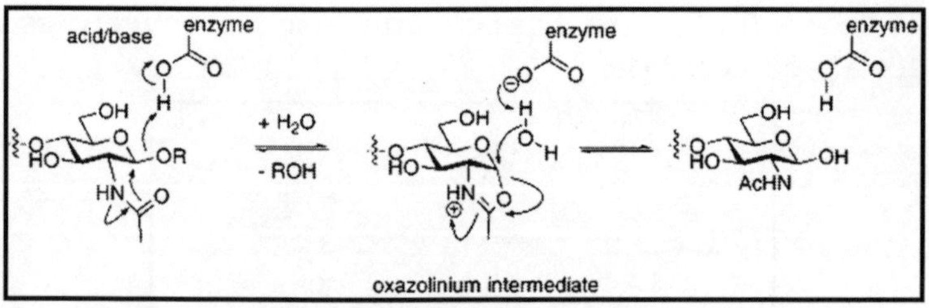

Fig. Glycoside Hydrolases

Nomenclature and Examples

Glycoside hydrolases are typically named after the substrate that they act upon. Thus glucosidases catalyze the hydrolysis of glucosides and xylanases catalyze the cleavage of the xylose based homopolymer xylan. Other examples include lactase, amylase, chitinase, sucrase, maltase, neuraminidase, invertase, hyaluronidase and lysozyme.

Uses

Glycoside hydrolases have a variety of uses including degradation of plant materials (eg cellulases for degrading cellulose to glucose, which can be used for ethanol production), in the food industry (invertase for manufacture of invert sugar, amylase for production of maltodextrins), and in the paper and pulp industry (xylanases for removing hemicelluloses from paper pulp). Cellulases are added to detergents for the washing of cotton fabrics and assist in the maintenance of colours through removing microfibres that are raised from the surface of threads during wear.

In organic chemistry, glycoside hydrolases can be used as synthetic catalysts to form glycosidic bonds through either reverse hydrolysis (kinetic approach) where the equilibrium position is reversed; or by transglycosylation (kinetic approach) whereby retaining glycoside hydrolases can catalyze the transfer of a glycosyl moiety from an activated glycoside to an acceptor alcohol to afford a new glycoside.

Mutant glycoside hydrolases termed glycosynthases have been developed that can achieve the synthesis of glycosides in high yield from activated glycosyl donors such as glycosyl fluorides. Glycosynthases are typically formed from retaining glycoside hydrolases by site-directed mutagenesis of the enzymic nucleophile to some other less nucleophilic group, such as alanine or glycine. Another group of mutant glycoside hydrolases termed thioglycoligases can be formed by site-drected mutagenesis of the acid-base residue of a retaining glycoside hydrolase. Thioglycoligases catalyze the condensation of activated glycosides and various thiol containing acceptors.

Inhibitors

Many compounds are known that can act to inhibit the action of a glycoside hydrolase. Nitrogen-containing, 'sugar-shaped' heterocycles have been found in nature, including deoxynojirimycin, swainsonine, australine and castanospermine. From these natural templates many other inhibitors have been developed, including isofagomine and deoxygalactonojirimycin, and various unsaturated compounds such as PUGNAc. Inhibitors that are in clinical use include the anti-diabetic drugs acarbose and miglitol, and the antiviral drugs oseltamivir and zanamivir. Some proteins have been found to act as glycoside hydrolase inhibitors.

Model Studies Reveal a Strain-Induced Destabilization of a Bound Substrate on Lysozyme

One of the premises of lysozyme models is that the native substrate would occupy the rest of the crevice or depression running across the surface of the enzyme, because there is room to fit three more sugar residues into the crevice and because the hexamer $(NAG)_6$ is in fact a good substrate for lysozyme. The model building studies refer to the six sugar residue-binding subsites in the crevice with the letters A through F, with A, B, and C representing the part of the crevice occupied by the $(NAG)_3$ inhibitor.

Modeling studies clearly show that NAG residues fit nicely into subsites A, B, C, E, and F of the crevice, but that fitting a residue of the $(NAG)_6$ hexamer into site D requires a substantial distortion of the sugar (out of its preferred chair conformation) to prevent steric crowding and overlap between atoms C-6 and O-6 of the sugar at the D site and Ile of the enzyme. This distorted sugar residue is adjacent to the glycosidic bond to be cleaved (between sites D and E), and the inference is made that this distortion or strain brings the substrate closer to the transition state for hydrolysis. This is a good example of strain-induced destabilization of an otherwise favourably binding substrate. Thus, the overall binding interaction of the rest of the sugar substrate would be favourable (DG<0), but distortion of the ring at the D site uses some of this binding energy to raise the substrate closer to the transition state for hydrolysis, an example of stabilization of a transition state (relative to the simple enzyme-substrate complex).

Chapter 6

Enzyme Kinetics

Thousands of chemical reactions are proceeding very rapidly at any given instant within all living cells. Virtually all of these transformations are mediated by enzymes, proteins (and occasionally RNA) specialized to catalyze metabolic reactions. The substances transformed in these reactions are often organic compounds that show little tendency for reaction outside the cell. An excellent example is glucose, a sugar that can be stored indefinitely on the shelf with no deterioration. Most cells quickly oxidize glucose, producing carbon dioxide and water and releasing lots of energy:

$$C_6H_{12}O_6 + 6\ O_2 \rightarrow 6\ CO_2 = 6\ H_2O + 2870 \text{ kJ of energy}$$

(–2870 kJ/mol is the standard free energy change [DG°′] for the oxidation of glucose. In chemical terms, 2870 kJ is a large amount of energy, and glucose can be viewed as an energy-rich compound even though at ambient temperature it is not readily reactive with oxygen outside of cells.

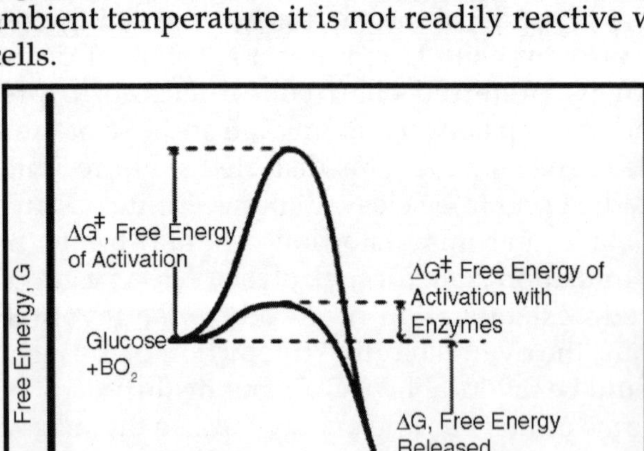

Fig. Reaction Profile Showing Large D G‡ for Glucose Oxidation, free Energy Change of -2,870 kJ/mol; Catalysts lower D G‡, Thereby Accelerating Rate.

Stated another way, glucose represents thermod-ynamic potentiality: its reaction with oxygen is strongly exergonic, but it just doesn't occur under normal conditions. On the other hand, enzymes can catalyze such thermodynamically favourable reactions so that they proceed at extraordinarily rapid rates. In glucose oxidation and countless other instances, enzymes provide cells with the ability to exert *kinetic control over thermodynamic potentiality*. That is, living systems use enzymes to accelerate and control the rates of vitally important biochemical reactions.

ENZYMES ARE THE AGENTS OF METABOLIC FUNCTION

Acting in sequence, enzymes form metabolic pathways by which nutrient molecules are degraded, energy is released and converted into metabolically useful forms, and precursors are generated and transformed to create the literally thousands of distinctive biomolecules found in any living cell.

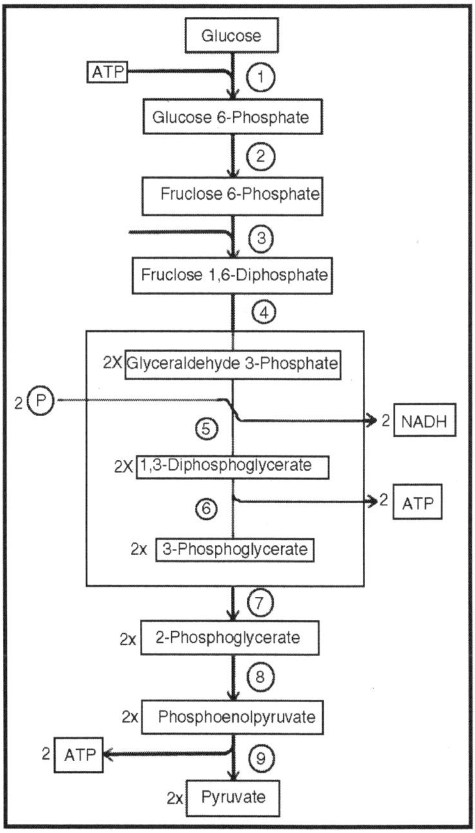

Fig. The Breakdown of Glucose by *Glycolysis* Provides a Prime Example of a Metabolic Pathway.

Situated at key junctions of metabolic pathways are specialized regulatory enzymes capable of sensing the momentary metabolic needs of the cell and adjusting their catalytic rates accordingly. The responses of these enzymes ensure the harmonious integration of the diverse and often divergent metabolic activities of cells so that the living state is promoted and preserved.

ENZYMES—CATALYTIC POWER, SPECIFICITY, AND REGULATION

Enzymes are characterized by three distinctive features: catalytic power, specificity, and regulation.

Catalytic Power

Enzymes display enormous catalytic power, accelerating reaction rates as much as 10^{16} over uncatalyzed levels, which is far greater than any synthetic catalysts can achieve, and enzymes accomplish these astounding feats in dilute aqueous solutions under mild conditions of temperature and pH. For example, the enzyme jack bean *urease* catalyzes the hydrolysis of urea:

$$H_2N-\overset{\overset{O}{\|}}{C}-NH_2 + 2H_2O + H \rightarrow 2NH_4^+ + HCO_3^-$$

At 20°C, the rate constant for the enzyme-catalyzed reaction is 3 x 10^4/sec; the rate constant for the uncatalyzed hydrolysis of urea is 3 x 10^{-10}/sec. Thus, 10^{14} is the ratio of the catalyzed rate to the uncatalyzed rate of reaction. Such a ratio is defined as the relative catalytic power of an enzyme, so the catalytic power of urease is 10^{14}.

Specificity

A given enzyme is very selective, both in the substances with which it interacts and in the reaction that it catalyzes. The substances upon which an enzyme acts are traditionally called substrates. In an enzyme-catalyzed reaction, none of the substrate is diverted into nonproductive side-reactions, so no wasteful by-products are produced. It follows then that the products formed by a given enzyme are also very specific.

This situation can be contrasted with your own experiences in the organic chemistry laboratory, where yields of 50% or even 30% are viewed as substantial accomplishments. The selective qualities of an enzyme are collectively recognized as its specificity. Intimate interaction between an enzyme and its substrates occurs through molecular recognition based on structural complementarity; such mutual recognition is the basis of specificity. The specific site on the enzyme where substrate binds and catalysis occurs is called the active site.

Regulation

Regulation of enzyme activity is achieved in a variety of ways, ranging from controls over the amount of enzyme protein produced by the cell to more rapid, reversible interactions of the enzyme with metabolic inhibitors and activators.

Because most enzymes are proteins, we can anticipate that the functional attributes of enzymes are due to the remarkable versatility found in protein structures.

ENZYME NOMENCLATURE

All enzymes contain a protein backbone. In some enzymes this is the only component in the structure. However there are additional non-protein moieties usually present which may or may not participate in the catalytic activity of the enzyme. Covalently attached carbohydrate groups are commonly encountered structural features which often have no direct bearing on the catalytic activity, although they may well effect an enzyme's stability and solubility. Other factors often found are metal ions (cofactors) and low molecular weight organic molecules (coenzymes). These may be loosely or tightly bound by noncovalent or covalent forces. They are often important constituents contributing to both the activity and stability of the enzymes. This requirement for cofactors and coenzymes must be recognised if the enzymes are to be used efficiently and is particularly relevant in continuous processes where there may be a tendency for them to become separated from an enzyme's protein moiety.

This enzyme commission assigned each enzyme a recommended name and a 4-part distinguishing number. It should be appreciated that some alternative names remain in such common usage that they will be used, where appropriate, in this text.

The enzyme commission (EC) numbers divide enzymes into six main groups according to the type of reaction catalysed:

1. Oxidoreductases which involve redox reactions in which hydrogen or oxygen atoms or electrons are transferred between molecules. This extensive class includes the dehydrogenases (hydride transfer), oxidases (electron transfer to molecular oxygen), oxygenases (oxygen transfer from molecular oxygen) and peroxidases (electron transfer to peroxide).

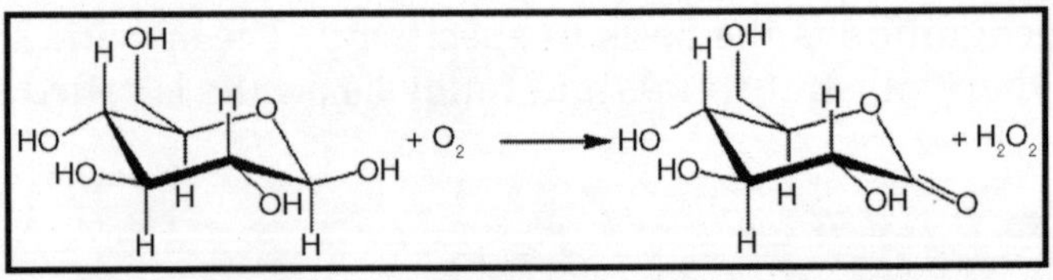

β-D-glucose + oxygen → D-glucono-1,5-lactone + hydrogen peroxide

2. Transferases which catalyse the transfer of an atom or group of atoms (*e.g.* acyl-, alkyl- and glycosyl-), between two molecules, but excluding such transfers as are classified in the other groups (*e.g.* oxidoreductases and hydrolases).

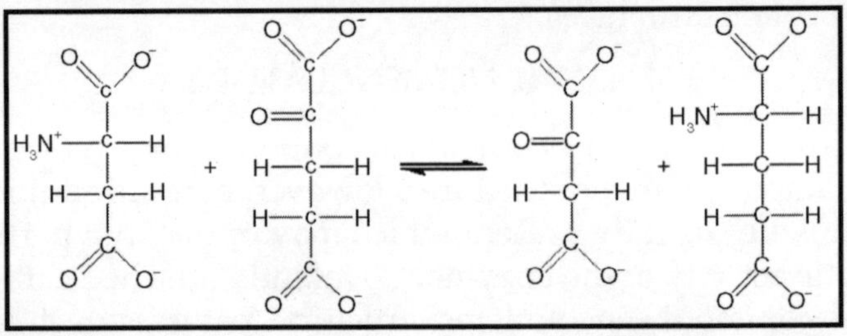

L-aspartate + 2-oxoglutarate □ oxaloacetate + L-glutamate

3. Hydrolases which involve hydrolytic reactions and their reversal. This is presently the most commonly encountered class of enzymes within the field of enzyme technology and includes the esterases, glycosidases, lipases and proteases. For example: chymosin (EC 3.4.23.4, no systematic name declared; also called rennin).

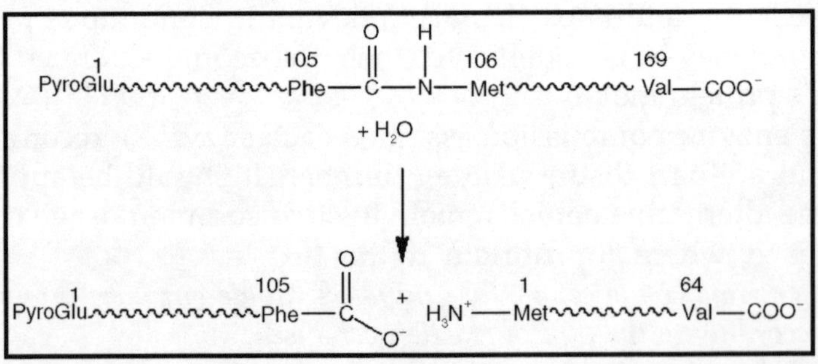

κ-casein + water → *para*-κ-casein + caseino macropeptide

4. Lyases which involve elimination reactions in which a group of atoms is removed from the substrate. This includes the aldolases, decarboxylases, dehydratases and some pectinases but does not include hydrolases. For example: histidine ammonia-lyase (EC 4.3.1.3, systematic name, L-histidine ammonia-lyase; also called histidase).

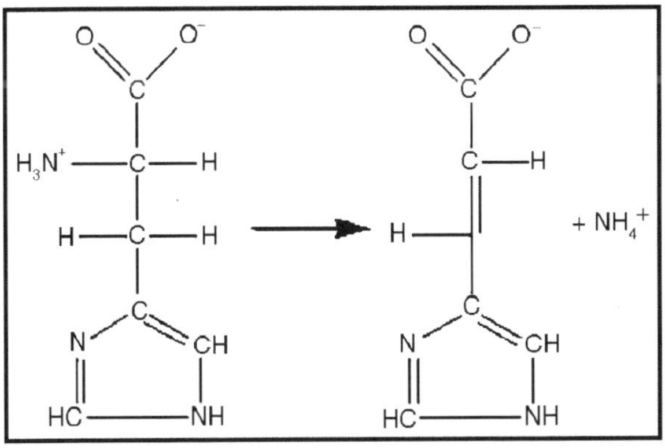

L-histidine → urocanate + ammonia

5. Isomerases which catalyse molecular isomerisations and includes the epimerases, racemases and intramolecular transferases. For example: xylose isomerase (EC 5.3.1.5, systematic name, D-xylose ketol-isomerase; commonly called glucose isomerase).

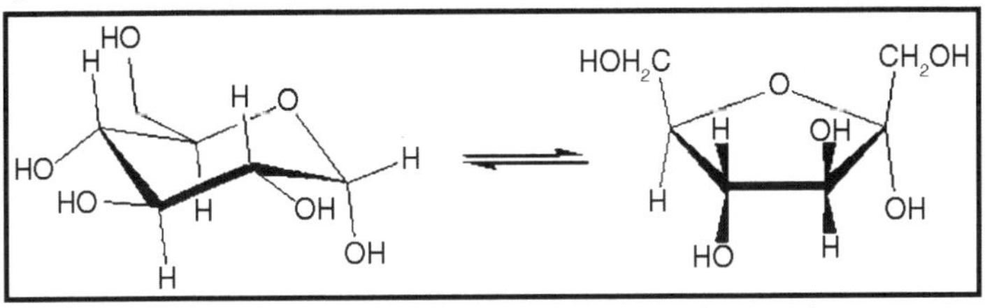

α-D-glucopyranose F□ α-D-fructofuranose

6. Ligases, also known as synthetases, form a relatively small group of enzymes which involve the formation of a covalent bond joining two molecules together, coupled with the hydrolysis of a nucleoside triphosphate. For example: glutathione synthase (EC 6.3.2.3, systematic name, g-L-glutamyl-L-cysteine:glycine ligase (ADP-forming); also called glutathione synthetase).

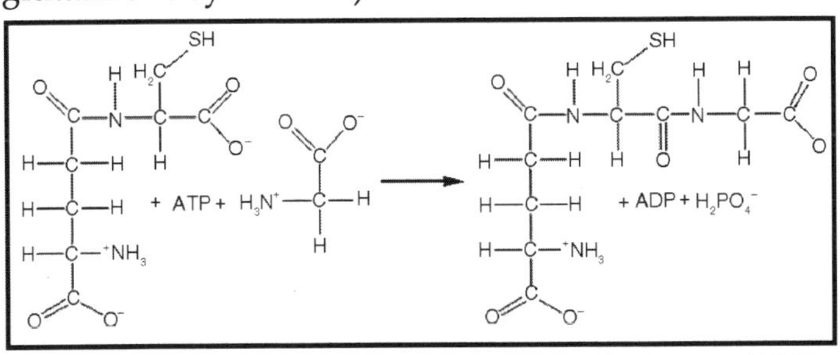

ATP + γ-L-glutamyl-L-cysteine + glycine →ADP + phosphate + glutathione

Coenzymes

Many enzymes carry out their catalytic function relying solely on their protein structure. Many others require nonprotein components, called cofactors. Cofactors may be metal ions or organic molecules referred to as coenzymes. Cofactors, because they are structurally less complex than proteins, tend to be stable to heat (incubation in a boiling water bath). Typically, proteins are denatured under such conditions. Many coenzymes are vitamins or contain vitamins as part of their structure. Usually coenzymes are actively involved in the catalytic reaction of the enzyme, often serving as intermediate carriers of functional groups in the conversion of substrates to products. In most cases, a coenzyme is firmly associated with its enzyme, perhaps even by covalent bonds, and it is difficult to separate the two. Such tightly bound coenzymes are referred to as prosthetic groups of the enzyme. The catalytically active complex of protein and prosthetic group is called the holoenzyme. The protein without the prosthetic group is called the apoenzyme; it is catalytically inactive.

ENZYME KINETICS

Kinetics is the branch of science concerned with the rates of chemical reactions. The study of enzyme kinetics addresses the biological roles of enzymatic catalysts and how they accomplish their remarkable feats. In enzyme kinetics, we seek to determine the maximum reaction velocity that the enzyme can attain and its binding affinities for substrates and inhibitors. Coupled with studies on the structure and chemistry of the enzyme, analysis of the enzymatic rate under different reaction conditions yields insights regarding the enzyme's mechanism of catalytic action. Such information is essential to an overall understanding of metabolism.

Significantly, this information can be exploited to control and manipulate the course of metabolic events. The science of pharmacology relies on such a strategy. Pharmaceuticals, or drugs, are often special inhibitors specifically targeted at a particular enzyme in order to overcome infection or to alleviate illness. A detailed knowledge of the enzyme's kinetics is indispensable to rational drug design and successful pharmacological intervention.

Review of Chemical Kinetics

Before beginning a quantitative treatment of enzyme kinetics, it will be fruitful to review briefly some basic principles of chemical kinetics. Chemical kinetics is the study of the rates of chemical reactions. Consider a reaction of overall stoichiometry,

$$A \rightarrow P$$

Although we treat this reaction as a simple, one-step conversion of A to P, it more likely occurs through a sequence of elementary reactions, each of which is a simple molecular process, as in,

$$A \rightarrow I \rightarrow J \rightarrow P$$

where I and J represent intermediates in the reaction. Precise description of all of the elementary reactions in a process is necessary to define the overall reaction mechanism for A → P. Let us assume that A → P *is* an elementary reaction and that it is spontaneous and essentially irreversible. Irreversibility is easily assumed if the rate of P conversion to A is very slow *or* the concentration of P (expressed as [P]) is negligible under the conditions chosen. The velocity, v, or rate, of the reaction A → P is the amount of P formed or the amount of A consumed per unit time, t. That is,

$$v = \frac{d[P]}{dt} \text{ or } v = \frac{-d[A]}{dt}$$

The mathematical relationship between reaction rate and concentration of reactant(s) is the rate law. For this simple case, the rate law is,

$$v - \frac{-d[A]}{dt} = k[A]$$

From this expression, it is obvious that the rate is proportional to the concentration of A, and k is the proportionality constant, or rate constant. k has the units of $(\text{time})^{-1}$, usually sec^{-1}. v is a function of [A] to the first power, or, in the terminology of kinetics, v is first-order with respect to A.

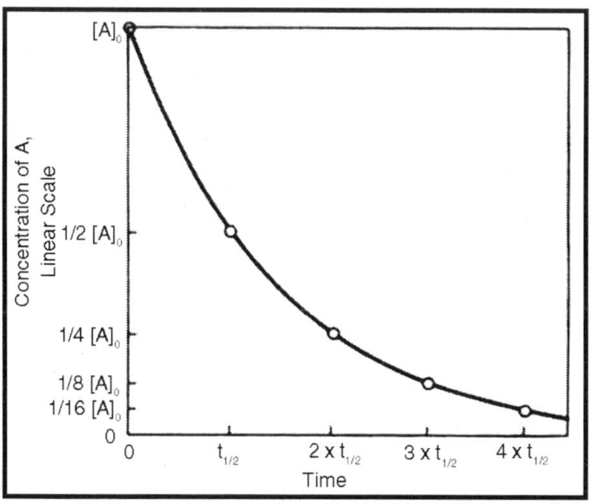

Fig. Plot of the Course of a First-order Reaction.

For an elementary reaction, the order for any reactant is given by its exponent in the rate equation. The number of molecules that must simultaneously interact is defined as the molecularity of the reaction. Thus, the simple elementary reaction of A $\rightarrow$ P is a first-order reaction. Portrays the course of a first-order reaction as a function of time. The rate of decay of a radioactive isotope, like ^{14}C or ^{32}P, is a first-order reaction, as is an intramolecular rearrangement, such as A $\rightarrow$ P. Both are unimolecular reactions (the molecularity equals 1).

Biomolecular Reactions

Consider the more complex reaction, where two molecules must react to yield products:

$$A + B \rightarrow P + Q$$

Assuming this reaction is an elementary reaction, its molecularity is 2; that is, it is a bimolecular reaction. The velocity of this reaction can be determined from the rate of disappearance of either A or B, or the rate of appearance of P or Q:

$$v = \frac{-d[A]}{dt} = \frac{-d[B]}{dt} = \frac{d[P]}{dt} = \frac{d[Q]}{dt}$$

The rate law is,

$$v = k[A][B]$$

The rate is proportional to the concentrations of both A and B. Because it is proportional to the product of two concentration terms, the reaction is second-order overall, first-order with respect to A and first-order with respect to B. (Were the elementary reaction 2 $\rightarrow$ P + Q, the rate law would be $v = k[A]^2$, second-order overall and Second-order with respect to A.) Second-order rate constants have the units of (concentration)$^{-1}$ (time)$^{-1}$, as in M^{-1} sec^{-1}.

Molecularities greater than two are rarely found (and greater than three, never). When the overall stoichiometry of a reaction is greater than two (for example, as in A+B+Cn or 2A+Bn), the reaction almost always proceeds via uni- or bimolecular elementary steps, and the overall rate obeys a simple first- or Second-order rate law.

At this point, it may be useful to remind ourselves of an important caveat that is the first principle of kinetics: *Kinetics cannot prove a hypothetical mechanism*. Kinetic experiments can only rule out various alternative hypotheses because they don't fit the data. However, through thoughtful kinetic studies, a process of elimination of alternative hypotheses leads ever closer to the reality.

Free Energy of Activation and the Action of Catalysts

In a first-order chemical reaction, the conversion of A to P occurs because, at any given instant, a fraction of the A molecules has the energy necessary to achieve a reactive condition known as the transition state. In this state, the probability is very high that the particular rearrangement accompanying the A → P transition will occur. This transition state sits at the apex of the energy profile in the energy diagram describing the energetic relationship between A and P.

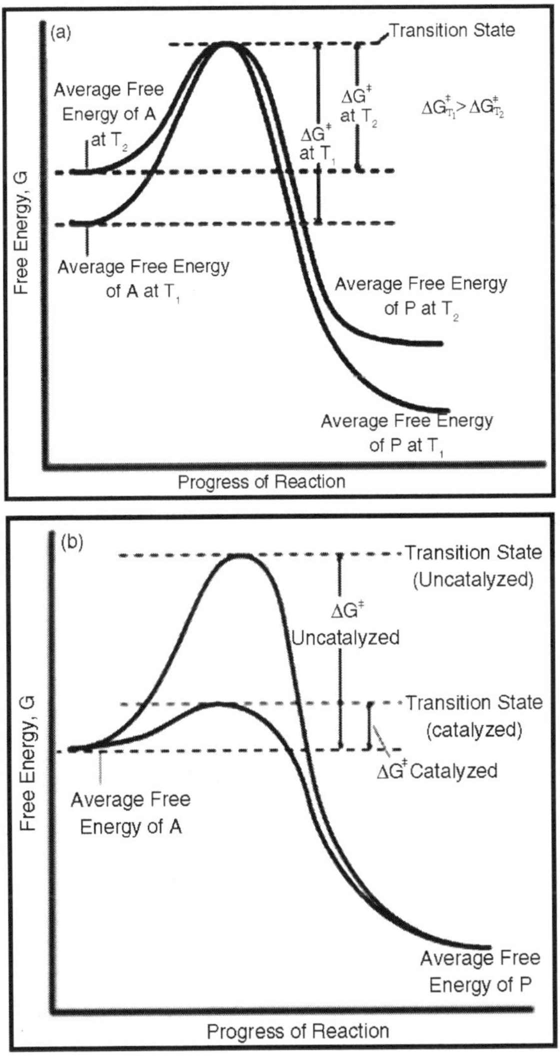

Fig. Energy Diagram for a Chemical Reaction (A → P) and the Effects of (a) Raising the Temperature from T_1 to T_2 or (b) Adding a Catalyst.

The average free energy of A molecules defines the initial state and the average free energy of P molecules is the final state along the reaction coordinate. The rate of any chemical reaction is proportional to the concentration of reactant molecules (A in this case) having this transition-state energy. Obviously, the higher this energy is above the average energy, the smaller the fraction of molecules that will have this energy, and the slower the reaction will proceed. The height of this energy barrier is called the free energy of activation, DG‡. Specifically, DG‡ is the energy required to raise the average energy of one mole of reactant (at a given temperature) to the transition-state energy. The relationship between activation energy and the rate constant of the reaction, k, is given by the Arrhenius equation:

$$k = Ae^{-DG‡/RT}$$

where A is a constant for a particular reaction (not to be confused with the reactant species, A, that we're discussing). Another way of writing this is $1/k = (1/A)e^{DG‡/RT}$. That is, k is inversely proportional to $e^{DG‡/RT}$. Therefore, if the energy of activation decreases, the reaction rate increases.

Decreasing DG‡ Increases Reaction Rate

We are familiar with two general ways that rates of chemical reactions may be accelerated. First, the temperature can be raised. This will increase the average energy of reactant molecules, which in effect lowers the energy needed to reach the transition state. The rates of many chemical reactions are doubled by a 10°C rise in temperature. Second, the rates of chemical reactions can also be accelerated by catalysts. Catalysts work by lowering the energy of activation rather than by raising the average energy of the reactants. Catalysts accomplish this remarkable feat by combining transiently with the reactants in a way that promotes their entry into the reactive, transition-state condition.

Two aspects of catalysts are worth noting:

- They are regenerated after each reaction cycle (A → P), and so can be used over and over again;
- Catalysts have *no* effect on the overall free energy change in the reaction, the free energy difference between A and P.

KINETICS OF ENZYME-CATALYZED REACTIONS

Examination of the change in reaction velocity as the reactant concentration is varied is one of the primary measurements in kinetic analysis. Returning to A → P, a plot of the reaction rate as a function of the concentration of A yields a straight line whose slope is k.

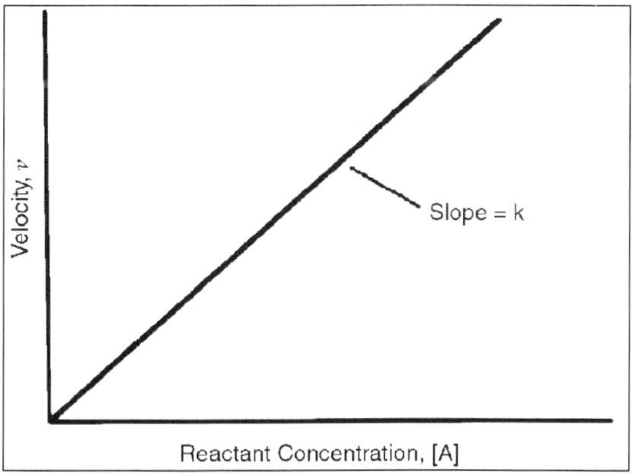

Fig. A Plot of v Versus [A] for the Unimolecular Chemical Reaction, A $\rightarrow$ P, Yields a Straight line Having a Slope Equal to k.

The more A that is available, the greater the rate of the reaction, v. Similar analyses of enzyme-catalyzed reactions involving only a single substrate yield remarkably different results. At low concentrations of the substrate S, v is proportional to [S], as expected for a first-order reaction. However, v does not increase proportionally as [S] increases, but instead begins to level off. At high [S], v becomes virtually independent of [S] and approaches a maximal limit. The value of v at this limit is written V_{max}. Because rate is no longer dependent on [S] at these high concentrations, the enzyme-catalyzed reaction is now obeying zero-order kinetics; that is, the rate is independent of the reactant (substrate) concentration. This behaviour is a saturation effect: when v shows no increase even though [S] is increased, the system is saturated with substrate. Such plots are called substrate saturation curves. The physical interpretation is that every enzyme molecule in the reaction mixture has its substrate-binding site occupied by S. Indeed, such curves were the initial clue that an enzyme interacts directly with its substrate by binding it.

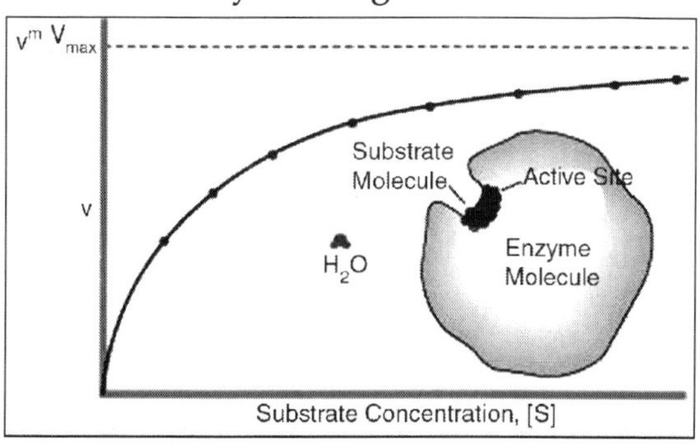

Fig. Substrate Saturation Curve for an Enzyme-catalyzed Reaction.

The Michaelis - menten Equation

Lenore Michaelis and Maud L. Menten proposed a general theory of enzyme action in 1913 consistent with observed enzyme kinetics. Their theory was based on the assumption that the enzyme, E, and its substrate, S, associate reversibly to form an enzyme-substrate complex, ES:

$$E + S \underset{k_{-1}}{\overset{k_1}{\rightleftharpoons}} ES$$

This association/dissociation is assumed to be a rapid equilibrium, and Ks is the *enzyme: substrate dissociation constant.*

At equilibrium:

$$k_{-1}[ES] = k_1[E][S]$$

and,

$$K_s = \frac{[E][S]}{[ES]} = \frac{k_{-1}}{k_1}$$

Product, P, is formed in a second step when ES breaks down to yield E + P.

$$E + S \underset{k_{-1}}{\overset{k_1}{\rightleftharpoons}} ES \xrightarrow{k_2} E + P$$

E is then free to interact with another molecule of S.

Steady-State Assumption

The interpretations of Michaelis and Menten were refined and extended in 1925 by Briggs and Haldane, by assuming the concentration of the enzyme-substrate complex ES quickly reaches a constant value in such a dynamic system.

That is, ES is formed as rapidly from E + S as it disappears by its two possible fates: dissociation to regenerate E + S, and reaction to form E + P. This assumption is termed the steady-state assumption and is expressed as,

$$\frac{d[ES]}{dt} = 0$$

That is, the change in concentration of ES with time, *t*, is 0. The time course for formation of the ES complex and establishment of the steady-state condition.

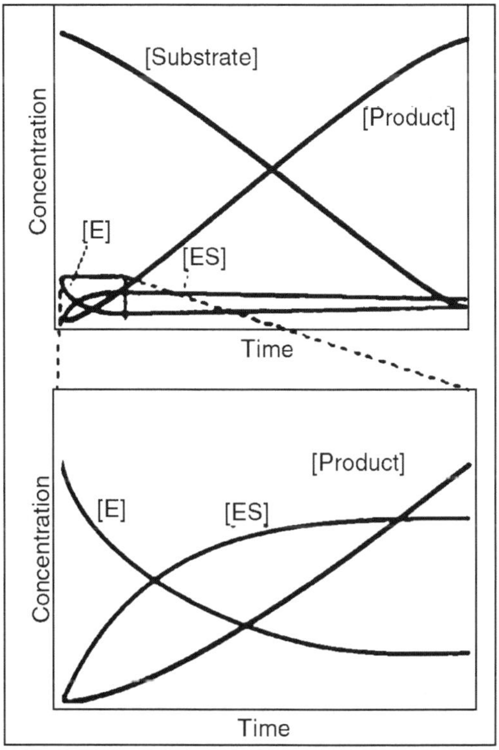

Fig. Consumption of Enzyme Substrate

Initial Velocity Assumption

One other simplification will be advantageous. Because enzymes accelerate the rate of the reverse reaction as well as the forward reaction, it would be helpful to ignore any back reaction by which E1P might form ES. The velocity of this back reaction would be given by $v = k_{-2}[E][P]$. However, if we observe only the *initial velocity* for the reaction immediately after E and S are mixed in the absence of P, the rate of any back reaction is negligible ecause its rate will be proportional to [P], and [P] is essentially 0. Given such simplification, we now analyse the system described by Equation in order to describe the initial velocity v as a function of [S] and amount of enzyme.

The total amount of enzyme is fixed and is given by the formula,

$$\text{Total enzyme, } [E_T] = [E] + [ES]$$

where [E] = free enzyme and [ES] = the amount of enzyme in the enzyme- substrate complex. From Equation, the rate of [ES] formation is,

$$v_f = k_1([ET] - [ES])[S]$$

where,

$$[ET] - [ES] = [E]$$

From Equation, the rate of [ES] disappearance is,

$$v_d = k_{-1}[ES] + k_2[ES] = (k_{-1} + k_2)[ES]$$

At steady state, $d[ES]/dt = 0$, and therefore, $v_f = v_d$.

So,

$$K_1([ET] - [ES])[S] = (k_{-1} + k_2)[ES]$$

Rearranging gives,

$$\frac{([E_T]-[ES])[S]}{[ES]} = \frac{(k_{-1}+k_2)}{k_1}$$

The Michaelis Constant, Km

The ratio of constants $(k_{-1} + k_2)/k_1$ is itself a constant and is defined as the Michaelis constant, K_m,

$$K_m = \frac{(k_{-1}+k_2)}{k_1}$$

Note from that K_m is given by the ratio of two concentrations (([ET] – [ES]) and [S]) to one ([ES]), so K_m has the units of molarity. From Equation, we can write

$$\frac{([E_T]-[ES])[S]}{[ES]} = K_m$$

which rearranges to

$$[ES] = \frac{[E_T]-[S]}{K_m+[S]}$$

Now, the most important parametre in the kinetics of any reaction is the rate of product formation.

This rate is given by:

$$v = \frac{d[P]}{dt}$$

and for this reaction

$$v = k_2[ES]$$

Substituting the expression for [ES] from Equation gives,

$$v = \frac{k_2[E_T][S]}{K_m+[S]}$$

The product $k_2[ET]$ has special meaning. When [S] is high enough to saturate all of the enzyme, the velocity of the reaction, v, is maximal. At saturation, the amount of [ES] complex is equal to the total enzyme concentration, E_T, its maximum possible value. From Equation, the initial velocity v then equals $k_2[E_T] = V_{max}$. Written symbolically, when $[S] >> [E_T]$ (and K_m), $[E_T] = [ES]$ and $v = V_{max}$. Therefore,

$$V_{max} = k_2[E_T]$$

Substituting this relationship into the expression for v gives the Michaelis-

Menten Equation

$$v = \frac{V_{max}[S]}{K_m + [S]}$$

This equation says that the rate of an enzyme-catalyzed reaction, v, at any moment is determined by two constants, K_m and V_{max}, and the concentration of substrate at that moment.

When $[S] = K_m$, $v = V_{max}/2$

We can provide an operational definition for the constant K_m by rearranging Equation to give

$$K_m = [S]\left(\frac{V_{max}}{v} - 1\right)$$

Then, at $v = V_{max}/2$, $K_m = [S]$. That is, K_m is defined by the substrate concentration that gives a velocity equal to one-half the maximal velocity.

Relationships Between V_{max}, K_m, and Reaction Order

The Michaelis-Menten equation describes a curve known from analytical geometry as a *rectangular hyperbola*.[1] In such curves, as [S] is increased, v approaches the limiting value, V_{max}, in an asymptotic fashion. V_{max} can be approximated experimentally from a substrate saturation curve and K_m can be derived from $V_{max}/2$, so the two constants of the Michaelis-Menten equation can be obtained from plots of v versus [S]. Note, however, that actual estimation of V_{max}, and consequently K_m, is only approximate from such graphs. That is, according to Equation, in order to get $v = 0.99\ V_{max}$, [S] must equal 99 K_m, a concentration that may be difficult to achieve in practice.

From Equation, when $[S] >> K_m$, then $v = V_{max}$. That is, v is no longer dependent on [S], so the reaction is obeying zero-order kinetics. Also, when $[S] < K_m$, then v » $(Vmax/K_m)[S]$. That is, the rate, v, approximately follows a first-order rate equation, $v = k'[A]$, where

$$k' = V_{max}/K_m.$$

K_m and V_{max},

once known explicitly, define the rate of the enzyme-catalyzed reaction, *provided*:

- The reaction involves only one substrate, *or* if the reaction is multisubstrate, the concentration of only one substrate is varied while the concentration of all other substrates is held constant.
- The reaction $ES \rightarrow E + P$ is irreversible, *or* the experiment is limited to observing only initial velocities where $[P] = 0$.
- $[S]_0 > [E_T]$ and $[E_T]$ is held constant.
- All other variables that might influence the rate of the reaction (temperature, pH, ionic strength, and so on) are constant.

Enzyme Units

In many situations, the actual molar amount of the enzyme is not known. However, its amount can be expressed in terms of the activity observed. The International Commission on Enzymes defines One International Unit of enzyme *as the amount that catalyzes the formation of one micromole of product in one minute.* (Because enzymes are very sensitive to factors such as pH, temperature, and ionic strength, the conditions of assay must be specified.) Another definition for units of enzyme activity is the katal. One katal is *that amount of enzyme catalyzing the conversion of one mole of substrate to product in one second*. Thus, one katal equals 6×10^7 international units.

Turnover Number

The turnover number of an enzyme, k_{cat}, is a measure of its maximal catalytic activity. k_{cat} is defined as the number of substrate molecules converted into product per enzyme molecule per unit time when the enzyme is saturated with substrate. The turnover number is also referred to as the molecular activity of the enzyme. For the simple Michaelis-Menten reaction under conditions of initial velocity measurements, $k_2 = k_{cat}$. Provided the concentration of enzyme, $[E_T]$, in the reaction mixture is known, k_{cat} can be determined from V_{max}. At saturating [S], $v = V_{max} = k_2 [E_T]$.

Thus,

$$k_2 = \frac{V_{max}}{[E_T]} = k_{cat}$$

The term k_{cat} represents the kinetic efficiency of the enzyme. Catalase has the highest turnover number known; each molecule of this enzyme can degrade 40 million molecules of H_2O_2 in one second!

At the other end of the scale, lysozyme requires 2 seconds to cleave a glycosidic bond in its glycan substrate.

Table. Values of k_{cat} (Turnover Number) for Some Enzymes

Enzyme	k_{cat} (sec^{-1})
Catalase	40,000,000
Carbonic anhydrase	1,000,000
Acetylcholinesterase	14,000
Penicillinase	2,000
Lactate dehydrogenase	1,000
Chymotrypsin	100
DNA polymerase I	15
Lysozyme	0.5

k_{cat}/K_m

Under physiological conditions, [S] is seldom saturating, and kcat itself is not particularly informative. That is, the *in vivo* ratio of $[S]/K_m$ usually falls in the range of 0.01 to 1.0, so active sites often are not filled with substrate. Nevertheless, we can derive a meaningful index of the efficiency of Michaelis-Menten-type enzymes under these conditions by employing the following equations. As presented in Equation, if

$$v = \frac{v_{max}[S]}{K_m [S]}$$

and

$$V_{max} = k_{cat}\,[E_T], \text{ then}$$

$$v = \frac{k_{cat}[E_T][S]}{K_m + [S]}$$

When [S] << K_m, the concentration of free enzyme, [E], is approximately equal to [ET], so that

$$v = \left(\frac{k_{cat}}{K_m}\right)[EIS]$$

That is, k_{cat}/K_m is an apparent Second-order rate constant for the reaction of E and S to form product. Because K_m is inversely proportional to the affinity of the enzyme for its substrate and k_{cat} is directly proportional to the kinetic efficiency of the enzyme, k_{cat}/K_m provides an index of the catalytic efficiency of an enzyme operating at substrate concentrations substantially below saturation amounts. An interesting point emerges if we restrict ourselves to the simple case where $k_{cat} = k_2$.

Then:

$$\frac{k_{cat}}{k_m} = \frac{k_1 k_2}{k_{-1} + k_2}$$

But k_1 must always be greater than or equal to $k_1k_2/(k_{-1} + k_2)$. That is, the reaction can go no faster than the rate at which E and S come together. Thus, k_1 sets the upper limit for k_{cat}/K_m. In other words, the catalytic efficiency of an enzyme cannot exceed the diffusion-controlled rate of combination of E and S to form ES. In H_2O, the rate constant for such diffusion is approximately $10^9/M \times$ sec. Those enzymes that are most efficient in their catalysis have k_{cat}/K_m ratios approaching this value. Their catalytic velocity is limited only by the rate at which they encounter S; enzymes this efficient have achieved so-called catalytic perfection. All E and S encounters lead to reaction because such *"catalytically perfect"* enzymes can channel S to the active site, regardless of where S hits E.

Linear Plots Can Be Derived from the Michaelis - Menten Equation

Because of the hyperbolic shape of v versus [S] plots, V_{max} can only be determined from an extrapolation of the asymptotic approach of v to some limiting value as [S] increases indefinitely; and K_m is derived from that value of [S] giving $v = V_{max}/2$. However, several rearrangements of the Michaelis-Menten equation transform it into a straight-line equation. The best known of these is the Lineweaver-Burk double-reciprocal plot: Taking the reciprocal of both sides of the Michaelis-Menten equation, Equation, yields the equality,

$$\frac{1}{v} = \left(\frac{K_m}{V_{max}}\right)\left(\frac{1}{[S]}\right) + \frac{1}{V_{max}}$$

This conforms to y5mx1b (the equation for a straight line), where $y = 1/v$; m, the slope, is K_m/V_{max}; $x = 1/[S]$; and $b = 1/V_{max}$. Plotting $1/v$ versus $1/[S]$ gives a straight line whose x-intercept is $-1/K_m$, whose y-intercept is $1/V_{max}$, and whose slope is K_m/V_{max}. The Hanes-Woolf plot is another rearrangement of the Michaelis-Menten equation that yields a straight line:

Multiplying both sides of Equation by [S] gives,

$$\frac{[S]}{v} = [S]\left(\frac{K_m}{V_{max}}\right)\left(\frac{1}{[S]}\right) + \frac{[S]}{V_{max}} = \frac{K_m}{V_{max}} + \frac{[S]}{V_{max}}$$

and,

$$\frac{[S]}{v} = \left(\frac{1}{V_{max}}\right)[S] + \frac{K_m}{V_{max}}$$

Graphing [S]/*v* versus [S] yields a straight line where the slope is $1/V_{max}$, the *y*-intercept is K_m/V_{max}, and the *x*-intercept is $-K_m$. The common advantage of these plots is that they allow both K_m and V_{max} to be accurately estimated by extrapolation of straight lines rather than asymptotes. Computer fitting of *v* versus [S] data to the Michaelis-Menten equation is more commonly done than graphical plotting.

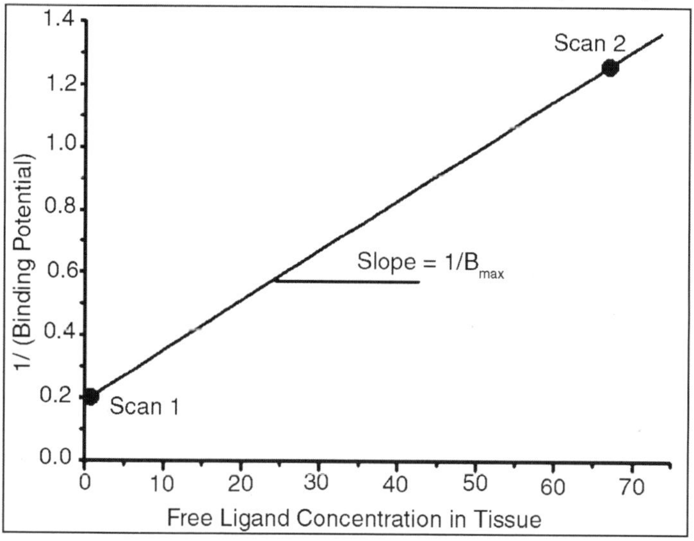

Fig. Hanes-Woolf Plot

Effect of pH on Enzymatic Activity

Enzyme-substrate recognition and the catalytic events that ensue are greatly dependent on pH. An enzyme possesses an array of ionizable side chains and prosthetic groups that not only determine its secondary and tertiary structure but may also be intimately involved in its active site. Further, the substrate itself often has ionizing groups, and one or another of the ionic forms may preferentially interact with the enzyme.

Enzymes in general are active only over a limited pH range and most have a particular pH at which their catalytic activity is optimal. These effects of pH may be due to effects on K_m or V_{max} or both. The relative activity of four enzymes as a function of pH. Although the pH optimum of an enzyme often reflects the pH of its normal environment, the optimum may not be precisely the same. This difference suggests that the pH-activity response of an enzyme may be a factor in the intracellular regulation of its activity.

Effect of Temperature on Enzymatic Activity

Like most chemical reactions, the rates of enzyme-catalyzed reactions generally increase with increasing temperature. However, at temperatures above 50° to 60°C, enzymes typically show a decline in activity.

Two effects are operating here:

1. The characteristic increase in reaction rate with temperature, and
2. Thermal denaturation of protein structure at higher temperatures.

Most enzymatic reactions double in rate for every 10°C rise in temperature (that is, $Q_{10} = 2$, where Q_{10} is defined as *the ratio of activities at two temperatures 10° apart*) as long as the enzyme is stable and fully active.

Some enzymes, those catalyzing reactions having very high activation energies, show proportionally greater Q_{10} values. The increasing rate with increasing temperature is ultimately offset by the instability of higher orders of protein structure at elevated temperatures, where the enzyme is inactivated. Not all enzymes are quite so thermally labile.

For example, the enzymes of thermophilic bacteria (*thermophilic* = "heat-loving") found in geothermal springs retain full activity at temperatures in excess of 85°C.

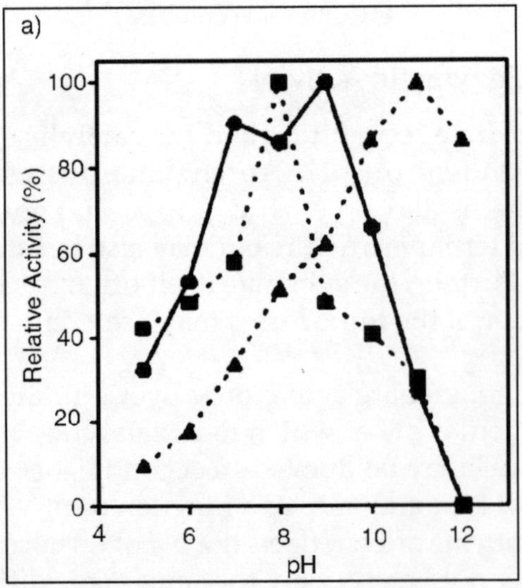

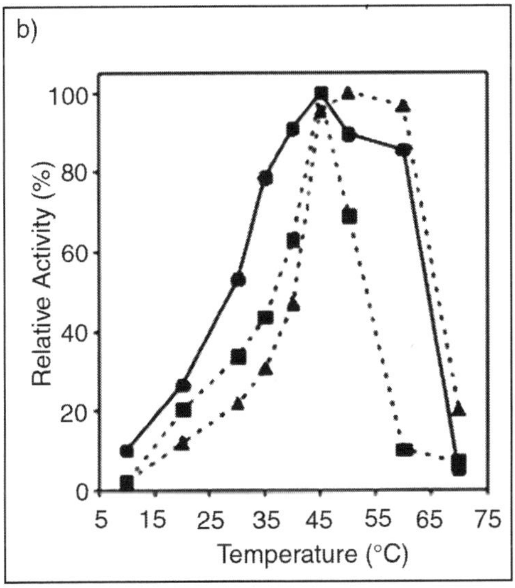

Fig. (a) Effect of pH (b) Effect of Temperature

ENZYME INHIBITION

If the velocity of an enzymatic reaction is decreased or inhibited, the kinetics of the reaction obviously have been perturbed. Systematic perturbations are a basic tool of experimental scientists; much can be learned about the normal workings of any system by inducing changes in it and then observing the effects of the change. The study of enzyme inhibition has contributed significantly to our understanding of enzymes.

The binding of an inhibitor can stop a substrate from entering the enzyme's active site and/or hinder the enzyme from catalysing its reaction. Inhibitor binding is either reversible or irreversible. Irreversible inhibitors usually react with the enzyme and change it chemically. These inhibitors modify key amino acid residues needed for enzymatic activity. In contrast, reversible inhibitors bind non-covalently and different types of inhibition are produced depending on whether these inhibitors bind the enzyme, the enzyme-substrate complex, or both.

Many drug molecules are enzyme inhibitors, so their discovery and improvement is an active area of research in biochemistry and pharmacology. A medicinal enzyme inhibitor is often judged by its specificity (its lack of binding to other proteins) and its potency (its dissociation constant, which indicates the concentration needed to inhibit the enzyme). A high specificity and potency ensure that a drug will have few side effects and thus low toxicity. Enzyme inhibitors also occur naturally and are involved in the regulation of metabolism. For example, enzymes in a metabolic pathway can be inhibited by downstream products. This type of negative feedback slows flux through a pathway when the products begin to build up and is an important way to maintain homeostasis in a cell. Other cellular enzyme inhibitors are proteins that specifically bind to and inhibit an enzyme target.

This can help control enzymes that may be damaging to a cell, such as proteases or nucleases; a well-characterised example is the ribonuclease inhibitor, which binds to ribonucleases in one of the tightest known protein–protein interactions. Natural enzyme inhibitors can also be poisons and are used as defences against predators or as ways of killing prey.

Reversible Inhibitors

Types of Reversible Inhibitors

Reversible inhibitors bind to enzymes with non-covalent interactions such as hydrogen bonds, hydrophobic interactions and ionic bonds.

Multiple weak bonds between the inhibitor and the active site combine to produce strong and specific binding. In contrast to substrates and irreversible inhibitors, reversible inhibitors generally do not undergo chemical reactions when bound to the enzyme and can be easily removed by dilution or dialysis.

There are four kinds of reversible enzyme inhibitors.

They are classified according to the effect of varying the concentration of the enzyme's substrate on the inhibitor:

1. In competitive inhibition, the substrate and inhibitor cannot bind to the enzyme at the same time. This usually results from the inhibitor having an affinity for the active site of an enzyme where the substrate also binds; the substrate and inhibitor *compete* for access to the enzyme's active site. This type of inhibition can be overcome by sufficiently high concentrations of substrate, *i.e.*, by out-competing the inhibitor. Competitive inhibitors are often similar in structure to the real substrate.
2. In uncompetitive inhibition, the inhibitor binds only to the substrate-enzyme complex, it should not be confused with non-competitive inhibitors. Both Vmax and Km decrease (maximum velocity decreases as a result of removing activated complex while binding efficiency increases as a result of Le Chatelier's principle).

3. In mixed inhibition, the inhibitor can bind to the enzyme at the same time as the enzyme's substrate. However, the binding of the inhibitor affects the binding of the substrate, and vice versa. This type of inhibition can be reduced, but not overcome by increasing concentrations of substrate. Although it is possible for mixed-type inhibitors to bind in the active site, this type of inhibition generally results from an allosteric effect where the inhibitor binds to a different site on an enzyme. Inhibitor binding to this allosteric site changes the conformation (*i.e.,* tertiary structure or three-dimensional shape) of the enzyme so that the affinity of the substrate for the active site is reduced.
4. Non-competitive inhibition is a form of mixed inhibition where the binding of the inhibitor to the enzyme reduces its activity but does not affect the binding of substrate. As a result, the extent of inhibition depends only on the concentration of the inhibitor.

Quantitative Description of Reversible Inhibition

Reversible inhibition can be described quantitatively in terms of the inhibitor's binding to the enzyme and to the enzyme–substrate complex, and its effects on the kinetic constants of the enzyme. In the classic Michaelis–Menten scheme below, an enzyme (E) binds to its substrate (S) to form the enzyme–substrate complex ES.

Upon catalysis, this complex breaks down to release product P and free enzyme.

The inhibitor (I) can bind to either E or ES with the dissociation constants K_i *or* K_i', *respectively*:

- Competitive inhibitors can bind to E, but not to ES. Competitive inhibition increases K_m (*i.e.,* the inhibitor interferes with substrate binding), but does not affect V_{max} (the inhibitor does not hamper catalysis in ES because it cannot bind to ES).
- Non-competitive inhibitors have identical affinities for E and ES ($K_i = K_i'$). Non-competitive inhibition does not change K_m (*i.e.,* it does not affect substrate binding) but decreases V_{max} (*i.e.,* inhibitor binding hampers catalysis).
- Mixed-type inhibitors bind to both E and ES, but their affinities for these two forms of the enzyme are different (K_i ‴ K_i'). Thus, mixed-type inhibitors interfere with substrate binding (increase K_m) and hamper catalysis in the ES complex (decrease V_{max}).

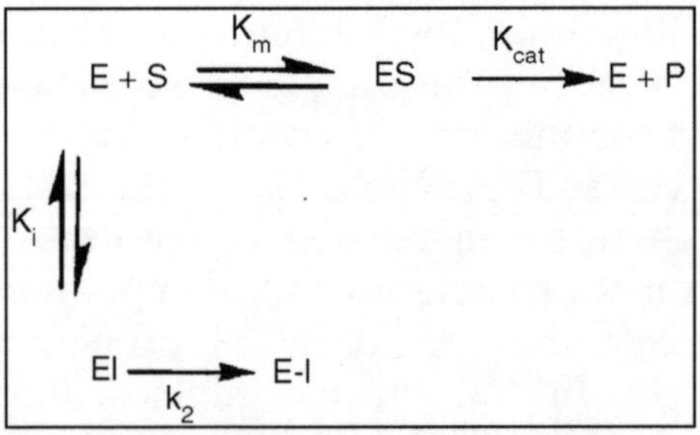

Fig. Kinetic Scheme for Reversible Enzyme Inhibitors

When an enzyme has multiple substrates, inhibitors can show different types of inhibition depending on which substrate is considered. This results from the active site containing two different binding sites within the active site, one for each substrate. For example, an inhibitor might compete with substrate A for the first binding site, but be a non-competitive inhibitor with respect to substrate B in the second binding site.

Measuring the Dissociation Constants of a Reversible Inhibitor

An enzyme inhibitor is characterised by its two dissociation constants, K_i and K_i', to the enzyme and to the enzyme-substrate complex, respectively. The enzyme-inhibitor constant K_i can be measured directly by various methods; one extremely accurate method is isothermal titration calorimetry, in which the inhibitor is titrated into a solution of enzyme and the heat released or absorbed is measured. However, the other dissociation constant K_i' is difficult to measure directly, since the enzyme-substrate complex is short-lived and undergoing a chemical reaction to form the product. Hence, K_i' is usually measured indirectly, by observing the enzyme activity under various substrate and inhibitor concentrations, and fitting the data to a modified Michaelis–Menten equation

$$V = \frac{V_{max}[S]}{\alpha K_m + \alpha'[S]} = \frac{(1/\alpha')V_{max}[S]}{(\alpha/\alpha')K_m + [S]}$$

where the modifying factors á and á′ are defined by the inhibitor concentration and its two dissociation constants,

$$\alpha = 1 + \frac{[I]}{K_i}$$

$$\alpha' = 1 + \frac{[I]}{K_i'}.$$

Thus, in the presence of the inhibitor, the enzyme's effective K_m

and V_{max} become $(\alpha/\alpha')K_m$ and $(1/\alpha')V_{max}$, respectively. However, the modified Michaelis-Menten equation assumes that binding of the inhibitor to the enzyme has reached equilibrium, which may be a very slow process for inhibitors with sub-nanomolar dissociation constants. In these cases, it is usually more practical to treat the tight-binding inhibitor as an irreversible inhibitor however, it can still be possible to estimate K_i' kinetically if K_i is measured independently.

The effects of different types of reversible enzyme inhibitors on enzymatic activity can be visualized using graphical representations of the Michaelis–Menten equation, such as Lineweaver–Burk and Eadie-Hofstee plots. For example, in the Lineweaver–Burk plots at the right, the competitive inhibition lines intersect on the *y*-axis, illustrating that such inhibitors do not affect V_{max}. Similarly, the non-competitive inhibition lines intersect on the *x*-axis, showing these inhibitors do not affect K_m. However, it can be difficult to estimate K_i and K_i' accurately from such plots, so it is advisable to estimate these constants using more reliable nonlinear regression methods, as described above.

SPECIAL CASES

- The mechanism of partially competitive inhibition is similar to that of non-competitive, except that the EIS complex has catalytic activity, which may be lower or even higher (partially competitive activation) than that of the enzyme–substrate (ES) complex. This inhibition typically displays a lower V_{max}, but an unaffected K_m value.
- Uncompetitive inhibition occurs when the inhibitor binds only to the enzyme–substrate complex, not to the free enzyme; the EIS complex is catalytically inactive. This mode of inhibition is rare and causes a decrease in both V_{max} and the K_m value.
- Substrate and product inhibition is where either the substrate or product of an enzyme reaction inhibit the enzyme's activity. This inhibition may follow the competitive, uncompetitive or mixed patterns. In substrate inhibition there is a progressive decrease in activity at high substrate concentrations. This may indicate the existence of two substrate-binding sites in the enzyme. At low substrate, the high-affinity site is occupied and normal kinetics are followed. However, at higher concentrations, the second inhibitory site becomes occupied, inhibiting the enzyme. Product inhibition is often a regulatory feature in metabolism and can be a form of negative feedback.
- Slow-tight inhibition occurs when the initial enzyme–inhibitor complex EI undergoes isomerisation to a second more tightly

held complex, EI*, but the overall inhibition process is reversible. This manifests itself as slowly increasing enzyme inhibition. Under these conditions, traditional Michaelis–Menten kinetics give a false value for K_i, which is time-dependent. The true value of K_i can be obtained through more complex analysis of the on (k_{on}) and off (k_{off}) rate constants for inhibitor association.

Examples of Reversible Inhibitors

As enzymes have evolved to bind their substrates tightly, and most reversible inhibitors bind in the active site of enzymes, it is unsurprising that some of these inhibitors are strikingly similar in structure to the substrates of their targets. An example of these substrate mimics are the protease inhibitors, a very successful class of antiretroviral drugs used to treat HIV. The structure of ritonavir, a protease inhibitor based on a peptide and containing three peptide bonds, is shown on the right. As this drug resembles the protein that is the substrate of the HIV protease, it competes with this substrate in the enzyme's active site.

Enzyme inhibitors are often designed to mimic the transition state or intermediate of an enzyme-catalysed reaction. This ensures that the inhibitor exploits the transition state stabilising effect of the enzyme, resulting in a better binding affinity (lower K_i) than substrate-based designs. An example of such a transition state inhibitor is the antiviral drug oseltamivir; this drug mimics the planar nature of the ring oxonium ion in the reaction of the viral enzyme neuraminidase.

Irreversible Inhibitors

Types of Irreversible Inhibition

Irreversible inhibitors usually covalently modify an enzyme, and inhibition cannot therefore be reversed. Irreversible inhibitors often contain reactive functional groups such as nitrogen mustards, aldehydes, haloalkanes, alkenes, Michael acceptors, phenyl sulfonates, or fluorophosphonates. These electrophilic groups react with amino acid side chains to form covalent adducts. The residues modified are those with side chains containing nucleophiles such as hydroxyl or sulfhydryl groups; these include the amino acids serine (as in DFP, right), cysteine, threonine or tyrosine.

Irreversible inhibition is different from irreversible enzyme inactivation. Irreversible inhibitors are generally specific for one class of enzyme and do not inactivate all proteins; they do not function by destroying protein structure but by specifically altering the active site of their target. For example, extremes of pH or temperature usually cause denaturation of all protein structure, but this is a non-specific effect. Similarly, some non-specific chemical treatments destroy protein structure: for example, heating in concentrated hydrochloric acid will hydrolyse the peptide bonds holding proteins together, releasing free amino acids.

Irreversible inhibitors display time-dependent inhibition and their potency therefore cannot be characterised by an IC_{50} value. This is because the amount of active enzyme at a given concentration of irreversible inhibitor will be different depending on how long the inhibitor is pre-incubated with the enzyme. Instead, $k_{obs}/[I]$ values are used, wherek_{obs} is the observed pseudo-first order rate of inactivation (obtained by plotting the log of% activity vs. time) and $[I]$ is the concentration of inhibitor. The $k_{obs}/[I]$ parametre is valid as long as the inhibitor does not saturate binding with the enzyme (in which case $k_{obs} = k_{inact}$).

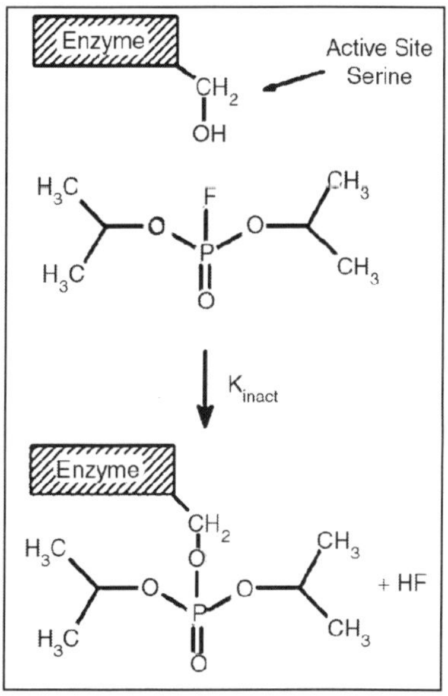

Fig. Reaction of the Irreversible Inhibitor

Analysis of Irreversible Inhibition

Irreversible inhibitors form a reversible non-covalent complex with the enzyme (EI or ESI) and this then reacts to produce the covalently modified "dead-end complex" EI*. The rate at which EI* is formed is called the inactivation rate or k_{inact}. Since formation of EI may compete with ES, binding of irreversible inhibitors can be prevented by competition either with substrate or with a second, reversible inhibitor.

This protection effect is good evidence of a specific reaction of the irreversible inhibitor with the active site.

The binding and inactivation steps of this reaction are investigated by incubating the enzyme with inhibitor and assaying the amount of activity remaining over time. The activity will be decrease in a time-dependent manner, usually following exponential decay. Fitting these data to a rate equation gives the rate of inactivation at this concentration of inhibitor. This is done at several different concentrations of inhibitor. If a reversible EI complex is involved the inactivation rate will be saturable and fitting this curve will give k_{inact} and K_i.

Another method that is widely used in these analyses is mass spectrometry. Here, accurate measurement of the mass of the unmodified native enzyme and the inactivated enzyme gives the increase in mass caused by reaction with the inhibitor and shows the stoichiometry of the reaction. This is usually done using a MALDI-TOF mass spectrometre. In a complementary technique, peptide mass fingerprinting involves digestion of the native and modified protein with a protease such as trypsin. This will produce a set of peptides that can be analysed using a mass spectrometre. The peptide that changes in mass after reaction with the inhibitor will be the one that contains the site of modification.

KINETICS OF ENZYME-CATALYZED REACTIONS

Involving Two or More Substrates

Thus far, we have considered only the simple case of enzymes that act upon a single substrate, S. This situation is not common. Usually, enzymes catalyze reactions in which two (or even more) substrates take part. Consider the case of an enzyme catalyzing a reaction involving two substrates, A and B, and yielding the products P and Q:

$$A + B \xrightarrow{\text{enzyme}} P + Q$$

Such a reaction is termed a bisubstrate reaction.

In general, bisubstrate reactions proceed by one of two possible routes:

1. Both A and B are bound to the enzyme and then reaction occurs to give P1Q:

   $$E + A + B \rightarrow AEB \rightarrow PEQ \rightarrow E + P + Q$$

   Reactions of this type are defined as sequential or single-displacement reactions.

   They can be either of two distinct classes:

   - Random, where either A or B may bind to the enzyme first, followed by the other substrate, or

– Ordered, where A, designated the *leading substrate,* must bind to E first before B can be bound.

Double-reciprocal Form of the Rate Equation: $\frac{1}{v} = \frac{1}{V_{max}}\left(K_m^A + \frac{K_s^A K_m^B}{[B]}\right)\left(\frac{1}{[A]} + \frac{1}{V_{max}}\right)\left(1 + \frac{K_m^B}{[B]}\right)$

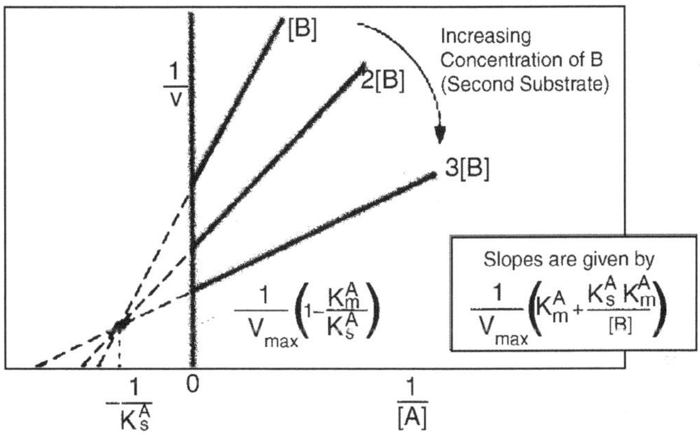

Fig. Single-displacement Bisubstrate Mechanism.

2. The other general possibility is that one substrate, A, binds to the enzyme and reacts with it to yield a chemically modified form of the enzyme (E′) plus the product, P. The second substrate, B, then reacts with E′, regenerating E and forming the other product, Q.

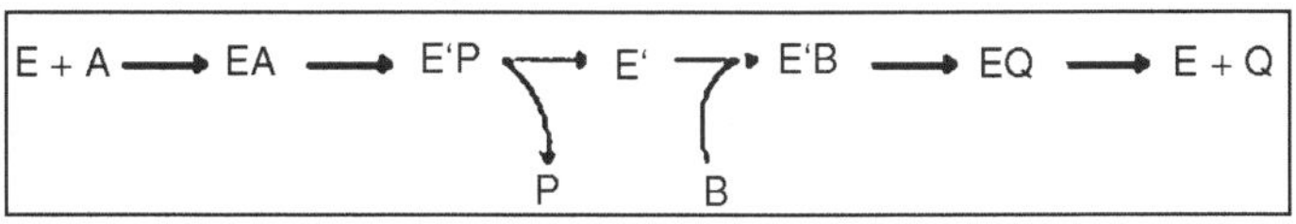

Reactions that fit this model are called ping-pong or double-displacement reactions. Two distinctive features of this mechanism are the obligatory formation of a modified enzyme intermediate, E′, and the pattern of parallel lines obtained in double-reciprocal plots.

Random, Single-displacement Reactions

In this type of sequential reaction, all possible binary enzyme: substrate complexes (AE, EB, QE, EP) are formed rapidly and reversibly when the enzyme is added to a reaction mixture containing A, B, P, and Q:

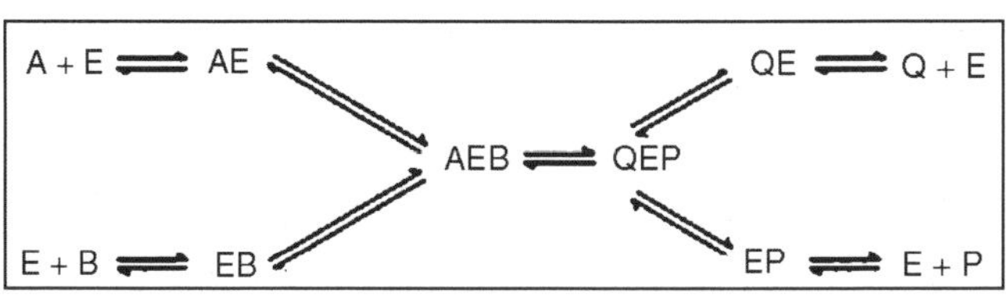

The rate-limiting step is the reaction AEB ® QEP. It doesn't matter whether A or B binds first to E, or whether Q or P is released first from QEP. Sometimes, reactions that follow this random order of addition of substrates to E can be distinguished mechanistically from reactions obeying an ordered, single-displacement mechanism, *if* A has *no* influence on the binding constant for B (and vice versa); that is, the mechanism is purely random.

Ordered, Single-displacement Reactions

In this case, the leading substrate, A (also called the obligatory or compulsory substrate), must bind first. Then the second substrate, B, binds. Strictly speaking, B cannot bind to free enzyme in the absence of A. Reaction between A and B occurs in the ternary complex, and is usually followed by an ordered release of the products of the reaction, P and Q. In the schemes below, Q is the product of A and is released last. One representation, suggested by W. W. Cleland, follows:

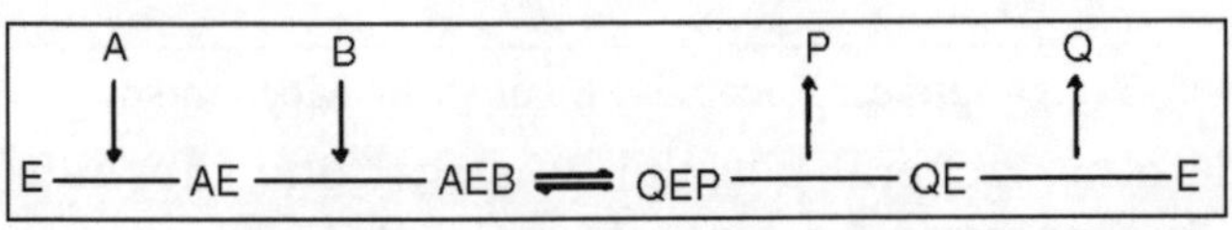

Another way of portraying this mechanism is as follows:

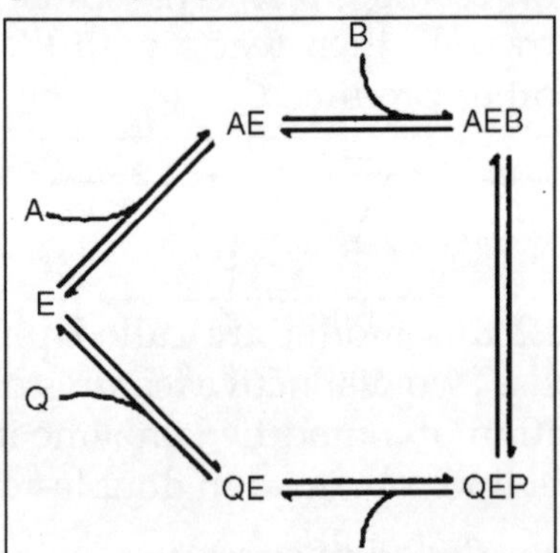

Note that A and Q are competitive for binding to the free enzyme, E, but not A and B (or Q and B).

NAD⁺-dependent Dehydrogenases Show Ordered

Single-Displacement Mechanisms

Nicotinamide adenine dinucleotide (NAD$^+$)-dependent dehydrogenases are enzymes that typically behave according to the kinetic pattern just described.

A general reaction of these dehydrogenases is:

$$NAD^+ + BH_2 \rightleftharpoons NADH + H^+ + B$$

The leading substrate (A) is nicotinamide adenine dinucleotide (NAD^+), and NAD^+ and NADH (product Q) compete for a common site on E.

A specific example is offered by alcohol dehydrogenase (ADH):

$$\underset{(A)}{NAD^+} + \underset{\substack{\text{ethanol}\\(B)}}{CH_3CH_2OH} \rightleftharpoons \underset{(Q)}{NADH} + H^+ + \underset{\substack{\text{acetaldehyde}\\(P)}}{CH_3CHO}$$

We can verify that this ordered mechanism is not random by demonstrating that no B (ethanol) is bound to E in the absence of A (NAD^+).

Double-Displacement (Ping-Pong) Reactions

Reactions conforming to this kinetic pattern are characterized by the fact that the product of the enzyme's reaction with A (called P in the following schemes) is released *prior* to reaction of the enzyme with the second substrate, B. As a result of this process, the enzyme, E, is converted to a modified form, E9, which then reacts with B to give the second product, Q, and regenerate the unmodified enzyme form, E:

$$E \overset{A\downarrow}{\rightleftharpoons} AE \rightleftharpoons PE' \overset{P\uparrow}{\rightleftharpoons} E' \overset{B\downarrow}{\rightleftharpoons} E'B \rightleftharpoons EQ \overset{Q\uparrow}{\rightleftharpoons} E$$

or,

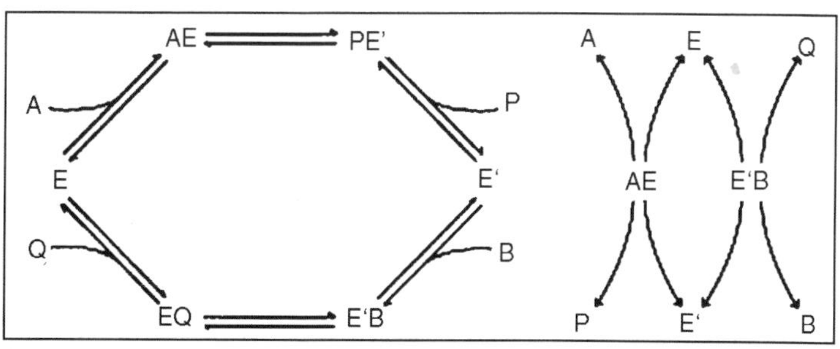

Note that these schemes predict that A and Q compete for the free enzyme form, E, while B and P compete for the modified enzyme form, E′. A and Q do not bind to E′, nor do B and P combine with E.

Aminotransferases Show Double-displacement Catalytic Mechanisms

One class of enzymes that follow a ping-pong-type mechanism are *aminotransferases* (previously known as transaminases). These enzymes catalyze the transfer of an amino group from an amino acid to an a-keto acid. The products are a new amino acid and the keto acid corresponding to the carbon skeleton of the amino donor:

$$\text{amino acid}_1 + \text{keto acid}_2 \rightarrow \text{keto acid}_1 + \text{amino acid}_2$$

A specific example would be *glutamate: aspartate aminotransferase*. The glutamate and aspartate are competitive for E, and that oxaloacetate and a-ketoglutarate compete for E′. In glutamate: aspartate aminotransferase, an enzyme-bound coenzyme, *pyridoxal phosphate* (a vitamin B_6 derivative), serves as the amino group acceptor/donor in the enzymatic reaction. The unmodified enzyme form, E, has the coenzyme in the aldehydic pyridoxal form, whereas the modified enzyme form, E9, is actually pyridoxamine phosphate. Not all enzymes displaying ping-pong-type mechanisms require coenzymes as carriers for the chemical substituent transferred in the reaction.

Diagnosis of Bisubstrate Mechanisms

Kineticists rely on a number of diagnostic tests for the assignment of a reaction mechanism to a specific enzyme. One is the graphic analysis of the kinetic patterns observed. It is usually easy to distinguish between single- and double-displacement reactions in this manner, and examining competitive effects between substrates aids in assigning reactions to random versus ordered patterns of S-binding. A second diagnostic test is to determine whether the enzyme catalyzes an exchange reaction. Consider as an example the two enzymes *sucrose phosphorylase* and *maltose phosphorylase*. Both catalyze the phosphorolysis of a disaccharide and both yield glucose-1-phosphate and a free hexose:

$$\text{Sucrose} + P_i \rightleftharpoons \text{glucose}-1-\text{phosphate} + \text{fructose}$$

$$\text{Maltose} + P_i \rightleftharpoons \text{glucose}-1-\text{phosphate} + \text{glucose}$$

Interestingly, in the absence of sucrose and fructose, sucrose phosphorylase will catalyze the exchange of inorganic phosphate, Pi, into glucose-1-phosphate. This reaction can be followed by using $^{32}P_i$ as a radioactive tracer and observing the appearance of ^{32}P into glucose-1-phosphate:

$$^{32}P_i + G{-}1{-}P \rightleftharpoons P_i + G{-}1{-}{}^{32}P$$

Maltose phosphorylase cannot carry out a similar reaction. The ^{32}P exchange reaction of sucrose phosphorylase is accounted for by a double-displacement mechanism where E′ = E-glucose:

$$\text{Sucrose} + E \rightleftharpoons E-\text{glucose} + \text{fructose}$$

$$E-\text{glucose} + P_i \rightleftharpoons E + \text{glucose}-1-\text{phosphate}$$

Thus, in the presence of just $^{32}P_i$ and glucose-1-phosphate, sucrose phosphory-lase still catalyzes the second reaction and radioactive P_i is incorporated into glucose-1-phosphate over time. Maltose phosphorylase proceeds via a single-displacement reaction that necessarily requires the formation of a ternary maltose: E: P_i (or glucose: E: glucose-1-phosphate) complex for any reaction to occur. Exchange reactions are a characteristic of enzymes that obey double-displacement mechanisms at some point in their catalysis.

RNA AND ANTIBODY MOLECULES AS ENZYMES: RIBOZYMES AND ABZYMES

RNA and DNA as Catalysts

DNA and RNA are usually seen as carriers of genetic information, with the occasional function of RNA as a scaffold, as for example in the ribosome. This picture has changed with the discovery of self-splicing of certain RNAs, most notably that of the Tetrahymena group I intron by Tom Cech, and the active role of RNA in the RNase P in the process of maturation of tRNAs by Sid Altman. These observations laid the foundation for the concept of catalytic RNA for which the Nobel prize of 1989 was awarded to these two colleagues.

Natural catalytic RNAs

These two first examples of catalytic RNA have been joined by others such as the group II intron, the hepatitis delta RNA and the plant viroid RNAs of the hammerhead and hairpin type. Although some of these reactions are facilitated by proteins *in vivo*, their reactions can be simulated *in vitro* without the aid of proteins, supporting the view that indeed the RNA by itself can do the job. With the exception of RNase P RNA, the RNAs undergo an intramolecular, or in *cis*, transesterification in these transformations.

These reactions are sequence specific, and more importantly, occur with rates orders of magnitude faster than expected from the reactivity of RNA. Thus there must be a catalytic element involved, and therefore these active RNAs are called ribozymes, in analogy to the classical protein enzymes. To study the mechanism of these ribozyme-catalysed reactions and their structural requirements, they were redesigned for intermolecular, or *in trans*, reactions. Thus substrate and ribozyme are localised on different RNA strands. This arrangement facilitates the determination of kinetic constants such as kcat and Km values. An example for an *in trans* arrangement is shown for the hammerhead ribozyme. A comparison of the efficiency of ribozyme- with enzyme-catalysed reactions shows that even though the reaction rate kcat is often smaller for the ribozymes, they compensate for it by a more favourable Km so that in the end efficiency is comparable for both catalysts.

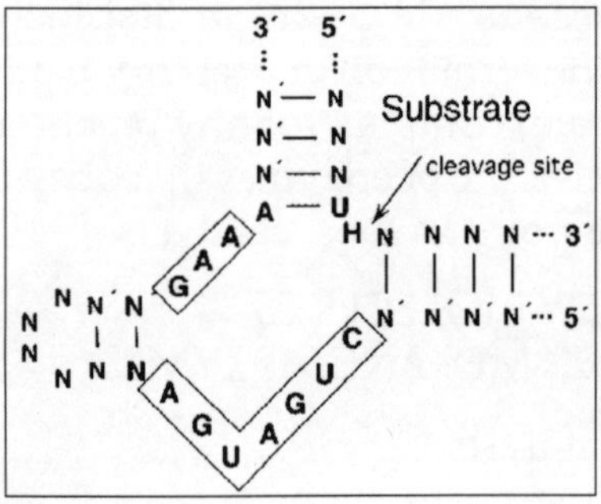

Fig. Generalised Two Dimensional Structure of the Hammerhead Ribozyme for Cleavage in Trans.

N, any nucleotide; N′, nucleotide complementary to N; H, any nucleotide except G; underlined, invariant core.

Mechanistic Aspects

Most of these RNA-catalysed reactions depend on the presence of a divalent metal ion such as Mg^{2+} but some, such as the hairpin and the hammerhead ribozyme, can also be coaxed into action with high concentrations of monovalent ions such as Na^{+}. Very interestingly, certain positively charged aminoglycosides can replace the metal ions for the hairpin ribozyme. Thus the question as to the role of the metal ions arises. Formation of a kinetically competent three-dimensional structure is quite certainly one of them. However, particularly in the presence of Mg^{2+} where most ribozymes work best, a more active role can be envisaged where it acts as a base and as a Lewis acid as originally proposed in the "two metal ion" mechanism for enzymes. Such mechanisms have been formulated particularly for the group I RNA and the hammerhead ribozymes.

In these cases, one metal ion is thought to abstract a proton from the nucleophilic hydroxy group to attack the phosphate group, and the other coordinates to the departing hydroxy group. Although there is experimental support for such a metal ion hydrate role, no metal ions which could take on this function have been detected in x-ray structures of these ribozymes. Thus, the final proof for the involvement of the metal ions as active participants in the chemical step of the reactions is still lacking.

This of course becomes a concern in light of some of the ribozymes not requiring metal ions. Looking for alternative functional groups one has to consider whether the nucleobases could not play such a role. The argument against this idea in the past has been that the pKa values of these groups are outside the pH optimum of the reactions. NMR studies have however shown that the pKa of adenosine, for example, can be shifted upwards by 2.5 units to approximately 6.5 as a consequence of a change in the micro-environment in the three-dimensional structure of the ribozymes. Thus it might well be that such groups can function as acid-base catalysts at neutral pH and that we have to reconsider our prejudice against nucleotides as active participants in such catalytic reactions.

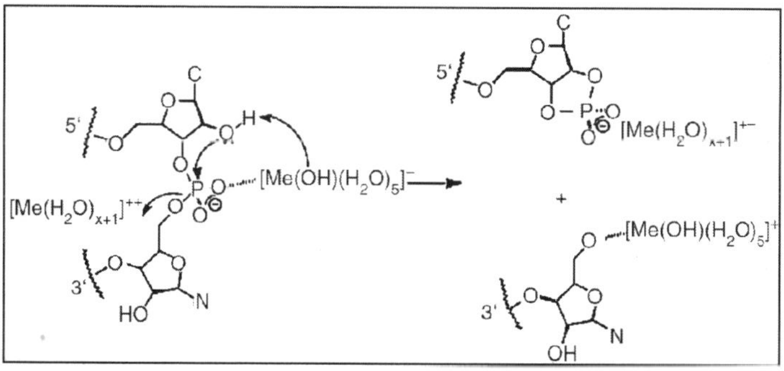

Fig. Suggested Hammerhead Ribozyme Cleavage Mechanism with Participation of Two Metal Ions.

New Ribozymes

The catalytic power of the naturally occuring ribozymes which all catalyse phosphoryl tranfers within RNA molecules, has stimulated the search for ribozymes with other activities. This has been made possible by the technique of *in vitro* selection. It involves the construction of double-stranded DNA containing a T7 promoter followed by a stretch of randomised nucleotides. Sequences necessary for binding of oligonucleotides as polymerase primers and for cloning are at both ends of such constructs. After transcription the RNA is incubated with the substrate and active RNA is separated from inactive RNA sequences. The success of the procedure depends crucially on the design of this selection step.

Often substrates with a biotin moiety are used which is transferred to the active RNA by the catalysed reaction. This facilitates separation of the active from the inactive RNA on a Streptavidin column. The active species are amplified by PCR, transcribed and incubated with the substrate again. This cycle is repeated between 6 and a dozen times until enough active RNA has accumulated for characterisation.. The method is extremely powerful, enabling the selection of active molecules from a large pool of sequences. Thus, for example, a DNA with 22 randomised positions has 4^{22} sequences which is equivalent to 1.7×10^{13} sequences. This represents a rather small degree of randomisation in the generally performed selection procedures. However, there are certain constraints created by Avogadro's number, i. e. the number of molecules in one mol which is 6×10^{23}.

It so happens that 39 randomisations is equivalent to 3.7×10^{23} sequences. Thus to cover the whole sequence space one would have to work with a 1 M DNA solution in one litre for the first selection. This becomes forbiddingly expensive in terms of enzyme required. Of course, larger degrees of randomisations can be used with the penalty that not the entire sequence space will be searched.

Such selections have yielded a most impressive number of ribozymes, catalysing the most diverse reactions. Examples are phosphoryl transfer like a kinase reaction, nucleoside triphosphate polymerisation, acyl transfer, N-glycosidic bond formation to obtain a nucleotide, and formation of a carbon-carbon bond in a Diels-Alder reaction.

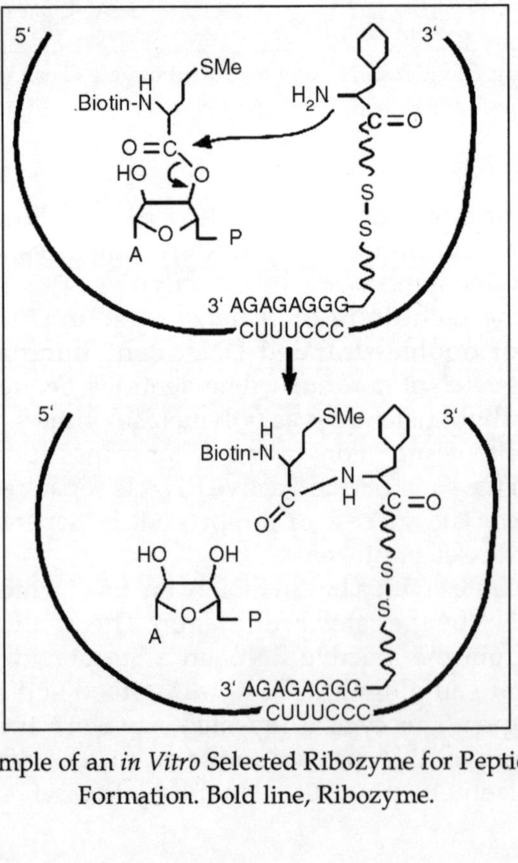

Fig. Example of an *in Vitro* Selected Ribozyme for Peptide bond Formation. Bold line, Ribozyme.

Surprisingly, the activity of ribozymes can be allosterically controlled. An example is the flavin-dependent activity of a hammerhead ribozyme to which a flavin-binding aptamer has been fused. In contrast to this positive effect, an ATP aptamer-ribozyme construct is inhibited in the presence of ATP. These examples obviously make ribozymes more akin to proteozymes. Besides the interest in mechanistic questions, such ribozymes might be very useful as therapeutic agents, as discussed below, as their activity can be modulated by small molecules.

DNAzymes

It was generally assumed that RNA was better suited as a catalyst than DNA because of its flexibility and the presence of the 2′-hydroxy group. It thus came as a big surprise when Joyce selected a DNAzyme capable of cleaving RNA, and this with somewhat better efficiency than a ribozyme. Thus our preconceived ideas have been proven wrong. DNA as a single-stranded molecule can apparently fold into three dimensional structures similarly to RNA, and the 2′-hydroxy group is dispensable for catalytic action. As ribozymes, DNAzymes can also be made, by selection, to depend on a cofactor. This has been demonstrated for a histidine-dependent DNAzyme for RNA hydrolysis. This narrows the gap to proteozymes and offers the possibility of extending the range of catalysed reactions. So far, selections have been performed with the aim in mind of mimicking reactions which are also catalysed by proteozymes. Besides the interesting mechanistic aspects of the RNA- or DNA-catalysed reactions, the ribozyme-catalysed reactions particularly are often considered as support for the prebiotic RNA world where RNA might have played the part of our present day enzymes. As the number of various types of reactions catalysed by ribozymes increases, this hypothesis gains in strength. Another aspect for the future is the development, by *in vitro* selection, of nucleic acid enzymes to catalyse reactions for which no natural enzymes exist. This could be of industrial importance.

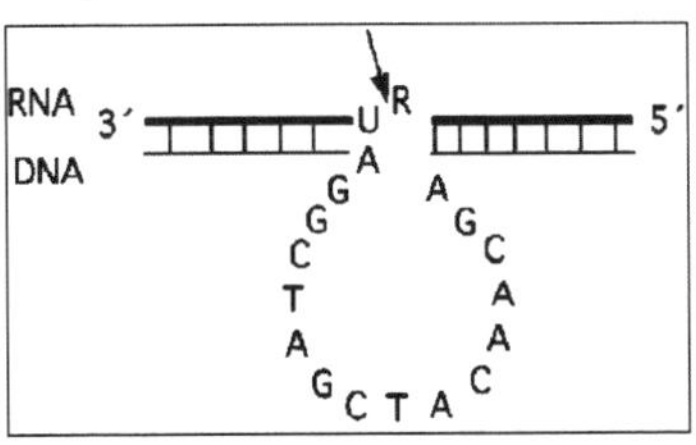

Fig. The first RNA-cleaving DNAzyme. R, Either Adenosine or Guanosine.

Catalytic Antibodies: Abzymes

Antibodies are *immunoglobulins*, which, of course, are proteins. Like other antibodies, catalytic antibodies, so-called abzymes, are elicited in an organism in response to immunological challenge by a foreign molecule called an antigen for discussions on the molecular basis of immunology). In this case, however, the antigen is purposefully engineered to be *an analog of the transition-state intermediate in a reaction*. The rationale is that a protein specific for binding the transition-state intermediate of a reaction will promote entry of the normal reactant into the reactive, transition-state conformation. Thus, a catalytic antibody facilitates, or catalyzes, a reaction by forcing the conformation of its substrate in the direction of its transition state. (A prominent explanation for the remarkable catalytic power of conventional enzymes is their great affinity for the transition-state intermediates in the reactions they catalyze.

One strategy has been to prepare ester analogs by substituting a phosphorus atom for the carbon in the ester group. The phospho-compound mimics the natural transition state of ester hydrolysis, and antibodies elicited against these analogs act like enzymes in accelerating the rate of ester hydrolysis as much as 1000-fold. Abzymes have been developed for a number of other classes of reactions, including COC bond formation via aldol condensation (the reverse of the aldolase reaction and the pyridoxal 5'-P-dependent aminotransferase reaction. In this latter instance, N_a-(5'-phosphopyridoxyl)-lysine coupled to a carrier protein served as the antigen. An antibody raised against this antigen catalyzed the conversion of d-alanine and pyridoxal 5'-P to pyruvate and pyridoxamine 5'-P. This biotechnology offers the real possibility of creating "designer enzymes," specially tailored enzymes designed to carry out specific catalytic processes.

Chapter 7

Enzyme in Organic Synthesis

Over the past few years the use of enzymes as catalysts for the preparation of novel organic molecules has received a steadily increasing amount of attention. Thus, until fairly recently the employment of enzymes and/or whole cell systems in organic chemistry was localised in a few pioneering laboratories—now many non-specialists are discovering the advantages of utilising these natural catalysts.

There are three easily recognised advantages that may be gained from utilising enzymes in organic synthesis:

- The proteins are able to promote reactions under very mild conditions of temperature, pH, and pressure.
- Enzymes are chiral catalysts and are often able to produce optically active molecules that can be used as building blocks (synthons) for the preparation of homochiral compounds.
- Biocatalysts can perform transformations that are difficult to emulate using more traditional organic chemistry.

USES OF HYDROLASE ENZYMES

Hydrolysis of Racemic Esters

Kinetic Resolutions

The enzyme-catalysed hydrolysis of alkyl carboxylates has been extensively investigated.

For example the acetate:

- Is easily prepared and is hydrolysed enantioselectively using *Mucor miehei* lipase or lyophilized yeast cells. The alcohol
- Is obtained [enantiomeric excess (e.e.) 80-100%] and this synthon is utilized in the preparation of 13-HODE (coriolic acid)
- A naturally occurring, biologically active compound.

While this chemo-enzymatic synthesis of 13-HODE is interesting and useful for the preparation of selected analogues, the preparation of the natural product itself is more efficiently accomplished using another enzyme-catalysed reaction namely the conversion of linoleic acid into 13-HODE (in 70% yield) using a lipoxygenase enzyme. We have developed this process to allow the production of gram quantities of the hydroxydienoic acid.

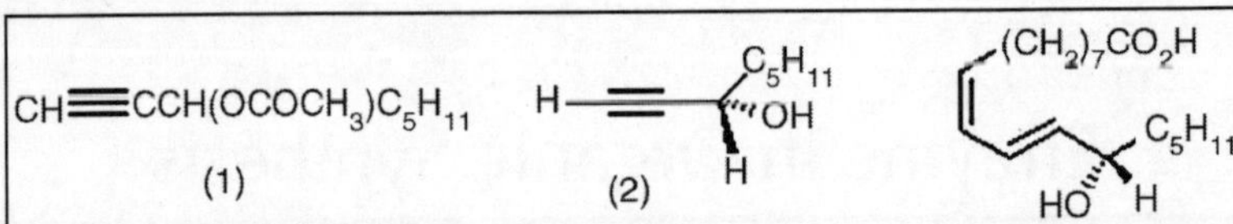

(1) (2) (3) 13- HODE Coriolic Acid

Kinetic resolutions of selected esters have been used in the preparation of a number of important compounds in these laboratories including a synthon for the hypocholestemic *delta*-lactones, anti-AIDS agents and the fascinating anti-fungal agent called brefeldin - A.

Asymmetrization of Meso-Esters

Meso-diacetates such as the compound (4) are interesting substrates in enzyme-catalysed transformations since "asymmetrization" can be observed. Thus compound (4) is converted into the half-ester (5) using porcine pancreatic lipase.

The yield of (5) is almost quantitative and the optical purity is very high (> 98% e.e.). Compound (5) has been converted into the nucleoside analogue (6) and the latter compound is being used in various enzyme studies.

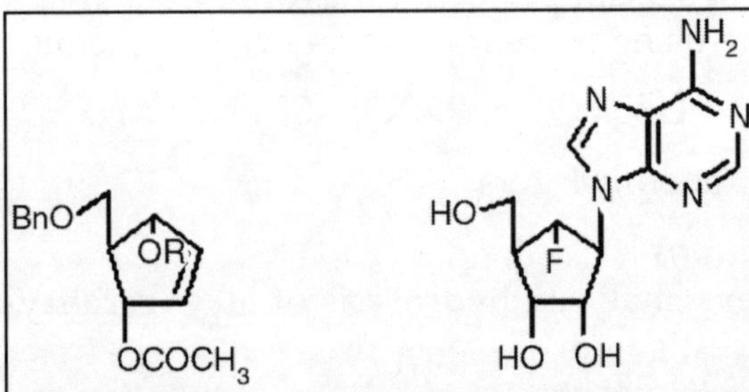

(4) R = $COCH_3$
(5) R = H
(6)

Similarly the dimethyl cyclopentane-dicarboxylate (7) is hydrolysed with exquisite selectivity using pig liver esterase to give the mono-ester (8) in a highly pure form (96% yield; 98% e.e.). Enzyme-catalysed reactions such as (4) => (5) and (7) => (8) give the possibility of forming optically pure products in quantitative yield from the chosen substrates, a process that is very difficult to emulate using conventional chemical catalysts.

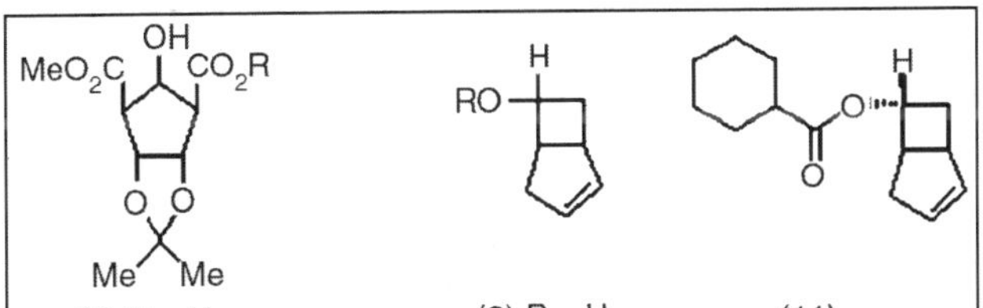

(7) R = Me
(8) R = H

(9) R = H
(10) R = $COCH_3$

(11)

Esterification of Racemic Acids

For example, the racemic alcohol (9) in hexane is converted into the optically active ester (11) (90% e.e.) and recovered optically enriched bicyclo[3.2.0]hept-2-en-6*endo*-ol using cyclohexane carboxylic acid and the catalyst Lipozyme® (*Mucor miehei* lipase attached to an inert solid support).

Inter-Esterification

Interestingly the ester (11) is obtained in even higher optical purity (> 99% e.e.) on employing an inter-esterification reaction involving the acetate (10) and cyclohexane carboxylic acid. The separated esters can be used in syntheses of prostaglandin-F2ã, a naturally occurring material with a plethora of biological activities (Scheme 1).

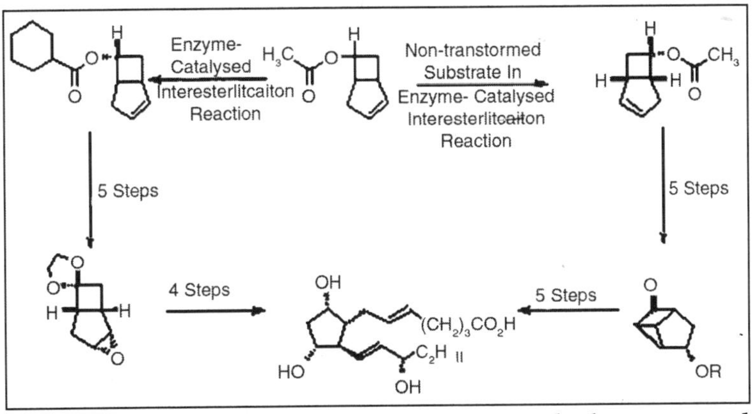

The inter-esterification process has recently been extended to provide an example of double enantioselection by using the ester (±)-10 and racemic 2-(*para*-chlorophenoxy)propanoic acid. One of the four possible diastereoisomeric esters, compound (12), is formed almost exclusively. Two stereoselective enzyme-catalysed processes are in operation, namely the preferential hydrolysis of the 6(R)-acetate (10), and selective acylation of the product alcohol by 2(-)-(*para*-chlorophenoxy)-propanoic acid.

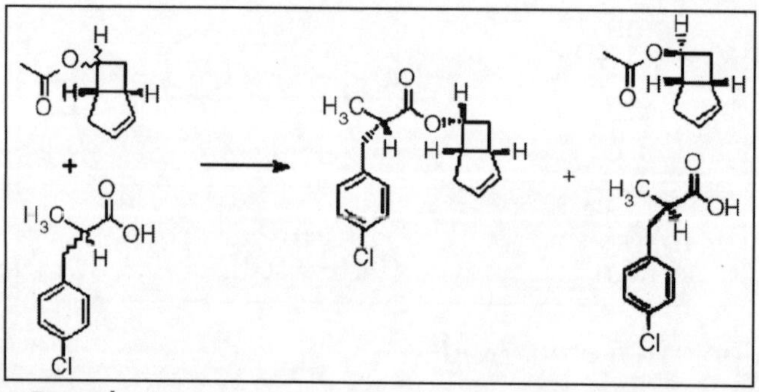

Dynamic Resolution

Lipozyme can also operate in organic solvents under almost anhydrous conditions as illustrated by the following example. Conversion of substituted racemic azlactones *e.g.* (13) to the corresponding optically active amino acid derivatives (14) occurs using toluene as the solvent in the presence of Lipozyme. In this process the azlactone starting material is able to racemise *in situ* leading to high yield (~ 94%) of the homochiral (99.5% e.e.) product (14). Subsequent two-step hydrolysis of (14) leads to L-(S)-tert-leucine (15), an important amino acid that is used in therapeutic peptides and as a chiral auxilliary.

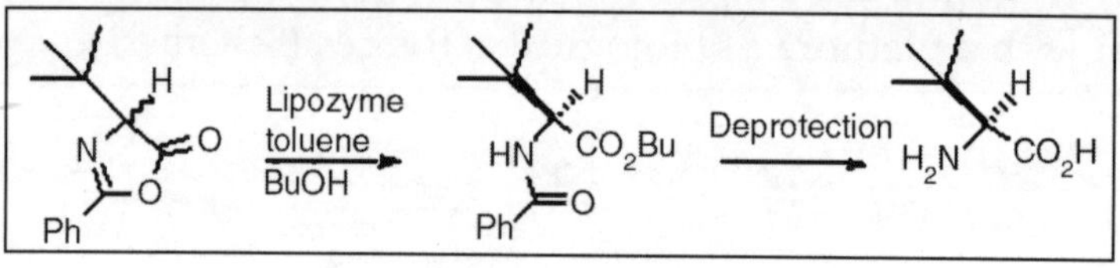

Polyesters

The use of lipases for the preparation of polyesters has been studied at Exeter and the condensation of adipic acid and butane-1,4-diol has been developed, by an industrial partner, to form part of a large scale process for the production of polyurethanes.

Amide Hydrolysis

The employment of acylases (such as hog kidney acylase) for the cleavage of amide bonds under mild conditions is well known. This type of hydrolysis is commercially important in the preparation of 6-amino-penicillanic acid and in the synthesis of some optically pure amino-acids. More recently it has been found that a microbial acylase can effect the enantiospecific hydrolysis of the lactam (16) to give the amino-acid (17) and recovered starting material. A second microorganism can effect the enantio-complementary hydrolysis to give the mirror image of compound (17) and recovered lactam. These optically active amino-acids and lactams can be used to prepare the important anti-AIDS agent carbovir (18) in homochiral form. The synthesis of the GABA-inhibitor (19) from both enantiomers of the lactam (16) in an enantio-convergent strategy exemplifies another important application of this useful enzyme-catalysed kinetic resolution.

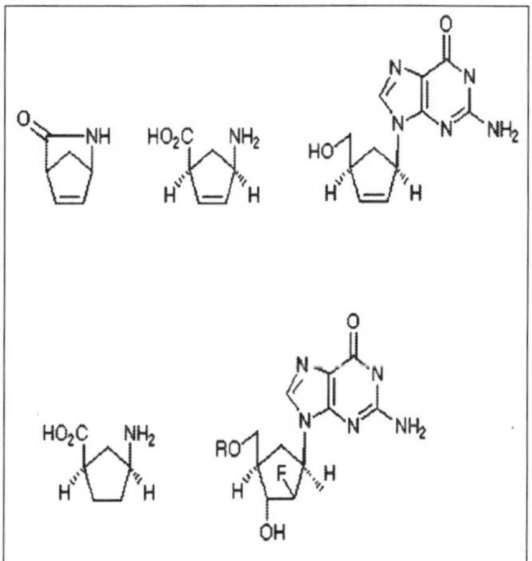

(20) R = H
(21) R = PO_3H_2

Phosphate Esters

The synthesis of phosphate esters can be accomplished by kinase enzymes using adenosine triphosphate (ATP) as the phosphate donor. The technique will work with unnatural substrates: for example the racemic nucleoside analogue (20) was converted into the racemic nucleotide (21) using thymidine kinase and ATP. Enantioselective hydrolysis of the (±)-phosphate (21) can be achieved using 5'-nucleotidase from snake venom to give the dextrorotatory carbocyclic nucleoside (+)-(20) which exhibited very powerful anti-Herpes activity.

Nitrile Hydrolysis

The hydrolysis of nitrile groups using hydratase enzymes is of great interest to many synthetic organic chemists, principally because the hydrolysis takes place under very mild conditions. The sequence outlined in Scheme 3 illustrates the selectivity of processes of this type.

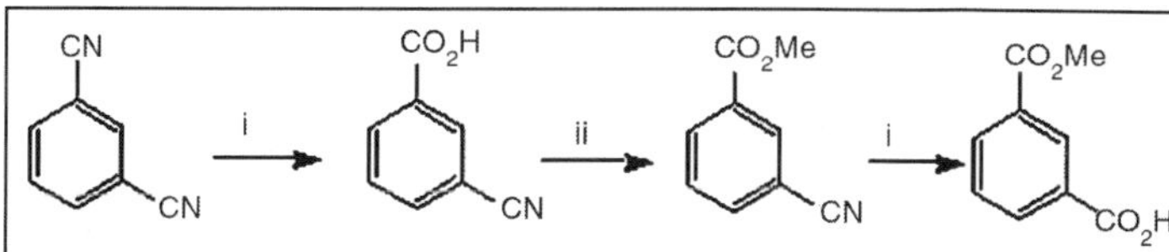

Reagents: i. Nitrile Hydratase SP361. buffer pH 7.0, 30°C. 12-24 h.
ii. CH_2N_2

Recently it has been shown that this reaction can be carried out in an enantioselective fashion in which the resolution occurs during the amide to acid conversion. For example racemic 2-phenylbutyronitrile yields (*R*)-2-phenylbutyramide and (*S*)-2-phenylbutyric acid, both with > 98% e.e. (Scheme 4). The asymmetric hydrolysis of 3-substituted glutaronitrile derivatives has also been achieved.

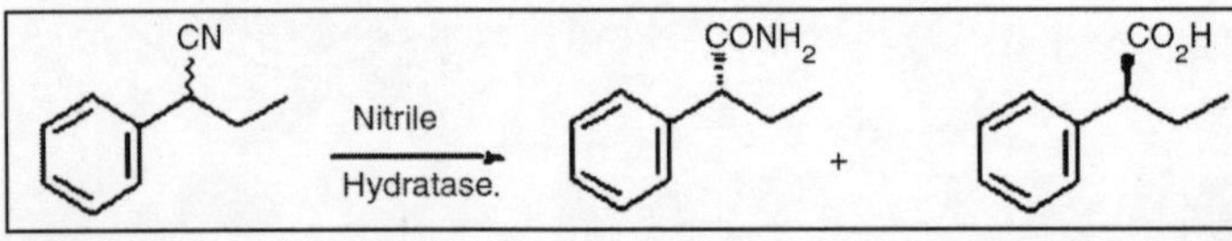

Epoxide Hydrolases

The substrate selectivity of epoxide hydrolase enzymes is being explored. To date most of the biotransformation systems have been derived from animal sources (*e.g.* rabbit or rat liver microsomes) and so the methodology is not at all useful to the non-specialist wishing to prepare optically active *vic*-diols on a reasonable scale.

Enzyme Models for Use by Non-Specialists.

The use of hydrolase enzymes by the non-specialist is aided by the availability of models of the active site of enzymes such as pig liver esterase, porcine pancreatic lipase, *Pseudomonas fluorescens* lipase and *Candida cylindracea* lipase.

Glycosidases

Finally, glycosidases are a diverse group of enzymes whose natural function is to catalyse the hydrolytic cleavage of glycosides. Recently we have exploited the inherent selectivity of one particular glycosidase, ß-glucuronidase, in the synthesis of the potent analgesic morphine-6-glucuronide (23). Treatment of morphine-3,6-diglucuronide (22) with ß-glucuronidase from *Patella vulgata* results in regioselective hydrolysis of the more reactive 3-glucuronide.

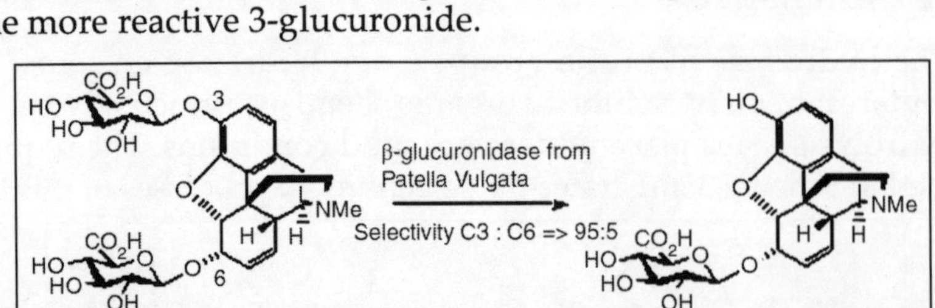

In addition it has long been known that these enzymes can also catalyse 'transglycosylation' reactions resulting in the formation of novel glycosides. By using certain amino acid derivatives as the nucleophilic alcohol component it is possible to prepare carbohydrate-amino acid linkages (*e.g.* 24) that are important constituents of many glycoproteins.

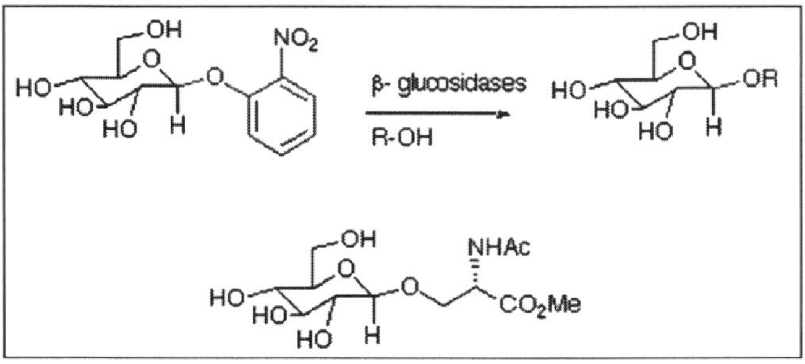

OXIDOREDUCTASE ENZYMES

Carbonyl Reductions

The employment of whole-cell systems (*e.g.* bakers' yeast) or isolated enzymes for the reduction of carbonyl groups to the corresponding alcohols is well researched. Bakers' yeast has been used to reduce a wide range of simple ketones, ß-ketoesters and cyclic diketones. The stereo-selectivity of the reduction process is predictable and very useful routes to homochiral anti-inflammatory steroids have been explored using this type of biotransformation. Other organisms, such as the fungus *Mortierella isabellina* can be used in place of bakers' yeast.

Partially purified dehydrogenase enzymes have also been utilised for the production of optically active secondary alcohols. During the purification of the protein the enzyme cofactor (nicotinamide adenine dinucleotide (phosphate) [NADPH]) is lost and so before the biotransformation is possible this cofactor must be re-coupled to the enzyme. The cofactors NADH and NADPH are expensive and cofactor recycling must be set up so that a less-than-stoichiometric amount of the dinucleotide can be employed. Cofactor recycling systems are now well defined thanks to the work of J. Bryan Jones and others, and are ready for use in non-specialist laboratories.

The ketone (25) has been reduced using *Mortierella* spp. and by 3ã,20ß-hydroxysteroid dehydrogenase/NADH. Thus the whole cell reduction gave an equimolar mixture of the 6*endo*-alcohol (26) and the 6*exo*-alcohol (27) (60% yield of diastereoisomeric alcohols of high optical purity (Cotterill *et al.*, 1991) while the dehydrogenase catalysed an enantioselective reduction of the ketone (25) by NADH, forming the alcohol (26) in an optically pure state and leaving optically active ketone. The alcohol (26) has been converted into eldanolide, the pheromone of an agricultural pest, the sugar cane borer (Scheme 6). The availability of the two enantiomers of the 7,7-dimethylbicyclo[3.2.0]heptane ring system allows the synthesis of biologically interesting leukotriene-B4.

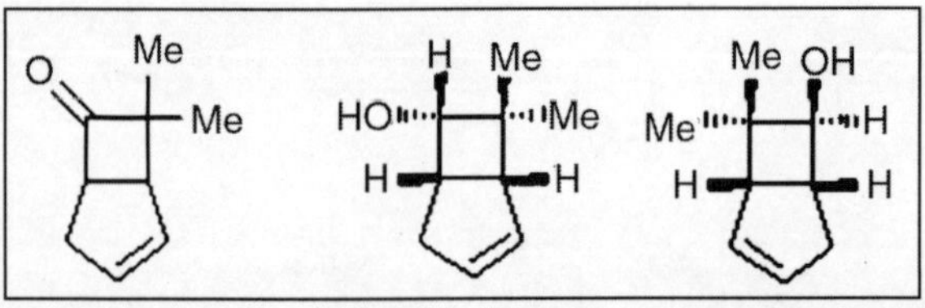

Note that one enantiomer of the ketone provides the C1-C6 section of the natural product (containing one chiral centre while the other enantiomer can provide the C7-C20 portion (containing the second chiral centre) (Scheme 6) (Cotterill *et al.*, 1990; Dormán *et al.*, 1989).

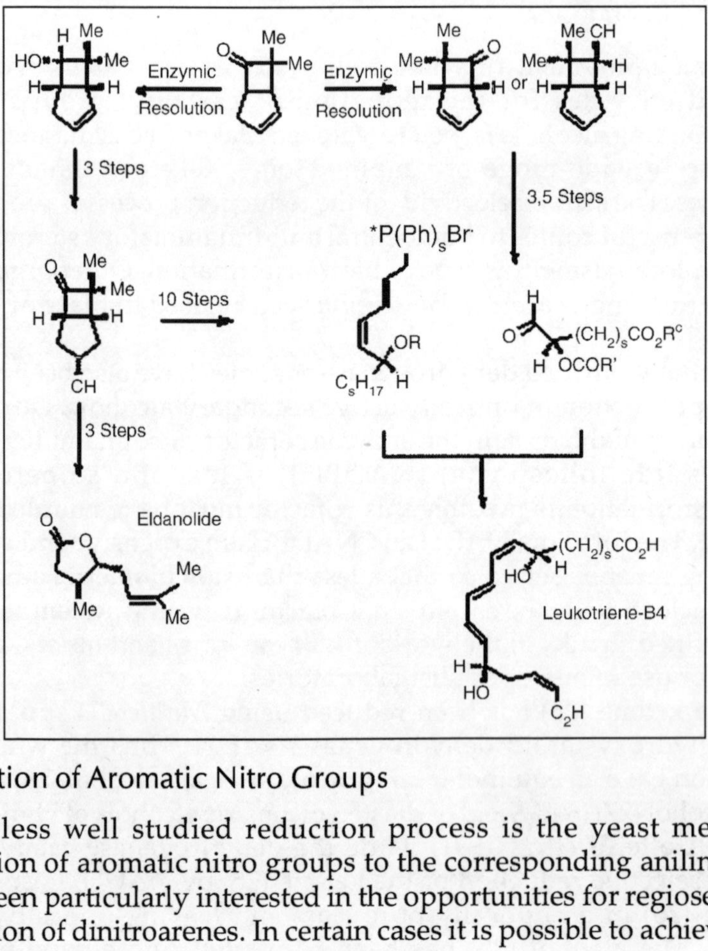

Reduction of Aromatic Nitro Groups

A less well studied reduction process is the yeast mediated reduction of aromatic nitro groups to the corresponding anilines. We have been particularly interested in the opportunities for regioselective reduction of dinitroarenes. In certain cases it is possible to achieve very high selectivity (*e.g.* reduction of the dinitroquinoline (28) to a single regioisomer of the product (29). On the basis of studying a large number of substrates we have formulated a series of rules to predict the outcome of the reduction.

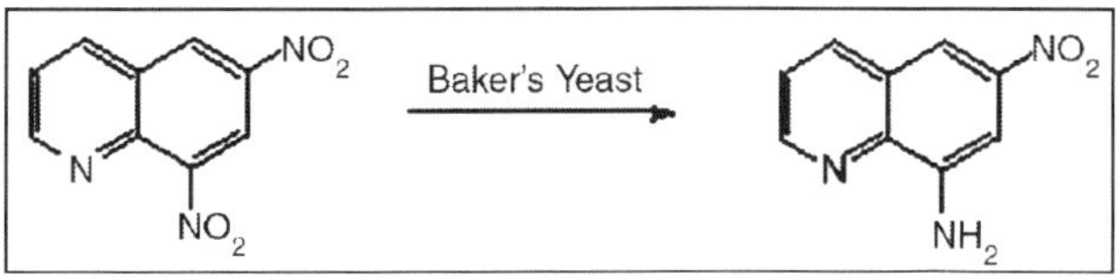

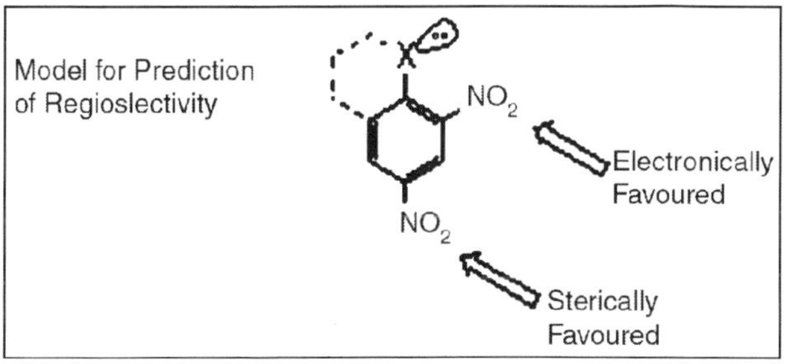

Oxidations

Oxidation of aromatic and alicyclic molecules using monooxygenase and dioxygenase enzymes has attracted a considerable amount of attention.

Dioxygenases

The oxidation of benzene to cyclohexa-3,5-diene-1,2-*cis*-diol is a reaction catalysed by *Pseudomonas putida* and substantial amounts of material can be obtained for further chemical studies.

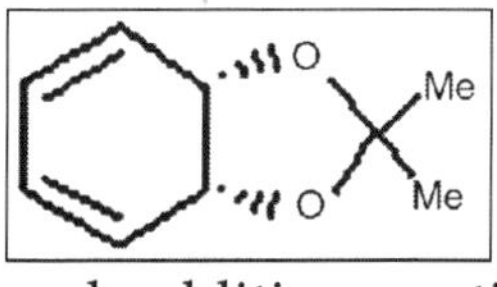

For example various cycloaddition reactions can be carried out on the acetonide (30), including Diels-Alder [4+2] reactions [to give, for example, the tricyclic compound (31)], [2+2] cycloaddition reactions [to give cyclobutanone derivatives such as (32)] and the more unusual [6+4] cycloaddition reactions.

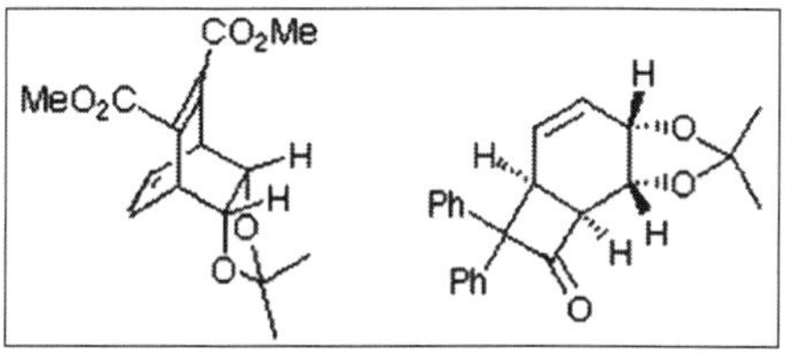

Not only is benzene biotransformed by *Pseudomonads* but so are many of its derivatives. Often the latter compounds are obtained in an optically pure form and it is established (at least for some of the compounds) that these diols have the 1(*S*),2(*R*)-configuration.

Trifluoromethylbenzene is biotransformed into the diol (33); the corresponding acetonide (34) can be converted into a wide range of derivatives through different chemical reactions.

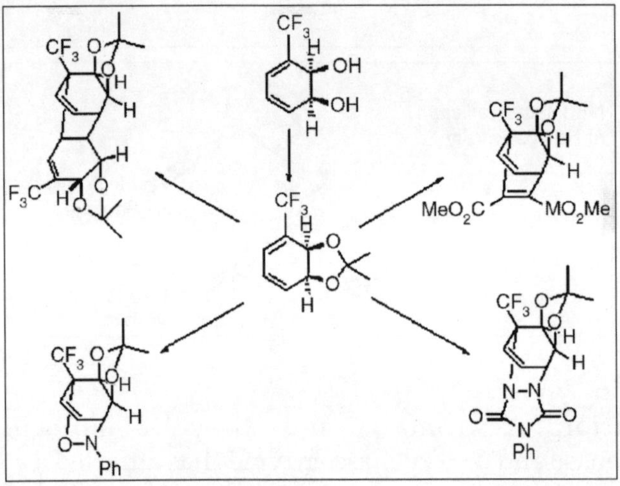

Interesting conversions involving 3-chlorocyclohexa-3,5-diene-1,2-diol and 3-methylcyclohexa-3,5-diene-1,2-diol have also been described.

Monooxygenases

Another oxidation reaction that is gaining an increasing amount of interest is the enzyme-catalysed conversion of a cyclic ketone into a lactone (Walsh and Chen, 1988). We have shown that (+/-)-7-*anti*-fluoro-5*endo*-bromobicyclo[2.2.1]heptan-2-one is oxidized in an enantioselective manner using the *Acinetobacter* NCIMB 9871 to give optically pure lactone (35) (30%) and recovered optically active ketone (36) (30%). Interestingly oxidation of the ketone (36) with the chemical oxidant, *meta*-chloroperoxybenzoic acid gave the 3-oxabicyclooctan-2-one (37) as the major product. The latter compound was converted into the carbocyclic nucleoside (38), a compound possessing anti-HIV activity.

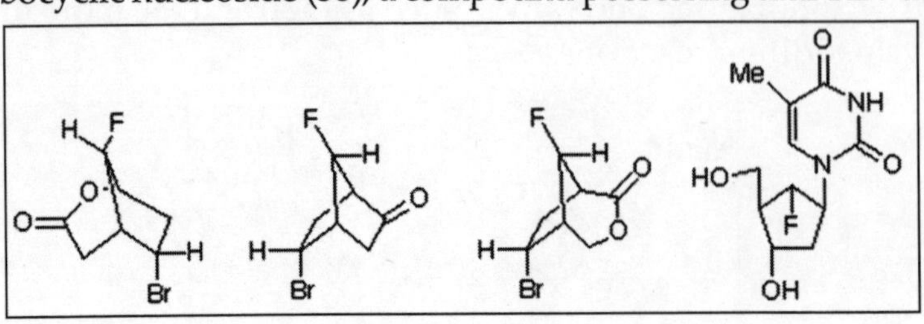

The monooxygenase from the *Acinetobacter* can be purified and Baeyer-Villiger oxidations can be performed using the enzyme in association with its cofactor NADPH. We have found that it is possible to convert a secondary alcohol into the corresponding lactone *via* the corresponding ketone by using a coupled enzyme system, that is *Thermoanaerobium brockii* dehydrogenase and *A. calcoaceticus* monooxygenase with *in situ* recycling of the cofactor. More recently a monooxygenase enzyme capable of using NADH as cofactor has been found, enabling cofactor recycling to be accomplished using formate dehydrogenase.

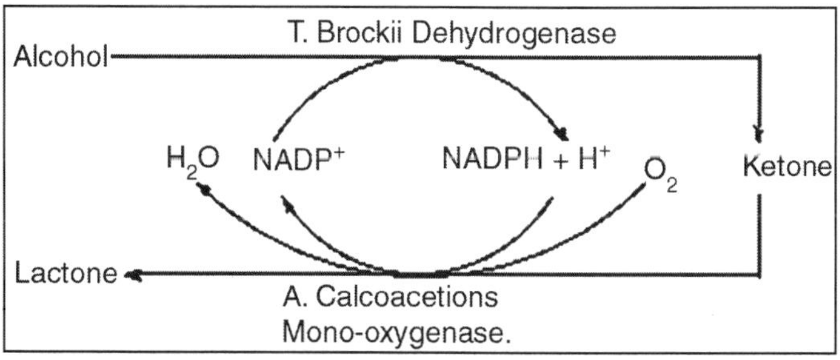

As adumbrated previously, immobilised soya bean lipoxygenase-1 efficiently converts linoleic acid into 13(*S*)-HODE (3) in 78% yield in a few hours. The immobilised enzyme can be recovered and reused. The chemistry of 13-HODE has been investigated: for example the bicyclic compound (39) has been obtained by macro-lactonization and a Diels-Alder reaction on the biotransformation product.

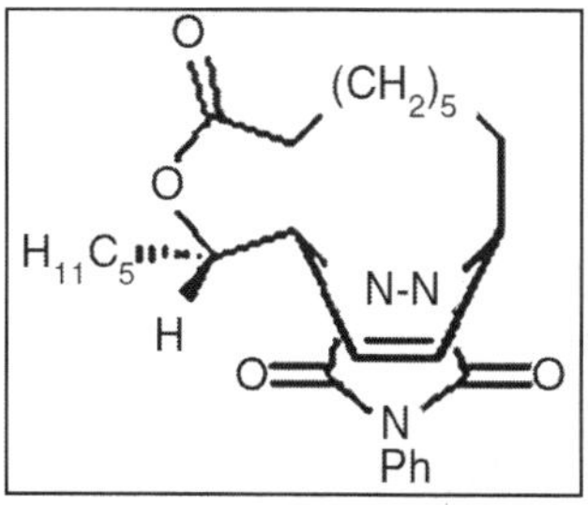

The ability to derivatize an organic molecule at a position distant from pre-existing functionality is an attractive goal for both organic chemists and microbiologists alike. Some progress has been made towards providing chemical methods for remote hydroxylation but, for the foreseeable future, microbiological methods seem to offer the most likely chances for success. The regioselective hydroxylation of steroids has been investigated in some depth, the formation of 11-hydroxyprogesterone from progesterone being one of the most noteworthy successes. The regio-selective (but not regiospecific)

oxidation of cyclohexylcyclohexane to give the diol (40) using *Cunninghamella blakesleeana* shows that even the most hydrophobic materials can be biotransformed. Spiro-heterocyclic compounds can be derivatized, often in very high yield, to produce single compounds using *Beauveria sulfurescens*: for example the piperidine derivative (41) furnishes the alcohol (42) in 70% isolated yield.

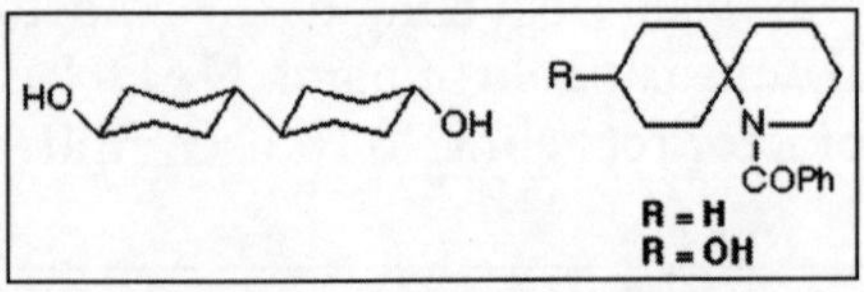

ENZYME-CATALYSED REACTIONS RESULTING IN THE FORMATION OF CARBON-CARBON BONDS

The stereocontrolled formation of carbon-carbon bonds is at the heart of organic synthesis. The aldol reaction is one popular way of joining together two synthons to give an advanced intermediate in a synthesis, for example, of biologically interesting compounds. Impressive progress has been made in controlling base-catalysed aldol reactions and the recent breakthroughs describe reactions that begin to approach the superb selectivity demonstrated by aldolase enzymes. Rabbit muscle aldolase has been studied in some detail and it has been shown, for example, that the triol monophosphate (43) is formed from phenylethanal and dihydroxyacetone monophosphate using this enzyme.

One restriction using this enzyme is the requirement for dihydroxyacetone mono-phosphate in this reaction. Recently we have begun to investigate an aldolase (transketolase) that does not require phosphorylated substrates. The enzyme has been successfully over-expressed resulting in large quantities suitable for biotransformation.

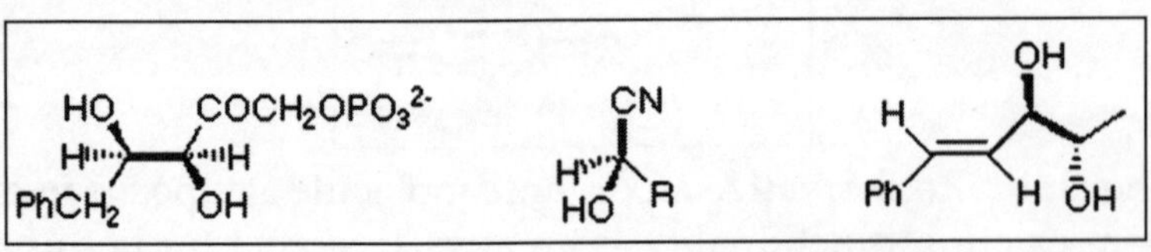

A less aesthetic process, but nevertheless a very useful transformation, involves the use of lyase enzymes which can catalyse the addition of HCN to aldehydes to give optically active cyanohydrins. Mandelonitrile lyase catalyses such an addition of HCN to a variety of aldehydes to afford the corresponding (*R*)-cyanohydrins (44). An enzyme that shows selectivity for the formation of (*S*)-cyanohydrins has been isolated.

Biotransformations

Current trends in the employment of lipases, esterases and amidases for the preparation of chiral compounds of high optical purity will continue. The use of these enzymes in non-specialist laboratories will become widespread and some of the newly-found processes will be scaled up to the multi-kilogram level. Other hydrolase enzymes, such as the nitrile hydratases will become increasingly used because of the mild hydrolysis conditions that can be employed using these enzymes. The use of microorganisms, such as bakers' yeast, for bio-reduction processes will also become more widespread. The rate of expansion of this area of biotransformations will be slower due to the inevitable problems associated with the use of whole-cell systems. The employment of dehydrogenase enzymes for the reduction or oxidation of organic substrates will also become more attractive to the synthetic organic chemist in the wake of further refinement of cofactor recycling systems.

More research work is needed on the use of enzymes for other oxidation reactions (Baeyer-Villiger type reactions, controlled hydroxylation *etc.*) before more extensive use is made of these catalysts. In five years time the situation should be different. Similarly the use of enzymes for the stereocontrolled formation of carbon-carbon bonds will need more attention in research laboratories before the full potential of these catalysts is realised. The exceptions are the enzymes that catalyse the stereoselective formation of cyanohydrins which are now available for more routine use.

The possibility of setting up a cascade of enzyme-catalysed reactions in the same "pot" is very attractive. The conversion of the racemic hydroxy acid (46) into the amino-acid (47) illustrates the sort of coupled process that is feasible note that co-factor recycling is accommodated within this sequence.

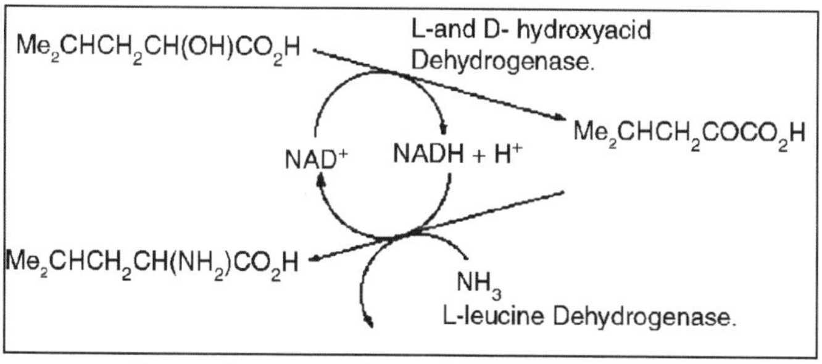

The preparation of unusual sugars, and sugar analogues, using enzyme-catalysed processes will gain prominence. The work of Card. The elegant way in which fairly complex end-products can be built up without the need for extensive protection/deprotection sequences.

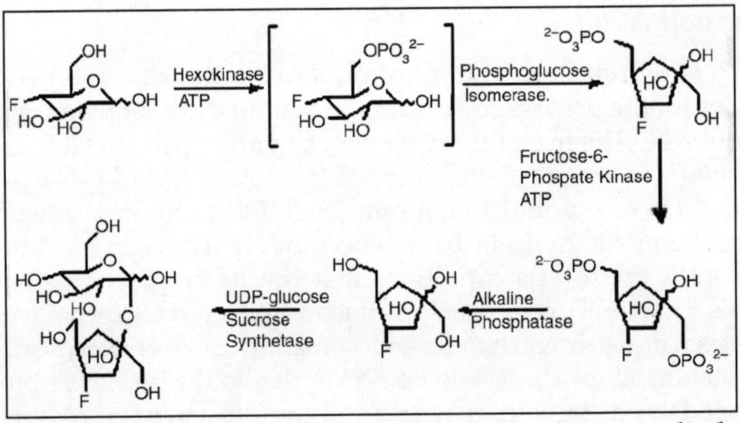

An analogous multi-enzyme approach has been applied to the synthesis of the nucleotide coenzyme GDP-mannose (48). In this reaction the GTP is initially prepared from GMP using cofactor recycling. The GTP thus formed can then be combined with mannose-1-phosphate (49) in the presence of GDP-mannose pyrophosphorylase to yield the product.

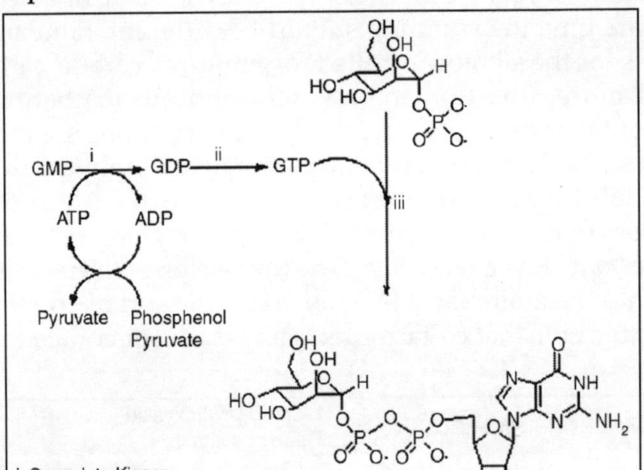

In an extension of this work, those enzymes that utilise GDP-mannose as a source of mannose are also being studied. These mannosyl transferases catalyse the transfer of mannose to lipid linked substrates thus generating complex mannosyl containing oligosaccharides that are important constituents of natural N-linked glycoproteins.

It has been shown that the novel lipid-linked substrate (50) can act as an efficient acceptor for mannose transfer thereby opening the way for the use of this approach to synthesise oligosaccharides on a preparative scale.

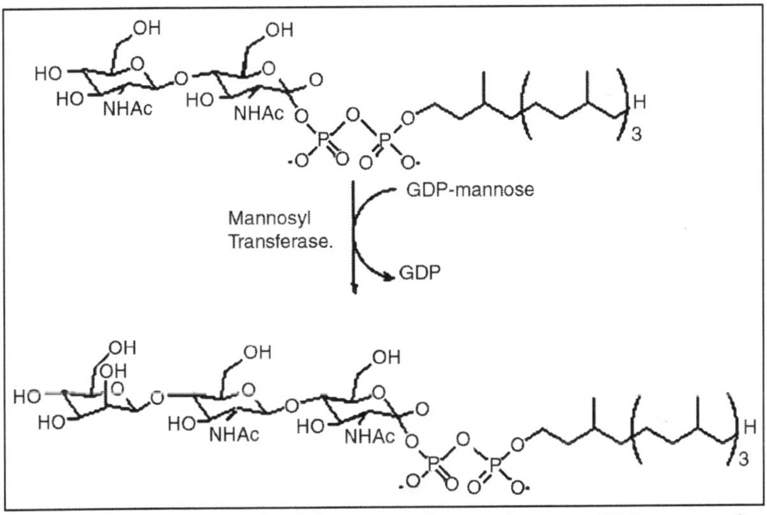

Progress in the field of biotransformations will be matched by equally exciting advances in the area of synthetic organic chemistry. The advances will often be complementary and the use of enzymes for the preparation of materials which are difficult to make by purely chemical methods will remain *in vogue* for the foreseeable future. Certainly the present problems with the disposal of waste organic solvents, with the consequent focus on reactions that can be conducted in an aqueous environment, concentrate attention on Nature's catalysts which are designed to operate in aqueous solution. Finally it is conceivable that "man-made" enzymes (*i.e.* catalytic antibodies) will also make a significant impact on research in this area and add to the range of organic reactions that can be catalysed by protein. Many of the forthcoming investigations will be aimed at producing bulk quantities of optically active, relatively simple molecules, while an equal volume of work will concentrate on the preparation of more complex, higher value materials in smaller quantities. It promises to be a fascinating decade for those scientists involved in the research and development of processes that have enzyme catalysed reactions as the key step(s).

BIOMIMETIC AND ORGANOCATALYTIC APPROACHES TO OXIDATION CATALYSIS

Catalytic Asymmetric Epoxidation and Cyclopropanation

Biomimetic oxidation catalysis aims at achieving the efficiencies and selectivities of enzymes such as monooxygenases or peroxidases with low-molecular-weight compounds. The design of biomimetic oxidation catalysts is based on the active-site structure and function of oxidative enzymes. In 1997, we reported the use of the chiral Ru-porphyrin 1a for the asymmetric epoxidation and cyclopropanation of olefins. For the electronic fine-tuning of the porphyrin ligand, electron-donating and -withdrawing substituents were attached to the dimethanoctahydroanthracene substituents present at the four *meso*-positions of the porphyrin ligand. We found that the introduction of the CF_3 substituent led to a particularly improved performance of the Ru-porphyrin catalyst 1b. High enantioselectivity was achieved in both epoxidation and cyclopropanation.

At the same time, remarkable catalyst stabilities resulted (>14.000 total turnovers in the asymmetric epoxidation of olefins under *Hirobe* conditions, and >7.500 total turnovers in the asymmetric cyclopropanation of olefins with diazo compounds). The epoxidation of olefins such as styrene requires electrophilic oxidants. On the other hand, the asymmetric epoxidation of, for example, enones calls for nucleophilic oxygen donors. The well-known Juliá–Colonna epoxidation of enones makes use of poly-amino acids as catalysts and of hydrogen peroxide as oxidant, and is a method of high practical utility. Unfortunately, the substrate spectrum is fairly narrow, with chalcone (2) and derivatives affording the best yields and enantiomeric excesses. On the other hand, cyclic enones and quinones are much less suitable substrates.

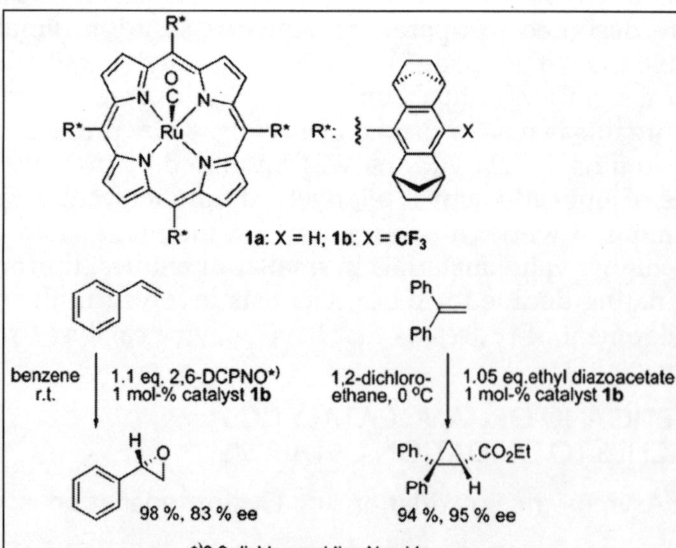

Our recent studies on the mechanism of the Julia–Colonna epoxidation revealed that this epoxidation is catalyzed by the *N*-terminus of the amino acid α-helix. The most important feature in this respect is the presence of three amide NH groups that are not involved in intra-helix hydrogen bonding, but protrude into the solution. Two of these NH moieties form hydrogen bonds to the oxygen atom of the enone carbonyl group, whereas the third one binds the hydroperoxide anion in the intermediate peroxyenolate. The induction of asymmetry thus rests on the helical chirality of the peptide backbone.

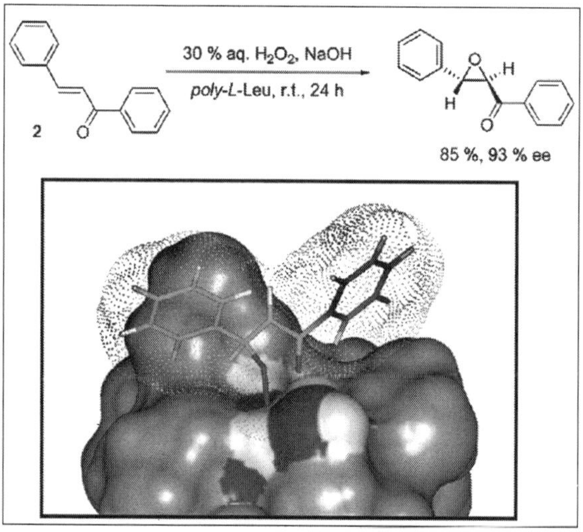

Fig. Poly-amino Acids do not Efficiently Catalyze the Asymmetric Epoxidation of Cyclic *Z*-enones

ASYMMETRIC C–C COUPLING

The synthesis of the DIANANE-salen ligand 5 (DIANANE: *endo,endo*-2,5-diamino- norbornane) and its application in the catalytic asymmetric Nozaki–Hiyama–Kishi coupling. The ligand 5 affords up to 92% ee in the addition of allylic electrophiles to aldehydes. Furthermore, we could show that vinylic iodides and triflates can be employed as well: In the addition to the PMB-protected 3-hydroxypropionic aldehyde, the resulting allylic alcohols were obtained with up to 75% ee. Our improved synthesis furnishes the DIANANE ligand 5 in large quantities.

Aldol additions undoubtedly belong to the most important C–C coupling reactions in organic synthesis. The recent years have seen the development of a number of proline-catalyzed asymmetric intermolecular aldol reactions. Proline is readily available at low cost and is thus an attractive and economic organocatalyst. However, one of its disadvantages is its poor solubility in organic solvents of low polarity. To overcome this hurdle, we synthesized the *N*-sulfonyl proline amides 6a–c.

These amino acid derivatives are of similar acidity as the amino acids themselves. In fact, the sulfonyl proline amides 6a–c turned out to be quite enantioselective catalysts: Enantiomeric excesses >95% ee were achieved for all three catalysts, with the nitro-compound 6c giving an optimum ee of 98%. Proline gave a maximum ee of 80% under the reaction conditions. Besides better enantioselectivity, catalyst loadings could be reduced from 30 mol% (typical for proline) to 5 mol% without significant loss in yield or enantioselectivity, in particular in the case of the tris-*iso*-propyl catalyst.

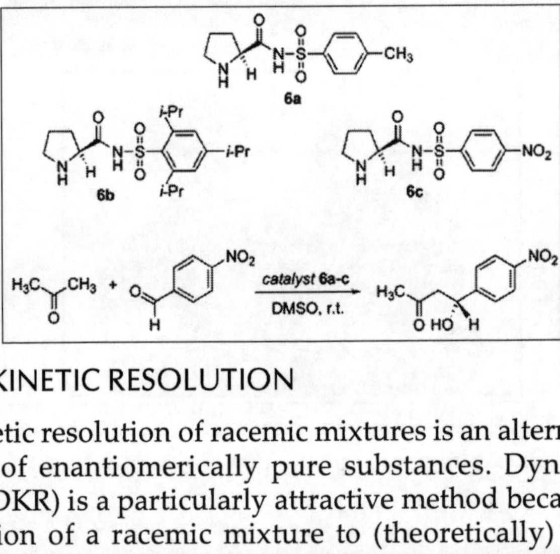

DYNAMIC KINETIC RESOLUTION

The kinetic resolution of racemic mixtures is an alternative for the production of enantiomerically pure substances. Dynamic kinetic resolution (DKR) is a particularly attractive method because it allows the conversion of a racemic mixture to (theoretically) 100% of the desired enantiomer. The azlactones 7 can be ring-opened with, for example, alcohol nucleophiles, and they readily racemize via enol formation. Consequently, they are ideal substrates for the production of enantiomerically pure a-amino acids by DKR.

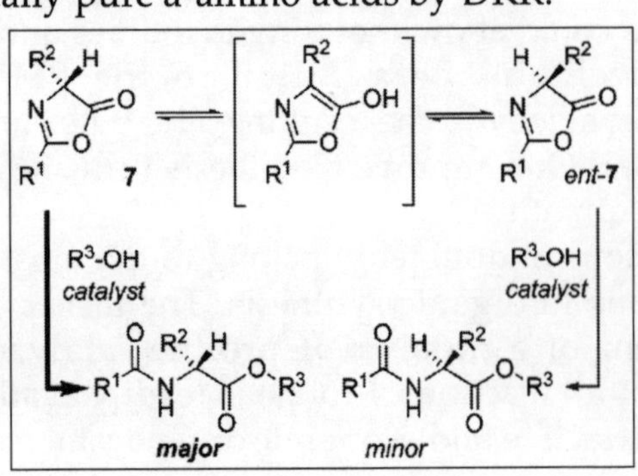

As catalysts for this type of transformation, we envisaged the bifunctional urea/thiourea–*tert*amine conjugates 8. The latter were used before as catalysts for asymmetric Michael additions to nitro-olefins and for the aza-Henry reaction.

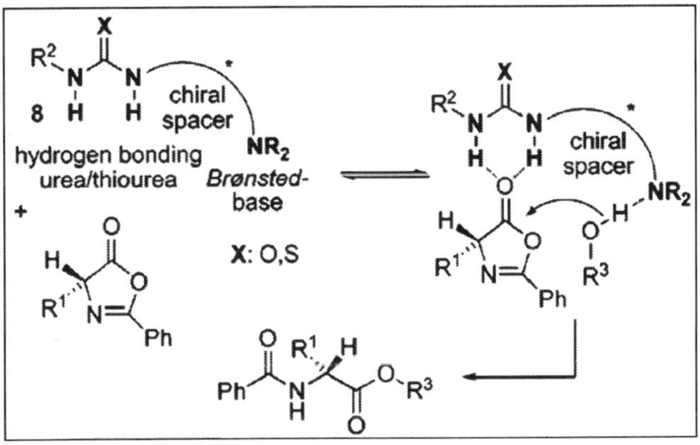

ENZYMES USED IN THE DAIRY INDUSTRY

In the field of biotechnology there are many food science applications that utilize enzymes. In the dairy industry, some enzymes are required for the production of cheeses, yogurt and other dairy products, while others are used in a more specialized fashion to improve texture or flavour.

Five of the more common types of enzymes and their role in the dairy industry are described below:

1. *Rennet*: Milk contains proteins, specifically caseins, that maintain its liquid form. Proteases are enzymes that are added to milk during cheese production, to hydrolyze caseins, specifically kappa casein, which stabilizes micelle formation preventing coagulation. Rennet and rennin are general terms for any enzyme used to coagulate milk.
2. *Other Proteases*: Milk contains a number of different types of proteins, in addition to the caseins. Cow milk also contains whey proteins such as lactalbumin and lactoglobulin. The denaturing of these whey proteins, using proteases, results in a creamier yogurt product. Destruction of why proteins is also essential for cheese production.
3. *Lactase*: Lactase is a glycoside hydrolase enzyme that cuts lactose into it's constituent sugars, galactose and glucose. Without sufficient production of lactase enzyme in the small intestine, humans become lactose intolerant, resulting in discomfort (cramps, gas and diarrhea) in the digestive tract

upon injestion of milk products. Lactase is used commercially to prepare lactose-free products, particularly milk, for such individuals. It is also used in preparation of ice cream, to make a creamier and sweeter-tasting product. Lactase is usually prepared from *Kluyveromyces sp.* of yeast and *Aspergillus sp.* of fungi.

4. *Catalase*: The enzyme Catalase has found limited use in one particular area of cheese production. Hydrogen peroxide is a potent oxidizer and toxic to cells. It is used instead of pasteurization, when making certain cheeses such as Swiss, in order to preserve natural milk enzymes that are beneficial to the end product and flavour development of the cheese. These enzymes would be destroyed by the high heat of pasteurization. However, residues of hydrogen peroxide in the milk will inhibit the bacterial cultures that are required for the actual cheese production, so all traces of it must be removed. Catalase enzymes are typically obtained from bovine livers or microbial sources, and are added to convert the hydrogen peroxide to water and molecular oxygen.
5. *Lipases*: Lipases are used to break down milk fats and give characteristic flavours to cheeses. Stronger flavoured cheeses, for example, the italian cheese, Romano, are prepared using lipases. The flavour comes from the free fatty acids produced when milk fats are hydrolized. Animal lipases are obtained from kid, calf and lamb, while microbial lipase is derived by fermentation with the fungal species *Mucor meihei*.

Enzymes for the Alcohol Industry

Specialty Enzymes and Biochemicals Co. is a worldwide leader in the production of industrial enzymes. Enzymes are eco-friendly solutions to industrial problems. They can improve manufacturing efficiency and lower costs. Specialty Enzymes and Biochemicals Co. prides itself on being one of the few manufacturers in the world that produces a full spectrum of industrial enzymes derived from all four natural origins: plant, animal, fungus and bacteria. And we use both solid state and submerge culture techniques of fermentation. Specialty Enzymes and Biochemicals Co. specializes in developing tailored enzyme solutions for individual customers. Our growing production base, and comprehensive advisory and after sales services, allow us to cater to our rapidly expanding global markets.

Products

Specialty Enzymes and Biochemicals' product portfolio consists of 160 unique enzyme solutions that address the needs of customers in 25 industries and over 30 countries. Our enzymes are all natural and non-GMO. They are a safe and effective alternative to the harsh chemicals often used in the manufacturing processes of many industries.

Industrial categories are:

- Alcohol
- Brewing
- Wine
- Food Processing
- Animals
- Baking
- Dairy
- Juice, Coffee and Tea
- Detergents
- Starch Processing
- Personal Care
- Pharmaceutical
- Textiles, Leather and Paper Pulp

Beverage Alcohol and Fuel Ethanol Industries

Enzymes are used in the alcohol processing industries to hydrolyze (break down) large starch molecules into simple sugars. Typically, the hydrolysis of starch to simple sugars is accomplished using a two-step, enzyme-catalyzed process called liquefaction and saccharifaction.

To save energy in alcohol production, starch can also be liquefied at lower temperatures using Specialty Enzymes' SEBamyl-BAL or fungal alpha-amylase SEBamyl-L. These alpha-amylases quickly reduce the viscosity of gelatinized starch. In distilleries with pre-saccharification systems, our glucoamylase blend SEBamyl-GL is used to convert liquefied-starch to fermentable sugars. SEBamyl retains excellent activity over a wide temperature range, making it suitable for use in simultaneous saccharification and fermentation distilleries.

Protease enzymes that breakdown proteins to their constituent amino acids are commonly added to whole-grain, alcohol fermentations to enhance yeast nutrition and alcohol yields. SEB-Neutral P increases free-amino nitrogen concentrations and the extraction of encapsulated starch when added to the fermentor.

Non-starch, polysaccharide-hydrolyzing enzymes, such as CelluSEB-L and ViscoSEB-L, are used in cereal-grain fermentations to reduce mash viscosity, improve alcohol yields and reduce fouling of distillation columns and spent-beer evaporators.

Brewing

Enzymes have proved to be useful for the brewing industry in many areas of beer production. They can be added to the beer after its fermentation to induce faster maturation. Enzymes also work as filtration improvers, reducing the presence of viscous polysaccharides such as xylans and glucans. Enzymes are often used to remove carbohydrates in the production of light beer and to induce chill proofing. But most importantly, enzyme supplementation is simpler and less expensive than the malting process.

Beer brewing involves the production of alcohol by allowing yeast to act on plant materials such as barley, maize, sorghum, hops and rice. Yeast converts simple sugars into alcohol and carbon dioxide. However, the sugars found in the plant materials are most often complex polysaccharides that yeast is unable to convert. The traditional method for breaking down these complex polysaccharides are called malting. This is the process whereby barley, for example, is allowed to partially germinate, producing enzymes that break down the complex polysaccharides into simple sugars that the yeast can utilize. However, the process of malting can be expensive and often difficult to control. By adding enzymes to unmalted barley the complex polysaccharides can be broken down to simple sugars and reduce or eliminate the costly and complicated process of malting.

Wine

Many of the biochemical reactions involved in wine production are enzyme-catalyzed. They begin during the ripening and harvesting of grapes, and continue through alcoholic and malolactic fermentation, clarification, and ageing. Winemakers often supplement naturally occurring grape enzymes with commercial enzymes to increase production capacity of clear and stable wines with enhanced body, flavour and bouquet. When added to grapes or musts, our LiquiSEB products increase free-run juice volume and extraction of colour, fermentable sugars and flavour components, as well as reduce pressing and fermentation time.

These pectinase or pectinase containing hemicellulase products can increase free-run juice volume by 20 to 30 per cent and lower fermentation time by 30 to 50 per cent by reducing grape-pectin viscosity. Rapid clarification and well-separated lees have a positive effect on finished wine flavour, texture and colour. Our purified ClariSEB pectinases depectinize grape-musts during fermentation or young wines prior to fining and filtration. Grape musts and wines treated with ClariSEB are less viscous. They ferment, settle and mature more quickly. ClariSEB, a beta-glucanase containing pectinase, is also used to degrade Botrytis-glucan. Wines made from overripe grapes infected with Botrytis cinerea mold are often difficult to clarify and filter due to high concentrations of Botrytis-produced glucan polysaccharides. The use of ClariSEB can speed up clarification and filtration. Our SEB-Acid and SEB-Prolase acid proteases clarify and stabilize some wines by reducing or removing naturally occurring and yeast synthesized, heat-labile proteins.

Animal Feed and Pet Care

We have blends that help provide natural relief to dry or scaly hair coats, skin problems, digestive disorders, joint difficulties, immune disorders, excessive shedding, weight problems, allergies, bloating, lethargy, flatulence, and coprophagia. We also have a proprietary blend called Exclzyme®Pet that promotes tissue, muscle and wound healing, helps along recovery from soreness and inflammation and is helpful for pets who are suffering from arthritis and pain. Exclzyme®Pet is ideal for active or athletic pets.

Baking

Enzymes are rapidly becoming very important to the Baking Industry. Enzyme supplements are used in baking to make consistently high-quality products by enabling better dough handling, providing anti-staling properties, and allowing control over crumb texture and colour, taste, moisture, and volume. Enzymes are now replacing chemical supplements that were used in the baking industry to account for these discrepancies. For instance, our enzyme called SEBake-GO can replace chemical oxidants, such as bromates, used to strengthen gluten.

Dairy

Specialty Enzymes' SEB-RMP, a microbial rennet used in cheese production, is an effective substitute for calf/cow, goat, sheep or swine-derived rennets. When milk is treated with milk-clotting proteases or microbial proteases, it causes casein to precipitate in the form of curd. The curd is separated from other milk ingredients (whey) and further processed to produce soft and hard cheeses. SEB-RMP is a superior option to animal derived rennets. Microbial lipases and proteases can be cost-effectively added to fresh milk or cheese curd to produce products with enhanced dairy and cheese flavours. Specialty Enzymes' fungal lipases SEB-ATlipase and SEB-CClipase are used to enhance the dairy-flavour in coffee whiteners, margarine and snack foods. These enzyme blends also accelerate the ripening of naturally aged cheeses and produce intensely-flavoured, enzyme-modified cheeses (EMC) used as a flavour in snack foods, soups, frozen dinners and cheese-sauces. Protease enzymes also play an important role cheese ripening and flavour development. The production of strong cheese-flavours is enhanced when fungal lipases and Specialty Enzymes' SEB-Npro, a neutral protease, or SEB-Pro, a fungal protease, are added to fresh cheese curd. It is possible develop a variety of intense cheese flavour-profiles in a few hours or days when using these protease and lipase enzymes together.

Specialty Enzymes offers SEB-Alkaline, SEB-Prolase, and SEBLipo, which are to be added with mild detergents and pH adjusting agents. These solutions will help remove milk protein and fat deposits from the filter, increasing filtration rates without using the harsh chemicals that add to waste effluent treatment costs.

Food Processing

SEB-Brolase and SEB-Prolase rapidly hydrolyze both animal and plant proteins to reduce or eliminate gel formation on heating. They are used to hydrolyze soy, pea, yeast, maize and wheat-flour proteins to produce non-bitter, protein hydrolysates. Pre-processed or separated plant and milk proteins such as soya isolates and concentrates, casein and whey can be easily hydrolyzed to produce non-bitter protein hydrolysates suitable for use as active ingredients in dietary supplements, functional beverages, fermented foods, baked goods and other foods.

SEB-Acid, SEB-Alkaline and SEB-Neutral are different types of protease enzymes used in food processing applications to break down and increase the solubility, dispersibility and digestibility of proteins. They also hydrolyze vegetable and animal proteins to peptides, polypeptides, and small amounts of amino acids.

SEBzyme-XL and CelluSEB is used for extraction, for liquefaction of plant materials and for downstream processing in the food, starch, and alcohol and brewing industries. In the food industry, these products are used to improve the processing and extraction of wheat and corn gluten, soybean protein, cereal starch, seed and nut oils, flavours and colours, coffee, tea and other botanical extracts.

Fruit and Vegetable Juice

Specialty Enzymes products, LiquiSEB and ExtractSEB, are pectinases that contain hemicellulase enzyme activities. These products not only increase juice yields, but also increase the colour and health-promoting antioxidants in fruit and vegetable juices extracted by pressing or decanter centrifuge. By reducing fruit and vegetable mash viscosity and improving solid/liquid separation, they increase colour extraction and juice volume by as much as 15 per cent.

ClariSEB and SEBamyl enzyme solutions speed up filtration and prevent storage or post-packaging haze formation by depectinizing and reducing starch in raw juices. Pectin and starch must be removed from freshly extracted juices prior to filtration, fining and concentration. Unripe fruits and vegetables can contain as much as 15 per cent starch. ClariSEB pectinase and SEBamyl amylase can reduce starch and pectin in raw fruits and juices, thus achieving clear and stable juices and juice concentrates.

Coffee and Tea

Bioclean TC has a blend of natural enzymes that gives manufacturing areas a 100 per cent safe-detergent and alkali-free cleansing. Bioclean TC is dissolved in water and sprayed in areas that come in contact with leaf material. The blend helps degrade biological material, which is then easily dislodged in cold water washing. Bioclean TC prevents bacterial growth by not providing any food to sustain the bacteria. And because it is 100 per cent natural, the blend doesn't impart any off-flavour or off-taste to the tea. Specialty Enzymes, a leading solution provider to the human healthcare, animal nutrition, and natural product processing industries, offers SPEEDOX. When added to the wash water or effluent stream in the coffee production process, SPEEDOX rapidly reduces organic loads in the wastewater that cause environmental pollution. SPEEDOX can reduce the pollution causing entities in coffee plantation wastewater. SPEEDOX is a 100% natural, high quality and eco-friendly product. It improves the clarity of water so the water can be used for irrigation. SPEEDOX also eliminates the foul odour near the effluent treatment tanks.

ENZYMES FOR INDUSTRIAL APPLICATIONS

Starch-Processing Industries

Specialty Enzymes' SEBstar-HTL and SEBamyl-BAL are blends used in the starch-liquefaction process for production of maltodextrins that are used in paperboard adhesives, paper sizing and in the manufacture of syrup-sweeteners.

Fungal alpha-amylase SEBamyl-L is used to saccharify low-DE, starch hydrolyzates in the production of crystalline maltose, low-DE maltose syrups and high-conversion, mixed maltose/glucose syrups. These stable sugar-syrups resist crystallization. They are also widely used as sweeteners in the beverage and food industries and as a source of fermentable sugars in the brewing and fermentation industries.

Detergents

Specialty Enzymes products, SEBrite-BP are a range of alkaline proteases that when added to surfactant-containing liquid and powder laundry detergents or pre-spotters, enhance the removal of protein-containing stains such as milk, egg, soya, blood, grass and body fluids of human origin. The use of Specialty Enzymes detergent alpha-amylases enzymes such as SEBrite-BA and SEBrite-A in liquid and powder laundry detergents solubilizes starch-containing stains such as baby food, spaghetti, mashed potatoes, oatmeal, gravy, chocolate, tomato sauce and other starch thickened foods. To enhance stain removal, brighten the colour, and soften cotton and cotton-containing garments, Specialty Enzymes offers a laundry-detergent compatible cellulase enzyme, SEBrite-COLOUR. SEBrite-COLOUR, effectively removes cotton cellulose microfibrils on wear and wash-damaged cotton garments. Repeated washing of cotton-containing garments with a laundry detergent containing SEBrite-COLOUR, improves soil removal, brightens fabric colour, softens and restores original garment appearance. Specialty Enzymes' protease and amylase enzymes, SEBalase-BP and SEBrite-BA can be used cost-effectively to improve the wash performance of reduced-pH, low temperature industrial and institutional laundry detergents used to remove blood, body fluids, organic material and food soils from hospital, restaurant and slaughterhouse textiles. Specialty Enzymes protease enzymes SEB-Prolase P or SEBalase-PB, carbohydrase enzyme, CelluSEB-T, alpha-amylase enzyme SEBamyl-B, Spectinase enzyme, ExtractSEB-R and lipase enzyme, SEBrite-L can be used to effectively clean filter membranes fouled with protein, starches, fats, gums, cellulose and pectin residues.

Leather

Specialty Enzymes' SEBsoak product, used during the soaking process, removes unwanted parts of hides and skins, provides an uniform soaking effect, speeds up the soaking process and better re-hydrates hides previously soaked in brine.

SEBlime is a blend that loosens hair from the corium layer of the skin, making hair removal easy. This method is faster, safer and more convenient than the method using sulphide. SEBlime offers eco-friendly, biodegradable, sulfide-free liming. It has optimum swell regulating properties, which result in excellent grain smoothness.

SEBbate Acid improves the hide's pliability and makes the grain smoother and more uniform for fabrication and dyeing. SEBbate Alkali helps stabilize the hide by ensuring the removal of all non-leather forming proteious matter.

Paper and Pulp

Specialty Enzymes' SEBrite-BB is custom made for bleach boosting in paper/wood pulping applications. It reduces pulp bleaching chemical consumption and effectively boosts the performance of kraft pulp bleaching systems.

Personal Care

Our special combinations of enzymes like SEBCareHR 413 L, SEBCareHR 413 P, SEBCareHR 414 L and SEBCareHR 414 P can be used to remove contaminants like dust, dead cells, and excessive oil, which often clog the hair follicle openings. Using our special blends in hair care formulations can clean the scalp, thereby increasing blood circulation, and leading to healthy keratin formation and less hair loss. Specialty Enzymes offers SEBCareOR 115L, which can be added to oral care products to break down excessive sugar in the mouth, thus preventing tooth decay, odour and discolouration. Specialty Enzymes also has proteolytic enzyme blends to break down the protein build up on teeth and gums - SEBCareOR 104 B, SEBCareOR 114 L and SEBCareOR 114 P. These are special blends of proteolytic enzymes, which can be used to prevent plaque development on teeth and gums and help whiten teeth. Our proteolytic enzyme blends like SoftSEB-SK 3131 L, SoftSEB-SK 3131 P, SoftSEB-SK 320 L and SoftSEB-SK 320 P, further help in this process by breaking down the keratin in dead cells, thus gently peeling off the dead outer layer without damaging the inner living cells. Our other blends, like SEBCareSK 309 L, are used in skin care products to remove excessive oil on skin, leaving it non-oily, soft and silky.

Pharmaceutical

Specialty Enzymes' bacterial blend Peptizyme SP is being used as a highly effective alternative to antibiotics in many countries. It treats inflammatory disorders, by not only fighting inflammation, but also relieving pain and swelling, improving recovery time and stimulating the immune system. Peptizyme SP has a "scavenging" effect. It helps chelate the heavy metals through which the body release toxins, and hence modulates the immune system, addresses hormonal imbalances and speeds wound and tissue repair time.

Textiles

Specialty Enzymes' biopolishers, such as Denisoft and Denistone PW help manufacturers achieve a permanent soft feel, increased absorbency and cool breathing nature of the fabric. More and more manufacturers are reducing the use of chemicals, like cationic softeners and silicones, and opting for enzyme-based biopolishing agents, particularly due to their ability to obtain permanent softness.

Chapter 8

Extraction of Enzymes

Fresh tissue is crushed into a paste with an extraction medium (often a buffer) in a mortar and pestle, or in a tissue homogenizer, or in a blender or by ultrasonic vibrations (sonication). The molarity and pH of the buffer is suitably adjusted (which may vary for different enzymes) to achieve maximum solubility and activity of the enzyme. EDT A (ethylene diamine tetra acetic acid) is often included in the extraction medium to remove heavy metals (which otherwise inhibit enzyme activity), and for disrupting the membranes of cells and cell organelles. Detergents such as Triton-X are also used sometimes to solubilise the membranes. Many enzyme proteins contain disulfide (S-S) bonds due to the presence of cysteine residues, which are easily broken during enzyme extraction leading -to loss of enzyme activity. To overcome this problem are added, thiols such as-mercaptoethanol whose sulfhydryl (-SH) group is able to maintain the S-S linkage in enzymes. If the extract is not homogeneous, the homogenate (extract) is filtered to remove cell debris, fibres, etc.; otherwise filtration may be avoided. All operations of extraction and purification are generally carried out in cold (0-4°C), since most of the enzymes get inactivated at higher temperatures.

PREPARATION OF CRUDE ENZYMES

CENTRIFUGATION

The enzyme extract, prepared as above, is centrifuged to remove cell debris, cell organelles and sometimes other molecular aggregates, leading to partial purification of enzymes. It also helps in characterization of an enzyme, since, depending upon its mass and shape, the enzyme will move through a solution at a definite speed and occupy a characteristic position in the centrifuge tube.

For most cytosolic enzymes, centrifugation at about 30,000g for 30 minutes is good enough to obtain a fair amount of activity in the supernatant. However, if the enzyme is located in a specific cell organelle, an extract rich in that organelle is prepared through preparative centrifugation. (Centrifugation for different durations at different velocities allows the cell organelles to sediment according to their sizes. All centrifugation operations are conducted in cold (0-4°C).

Table. Cell Fractions by Different Centrifugation Protocols

Centrifugation speed (x g)	Time (min)	Cell fraction to settle
1,000	5	Eukaryotic cells
4.00	10	Chloroplasts, celldebris, cell nuclei
15,000	20	Mitochondria, bacteria
30,000	30	Lysosomes, bacterialcell debris
100,000	180-600	Ribosomes

Precipitation

Enzymes and other proteins are highly charged molecules, and can be precipitated with appropriate charge neutralizing chemicals. Once their charges are broken they form aggregates- and settle down as precipitate. When, an acid or base is added, the enzyme protein can be brought to its isoelectric pH. At this pH, there is no net charge on enzyme molecules and electrostatic repulsion between them is low so that they tend to aggregate. Therefore, adjusting the pH to isoelectric point of a protein, causes its precipitation.

Acids and bases, however, often inactivate the enzyme, so that their use for precipitation is not recommended in most cases. Instead ammonium sulphate and other salts are used for precipitation in a process called salting out. Salts can change the structure of the solvent, which can lead to large changes in protein conformation by altering the electrostatic interaction between charged groups on the protein surface. The salt also competes with the protein for solvent molecules and thereby lowers its solvation. In large scale enzyme precipitation, use of many other neutral salts is preferred over ammonium sulphate, which is corrosive and releases NH_3 at higher temperatures.

Some organic solvents like acetone, methanol and ehtanol are also used for enzyme precipitation, since water miscible solvents decrease the solubility of proteins, leading to precipitation. They are cooled upto 40-60°C before their use, and precipitation is carried out at 0°C, because precipitation at room temperature causes denaturation of the enzyme, in most cases. Organic solvents are added drop' by drop to avoid local concentration.

Water soluble non-ionic polymers such as polyethylene glycol, alginate, pectinate, carboxy-methyl cellulose, polyacrylic and polymetaacrylic acids, etc. also cause enzyme precipitation. Poly-ethyleneimine is also widely used as protein precipitant at large scale. They primarily act through the removal of solvent sphere of the enzyme protein.

Purification of Enzymes

Enzyme purification involves three steps, *i.e.*, dialysis, chromatography and electrophoresis. These three techniques and their variants are briefly described in the following text.

Dialysis

Dialysis is the process that is used to remove small molecules from crude enzyme. For this, enzyme precipitate obtained in previous step is dissolved in a small quantity of buffer solution in which the enzyme was originally extracted. The solution is taken in a dialysis bag (may be a cellophane tube) and after sealing securely, the bag is suspended in either distilled water or a buffer of known molarity and ionic composition. Some other salts or chemicals may have to be added sometimes in the outer solution, to prevent the loss of enzyme activity during dialysis. The dialysis is carried out for a few hours with regular change of the outer solution or distilled water. At large scale enzyme purification dialafiltration instead of dialysis is used. The enzyme solution is filtered for small molecules through a membrane generally mounted on a fibrous support, by pressure driven operation.

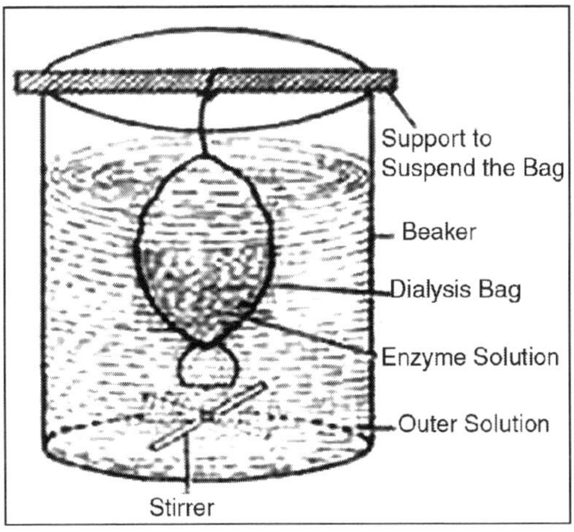

Fig. An Arrangement for Dialysis to Purify an Enzyme

Chromatography

Chromatographic separation of proteins is the most common method of enzyme purification.

Following four types of chromatography are available for this purpose:

1. Adsorption or column chromatography;
2. Ion exchange chromatography;
3. Filtration chromatography
4. Chromatography.

Adsorption Chromatography

In this chromatography, the protein or enzyme solution suspected to contain other proteinaceous impurities, is passed through a column of inert material packed in a glass or steel tube. Most commonly used column materials include finely divided solids such as charcoal, silica, alumina, calcium phosphate, hydroxyapatite, etc. The effluent solution is continuously collected in small fractions of 1.0 to 2.0 ml. The protein in each fraction is estimated by measuring the absorption at 280 nm using a UV spectrophotometre. The enzyme is also assayed in each fraction.

Various spleen enzymes such as basic RNAase, acidic RNAase, acidic DNAase, phosphodiesterase, phosphomonoesterase, etc. are often separated from each other using adsorption chromatography. For large scale chromatographic. separation of enzymes, the process is accelerated by using motors and other mechanical devices for packing the column, for loading the enzyme on the column and for eluting the enzyme.

Ion Exchange Chromatography

In ion exchange chromatography, generally a cellulosic ion exchanger is taken in the column. The proteins are separated according to their charges. The resolution is quite high and the technique can facilitate large scale protein purification. This has been successfully employed for insulin purification, plasma fractionation and for purification of many enzymes.

Gel Filtration Chromatography

In this chromatography, various proteins are separated on the basis of differences in their molecular size. This technique of gel filtration chromatography is also known as molecular exclusion chromatography or molecular sieve chromatography. The basic arrangement for gel filtration chromatography is similar to that for adsorption chromatography. A column made up of glass or steel is taken and packed with a gel.

An Arrangement for Gel Filtration Chromatography to Purify an Enzyme,

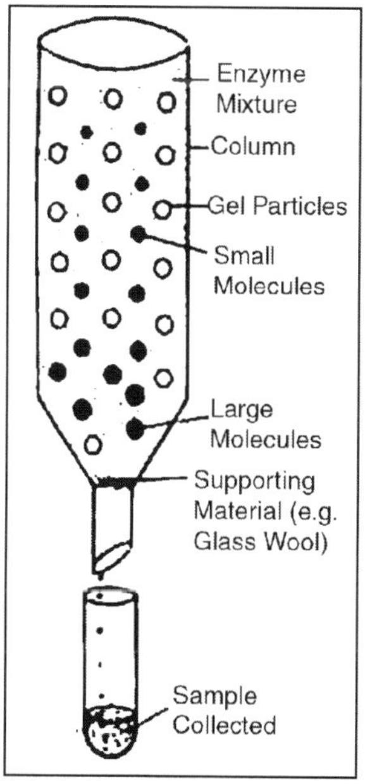

Fig. Gel Filtration Chromatography

The most commonly used gel- is sephadex, which is a cross-linked dextran produced by certain strains of bacteria. Several types of sephadex, namely G-10, G-30, G-50, G-100, G 150. G-200, etc. are available, which differ according to their pore size. When a mixture of enzymes or proteins is poured on top of the column different proteins move downwards according to their molecular sizes and come out from the column in order of decreasing sizes; the larger molecules are eluted first. The elution volume is logarithmically proportional to the molecular size. Gel filtration chromatography can be used for determining the molecular weight of the protein by calibrating the column with proteins of known molecular weights.

Affinity Chromatography

In this method, enzymes are purified according to their specificity for a particular substrate or cofactor. One component of the mixture, containing the enzyme, binds covalently to the solid support of the column, and the other components percolate down through the column. The basic requirements for affinity chromatography are the same as in adsorption or gel filtration chromatography but the packing gel must have some component which can bind with one component of the mixture. Many commercial gels available for affinity chromatography contain functional groups attached to the 'spacer arms' of the gel. The spacer arm is a chemical linkage between the functional group and the gel or matrix proper, so that the binding between Functional group and the enzyme is kept away from the gel. Thus the steric hindrance will be unlikely to prevent binding of the specific enzyme to the column. An Arrangement for Affinity Chromatography, Showing the Role of Functional Group having Affinity with the Enzyme,

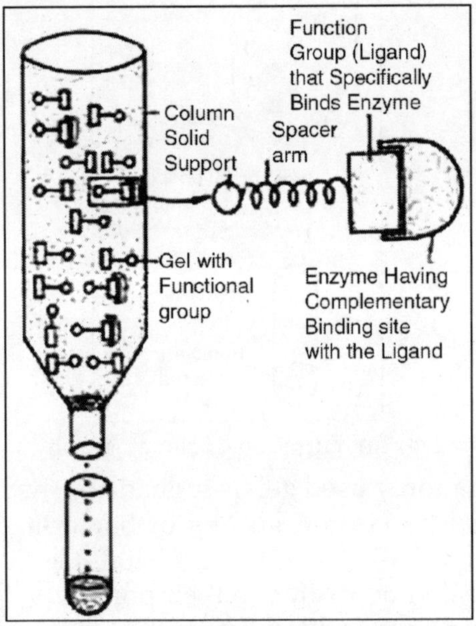

Fig. Affinity Chromatography

Electrophoresis

Electrophoresis is a technique in which molecules (enzymes, proteins, amino acids, nucleotides and nucleic acids) are separated by differences in their net charge in the presence of an externally applied electric field. This technique is routinely used in enzyme purification and isozyme separation in the laboratories, although it has found only limited application at large scale, since the technique is time consuming and is a bit expensive. Various types of instrumental approaches have been used to separate and purify charged molecules using electrophoresis. However, the most common method for purifying enzymes, is through electrophoresis on polyacrylamide gel. Polyacrylamide is a polymer of acrylamide and methylene bisacrylamide, and when prepared as a gel it is transparent, thermostable, non-ionic and extremely regular in structure. The gel may be taken either in the form of a column or a slab, although the latter is preferred over the former.

An Arrangement for Vertical Polyacrylamide Gel Electrophoresis Technique for Separation of Macromolecules,

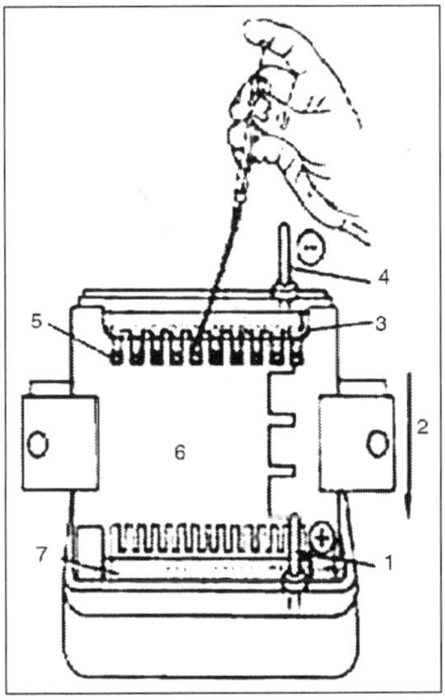

Fig. Electrophoresis

1. Electrode
2. Direction of Electrophoresis
3. Buffer
4. Electrode
5. Sample
6. Gel (Covered by a Plate)
7. Buffer

The protein mixture is loaded on the gel and the components are separated under a direct current of constant voltage. The migration rate of the various components of the mixture is dependent upon their charge and molecular weight. A variation of the above polyacrylamide gel electrophoresis, is the sodium dodecyl sulfate-polycrylamide gel electrophoresis (SDS-PAGE), which is used to determine the molecular weight of proteins. In this method, the separation is caused by the seiving action of the gel. The proteins migrate through the gel depending on their shapes and mass/charge (m/z) ratio. Gel electrophoresis is also used to separate various isozymes of a given enzyme. Isozymes perform the same catalytic function but differ in their regulatory and some kinetic aspects.

Final Processing of Enzymes

Most of the commercially available enzyme preparations, purified as above, are concentrated and sterile filtered, after purification. This is done to reduce both, the volume and the microbial contamination of the sample. Often, before storage and transport, the sample is freeze-dried with additives such as sugar substrates and dextrans.

Immobilization of Enzymes or Whole Cells

In recent years, the technique of enzyme or whole cell immobilization has revolutionized the prospects of enzyme application in industry. Immobilization is defined as the imprisonment of a biocatalyst in distinct phase that allows exchange with, but is separated from, bulk phase in which substrate, effector or inhibitor molecules are dispersed and monitored'. In other words, an immobilized enzyme is physically entrapped or covalently bonded by chemical means to an inert and usually insoluble matrix, where it can act upon its natural substrate. The matrix is usually a high molecular' weight polymer such as polyacrylamide, cellulose, starch, glass, beads, etc.

The concept of enzyme immobilization was first evolved, when difficulties were experienced during the use of crude enzyme preparations in the production of wine, cheese or in tanning. The major limitation that was due to limited availability, instability and the consequent high cost of the enzyme, was largely overcome by the immobilization of enzyme on a support, a phenomenon reported for the first time by J.M. Nelson and *E.G.* Griffin (1916). They reported an adsorption (immobilization) of invertase on charcoal/alumina without loss of activity. However, the technique of enzyme immobilization could be established only after a lapse of about 40 years, when between 1954 and 1961 many researchers developed relevant procedures and the equipments.

Some of the advantages of using immobilized enzyme over free enzyme include the following:

- Because of its binding with a matrix, the immobilized enzyme has better stability in many cases.
- Efficiency of immobilized enzyme is better.
- The enzyme can be recovered at the end of reaction and can be used repeatedly; also the processed product is not contaminated with enzyme.
- Some manipulations of enzyme catalysed reactions are better and easier, when the enzyme is used in the immobilized form. For example, the reaction can be stopped instantly by removing the immobilized enzyme from the reaction solution.

The immobilization of enzyme is performed under mild and controlled condition, so that the enzymes retain their tertiary and quaternary structures', which are necessary for their activity.

Immobilized enzymes are preferred over immobilized cells or tissues, since they have a high specificity and yield pure products.

Techniques for Immobilization

Various methods developed for immobilizing enzymes, which are described in this section, can be classified into the following five general categories:

1. Adsorption method,
2. Covalent binding method,
3. Cross-binding method,
4. Entrapping method, and
5. Entrapping by micro-encapsulation.

Adsorption Method

An enzyme may be immobilized by adsorption to several types of materials (adsorbents with charged or neutral surfaces). Since the activity of enzyme may be significantly reduced or lost during adsorption and subsequent release, the adsorbents should be chosen such that enzymes are bound firmly with minimum inactivation. Following are some of the adsorbents often used for immobilization: alumina, amberlite CG-50, bentonite phosphate gels, carbon, carboxymethyl cellulose, carboxymethyl sephadex, collagen, DEAE-cellulose, DEAE-sephadex, glass, silica gel and titania (ceramics). The pH and ionic conditions of enzyme and adsorbent solution should be carefully controlled during immobilization. If adsorption involves predominantly ion-ion interaction with very little hydrogen bonding, then a simple shift in pH or ionic strength could exchange the protein ion for another ion causing desorption. In a typical adsorption method for immobilization, adsorbent is packed in a water jacketed column, which is washed with a preconditioning solution. For instance, if we take titania as an adsorbent, it is washed with 0.5 $NaHCO_3$ and internal surfaces of the adsorbent are rendered air free. Enzyme solution is buffered fairly close to isoelectric point of the enzyme with a low ionic strength (below 0.01 μM), and is then circulated through the column at a desired temperature for several hours. After the unabsorbed enzyme solution is drained from the column, the column is washed with water, then with 0.5M NaCI and finally again with water. The immobilized enzyme column is now ready for the delivery of substrate solution and evaluation of performance.

Covalent Binding Method

An enzyme can be covalently bound to' support materials by different methods.

The enzyme forms a covalent link with active groups of support material either:

- Through the reactive groups on side chains of its amino acids like lysine, arginine, histidine, tyrosine, cysteine, serine, aspartic acid and glutamic acid.
- With terminal amino and carboxyl groups of the polypeptide chains. Some examples of covalent linkage between the support and the enzyme are as follows.

Support with - OH Group

Such supports can be activated for covalent linking by treating with either triazines or cyanogen bromide. The reaction with enzyme protein in each case involves the $-NH_2$ group of the lysine. Immobilization of Enzymes, Using Supports Having OH Groups, that are Activated by Covalent Bond Formation with (A) Triazines or With,

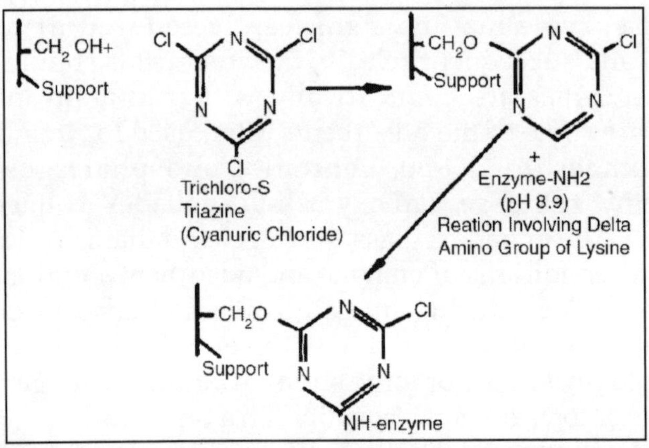

Immobilization of Enzymes, Using Supports Having OH Groups, that are Activated by Covalent Bond Formation with (B) Cyanogen Bromide,

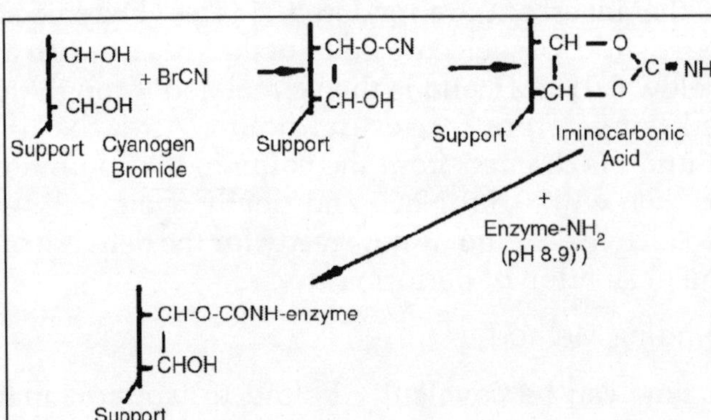

Support with -COOH Group

Such supports including carboxymethyl cellulose can be activated via azide derivative or acyl isourea formation. The reaction again involves $-NH_2$ group of lysine. However, other amino acids such as tyrosine, cysteine and serine are also implicated.

Immobilization of Enzymes, Using Supporths Having-COOH, with Hydrazine (NH_2 - NH_2) or with - NH_2 Groups Through Covalent Linkage Involving (A) Azide Derivative

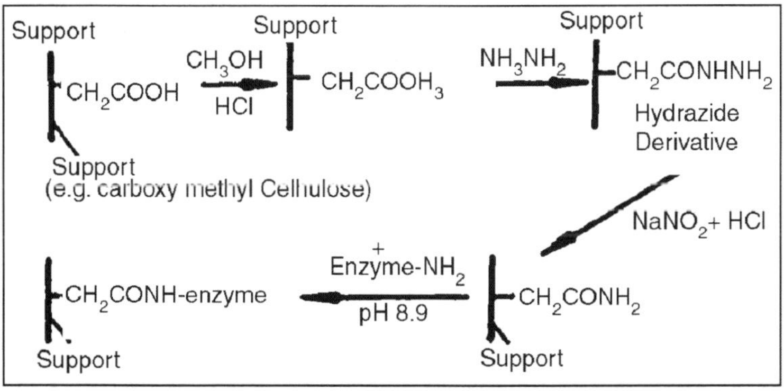

Immobilization of Enzymes, Using Supporths Having-COOH, with Hydrazine (NH_2 - NH_2) or with - NH_2 Groups Through Covalent Linkage Involving (B) Acyl Isourea

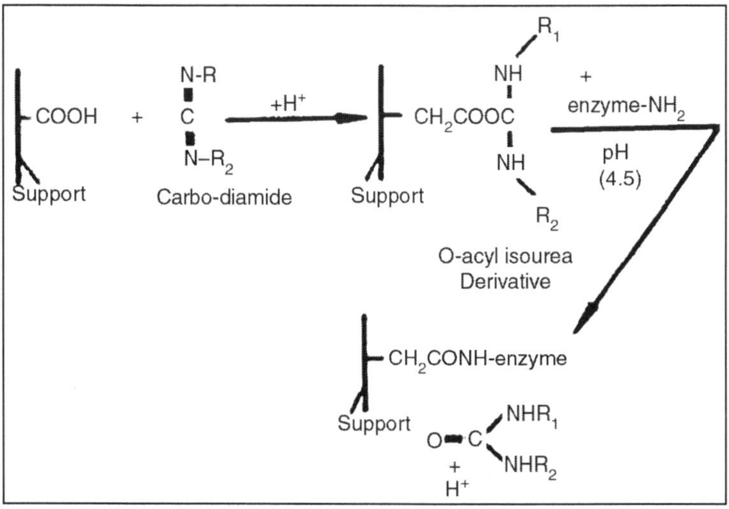

Support with -NH_2 Group

Support containing amino group can be converted to a diazonium chloride by treating with $NaNO_2$+HCl. The enzyme protein links with this derivative forming an azo- linkage involving the tyrosine residue of enzyme protein. Immobilization of Enzymes Using Supports Having $-NH_2$ Group Involving (A) Formation of Diazonium Chloride

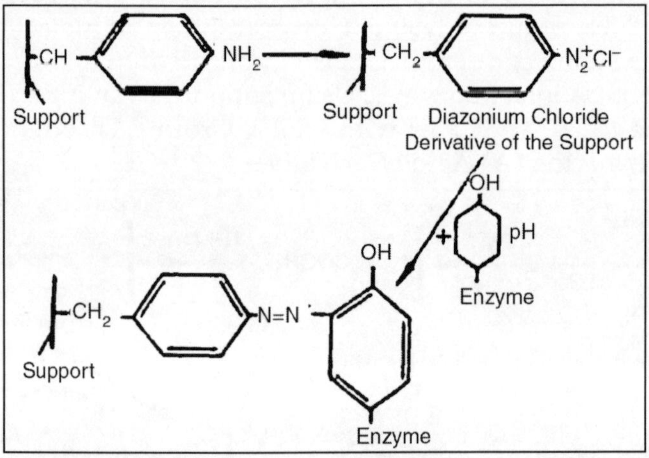

Glutaraldehyde is also used sometime to activate the support containing $-NH_2$ group. The reaction involves a Schiffs - base formation between amino group of support and amino group of one of the amino acids of protein. Immobilization of Enzymes Using Supports Having - NH_2 Group Involving (B) Activation with Glutaraldehyde

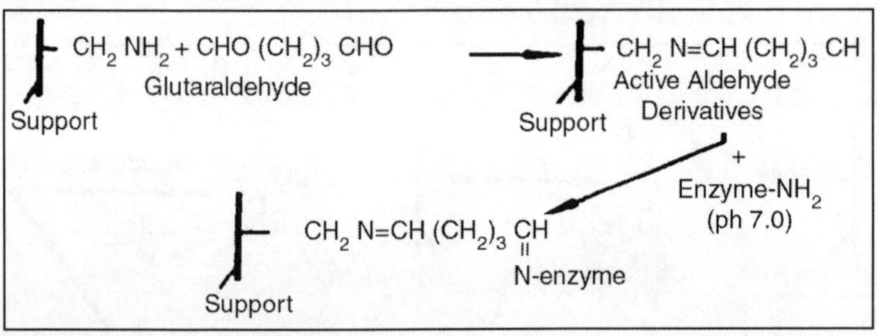

Number of bonds between support and the enzyme molecule during covalent linkage is variable. There are, for example, 17 bonds per papain molecule and 8 for subtilopeptidase immobilized on semi aldehyde starch derivative. Papain bound to porous glass is linked through three azo links per enzyme molecule.

Cross Binding Method

The enzyme can be immobilized by cross-linking to a multifunctional reagent without involving a solid support. Alternatively, a chemical bridge (analogous to the spacer arm in affinity chromatography) of some other molecule can be established between the enzyme and a chemical support. The reagent or the bridge chemical, reacts with the enzyme and forms a linkage. Some of the commonly used reagents for cross-linking enzymes.

There types of basic approaches have been used in immobilizing enzymes by cross-linking:

1. Cross-linking of enzyme with glutaraldehyde to form an insoluble aggregate; *e.g.* papain.
2. Adsorption of enzyme onto a surface followed by cross-linking; for instance, cross-linking trypsin adsorbed to the surface of colloidal silica particles.
3. Impregnation of porous material with the enzyme followed by cross-linking of the enzyme in the pores; for instance, papain in collodion membrane.

Some Reagents Used to Immobilize Enzymes by Cross-Linking:

- Bisdiazobenzidine 2, 2, -disulphonic acid
- Diazobenzidine.
- Diazobenzidine 3,3, -dianisidine
- 4,4' - Difluoro 3,3' dinitrophenyl sulphone
- Diphenyl 4,4' -diisothiocyanate 2,2' -disulphonic acid
- Glutaraldehyde
- N,N′ -hexamethylene bisisodiacetamide
- Hexamethylene di-isocyanate
- 3-methoxydiphenyl methane 4,4' -di-isocyanate
- Toluene 2. 4-di-isothiocyanate
- Toluene 2-isocyanate 4-isothiocyanate

Entrapping Method

Enzymes can be entrapped inside a cross-linked gel matrix by allowing the gel to be formed in an aqueous solution containing one or more enzymes. The polymerization of the gel is carried out in the presence of enzyme(s). The enzyme is physically entrapped within the matrix and cannot escape by permeation.

Entrapping of Enzyme Inside a Polymerized Gel,

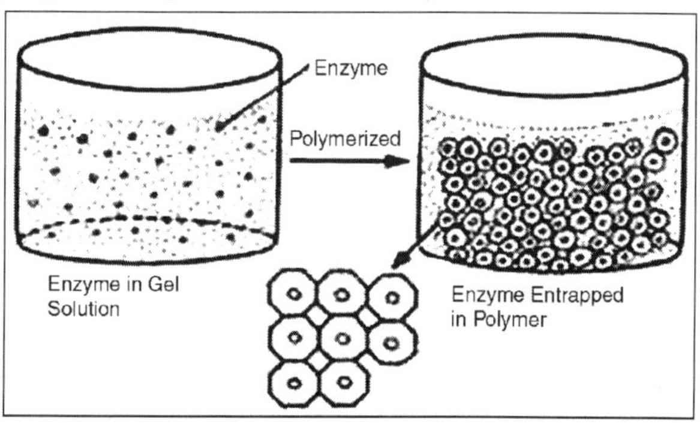

However, substrate molecules, being smaller can defuse in, can be acted upon by enzyme and can then diffuse out of the matrix. The important matrices for entrapping enzymes are polyacrylamide, starch and silicone rubber gel.

However, immobilization of enzymes by such an entrapment method is often associated with following three types of difficulties:

1. There is continuous leakage of enzyme because of wide pore size distribution in the gel.
2. There is reduced substrate accessibility to the enzyme.
3. There is some loss of enzyme activity due to free radicals produced during polymerization of the gel.

Entrapping by Microencapsulation

The microencapsulation is another approach in enzyme immobilization by entrapping method. In this approach enzyme can be immobilized within microcapsules prepared from organic polymers, so that the enzyme can not escape, although low molecular weight substrates and products can enter and leave the capsule by diffusion through the membrane.

There are two general methods for preparing the membranous capsules for enzyme entrapping:

1. The membranes are prepared by the process of phase separation, which is similar to homogenization of water in oil. One phase is not miscible with the other but forms a droplet or coacervate with the.other phase, when mixed. The enzyme is entrapped within this droplet or coacervate.
2. A water insoluble membrane is prepared by chemical polymerization and the enzyme is entrapped during this polymerization.

Table. Immobilization of Aminocyclase.
(from 'Fermentation Technology Today', 1972)

Support	Method	Retention of enzyme activity (%)
Polyacrylamide	Entrapment	53
Nylon	Encapsulation	36
DEAE cellulose	lonic binding	55
DEAE sephadex A-59	Ionic binding	56
CM sephadex C-50	Ionic binding	0
Iodoacetyl cellulose	Covalent binding	39
CHBr activatedsephadex	Covalent binding	1.0
AE-cellulose	Cross linked with glutaraldehyde	0.6

Spheres made up of semipermeable nylon or collodion membranes are often used for microencapsulation of an enzyme. Enzymes are also entrapped inside fibres of cellulose triacetate. These fibres may be packed in columns or woven into fabric. The choice of the method to be employed for immobilization would certainly vary according to the enzyme involved and its application. Sufficient variation has been reported in the level of retention of enzyme activity on immobilization to different supports.

Requirements for Commercial use of Immobilized Enzyme Systems

The use of immobilized enzymes in industry attracted increased emphasis during 1970-1980, so that by late 1970s immobilized isomerase became a major commercial success, making possible the production of high-fructose com syrup, so vital for soft drinks industry. Immobilized enzymes and whole cells are today utilized in a variety of industrial processes.

It should however, be appreciated that immobilized enzymes make only a fraction of the total cost. In developing an immobilized system, once the cost: benefit ratio is favourable, immobilized system may be developed and improved for commercial purposes. For this purpose, the following four areas should receive adequate attention of the decision makers for the use of an immobilized enzyme system.

Enzyme Carrier Systems

The choice of an immobilized enzyme system depends on the following two factors:

1. *Enzyme purification and stability*: The manufacturers have to make a choice about the purity of enzyme. The less purified enzyme requires a larger bulk of enzyme, while highly purified is more expensive.

   The higher purity may also avoid undesirable side reactions due to:

   - Absence of other enzymes
   - Low duration, for which the substrate has to be in contact with the enzyme.

   Since, long operating life or maximum yield of product per unit weight of immobilized enzyme is critical for its commercial success, stability of enzyme system is important. Highest total productivity is achieved at lowest temperature, although contamination and capital costs may be higher at low temperature. The stability of the support material as well as the activity of enzyme are also influenced by pH, so that pH profile of enzyme system should also be known.

2. *Nature and composition of carrier system*: The charge on the carrier also influences pH. For instance, if carrier is negatively charged. H^+ ions will accumulate at its boundary leading to drop in pH and if positively charged, there will be increase in pH. Such changes in there will reduce stability of the system and may be overcome by incorporation of the ions of interest into the carrier. For example, magnesia incorporated into porous alumina support (developed by Corning) for glucose isomerase increased the observed activity of enzyme by more than 50%.

APPLICATION OF IMMOBILIZED ENZYMES AND WHOLE CELLS

Reactor Types

After the enzyme and the carrier system have been chosen, decision needs to be taken on reactor design, which can be any one of the following four types:

1. *Batch reactor*: It is the simplest type of reactor. The immobilized enzyme is placed in a container with the reactants, and the reaction is allowed to proceed until the desired level of conversion is reached. Some stirring or agitation of reaction mixture is also required. Many modification of these reactors have been designed to simplify recovery and reuse of the enzyme composite.
2. *Continuously stirred reactor*: This type of reactors employ a stirred rank, to which reactants are continuously added and products continuously withdrawn.
3. *Fluidized-bed reactor*: In these reactors, enzyme system is fluidized by the upward flowing of substrate solution. This helps in eliminating any plugging of enzyme system, although small duration of contact may be insufficient for the desired conversion. This can be overcome by decreasing the velocity of solution by different methods. Glucose isomerase and lactase have been shown to double their efficiency due to fluidization when compared with fixed beds.
4. *Fixed bed reactor*: These are the most widely used reactors for large scale commercial operations. Different companies use these reactors for enzymes like:
   - Aminocyclase,
   - Glucose isomerase
   - Lactase.

These reactors will keep on dominating the large scale commercial application due to their high efficiency, ease and simplicity of operation. There are many types of fixed bed reactors including those with packed bed of particular material to which enzyme is coupled. A packed bed of immobilized enzyme has clear advantages, since the void space is reduced to 30%, as compared to 80-90% in stirred tank reactors.

However, in the packed bed reactors, a concentration gradient of substrate and the product are often generated, so that a low K_M for the enzyme and substrate inhibition become critical to capitalize on this concentration gradient. Moreover in a packed bed reactor, pH adjustment is not possible, making it necessary that the enzyme has a sufficiently broad range of pH to accommodate any changes in pH.

Operating Strategy

An operating strategy also need to be prepared for the use of immobilized enzymes to have maximum production at minimum cost. Since the enzyme activity falls down in due course of time, the production rate also falls.

This is minimized by:

- Using a multiple reactor system, in which start-up and reloading time for each reactor is staggered with other reactors in series or parallel (as done in case of glucose isomerase),
- Raising temperature to compensate for loss in activity (as has been done for lactase).

The operating costs have to be kept in mind, while deciding about the operating mind, while deciding about the operating strategies and will include cost of the following:

- Immobilized enzymes (including carrier system, immobilization equipment, etc.);
- Labour (including fringe benefits, supervision and other overhead expenses);
- Depreciation charges of the equipment used, which may be upto 20% of the capital annually
- Raw material losses at each processing step.

Regulatory Considerations

For using immobilized enzymes, particularly for producing food or pharmaceuticals, it is necessary to get approval of appropriate authority (*e.g.* Food and Drug Administration or FDA in USA) for the use of a particular fixing agent or a carrier system. These fixing agents have to be declared safe by these authorities.

Only two fixing agents for glucose isomerase are approved and these are:

1. Diethylaminoethyl cellulose
2. Glutaraldehyde. Many of the fixing agents reported in literature are not 'approved for use in food industry.

Stability of Immobilization and Protection of Immobilized Enzymes/Cells

Strategies have also been developed to achieve stabilization of immobilized enzymes, using multipoint covalent attachment to gels. This also provides resistance to conformational changes due to heat and altered pH, although very strong support may also lead to adverse conformational changes. As will be shown in the next chapter, a large number of bioprocesses in future will be carried out in organic solvents and stability in these solvents may be needed. Although, immobilization itself provides stability in organic solvents, cross-linking with a bifunctional reagent, such as glutaraldehyde proves useful. Stability of immobilized cells has also been achieved by cross-linking, particularly in case of halophilic cells used for bioconversions under high salt conditions.

Some Example of the Use of Immobilized Enzymes/Cells

Immobilized enzymes/cells have already been used for a variety of purposes including bioprocess/fermentation industry, agriculture and environment protection. In India, ressearch in this area is in progress at BARC, Mumbai under the leadership of S.F.D Souza. Some uses of immobilized enzymes/cells will be discussed.

Immobilized enzymes/cells for Bioprocess Industry (Including Fermentation)

A number of examples of the use of immobilized enzymes/cells in a variety of bioprocesses. In addition to a variety of immobilized enzymes used in industry, yeast cells have been extensively used for the production of fuel ethanol, alcoholic beverages (including beer) and for removal of lactose from milk. Immobilized cells are also used for the continuous production of enzymes in several bioprocesses.

Immobilized Cells for Agriculture

Microbial inoculants are often used in agriculture for:

- Enhancement of symbiotic/associative nitrogen fixation;
- For biological control of soil-borne plant pathogens.
- Reduction of aflatoxins

- Biodegradation of xenobiotic compounds.

However, in these practices, survival of microbes under biotic/abiotic stresses in the soil is a major limitation. Immobilization of microbial cells (encapsulation or entrapment) in these cases has been shown to provide protection against these stresses. For instance. microbial cells entrapped in alginate were found to survive, while the survival of free cells was greatly reduced under drying/wetting cycles in soil. In case of algal biofertilizers also, immobilization provided a definite advantage.

Immobilized Cells for Bioremediation

Immobilized cells (bacteria, yeast fungi, algae, etc.) have also been used during bioremediation operations for removing heavy metals and radionuclide waste from the atmosphere. Bioresins containing a variety of microbes are also being prepared for the treatment of radionuclide and allied wastes.

Industrial Biocatalysis

Industrial uses of enzymes were described earlier in this chapter. However the use of enzymes in industry need to be economical and costeffective. Both isolated enzymes and whole cells are used in industry today, making this an active area of research.

To be economically viable, most industrial processes should involve:

- High product concentration and productivities;
- No undesirable by-products
- Enzymes that do not require expensive cofactors.

The economic feasibility of a biocatalytic process at Industrial scale depends on several factors depicted in diagram. The Biocatalytic Cycle Showing Different Factors, which Determine the Economic Feasibility of an Industrial Biocatalytic Process

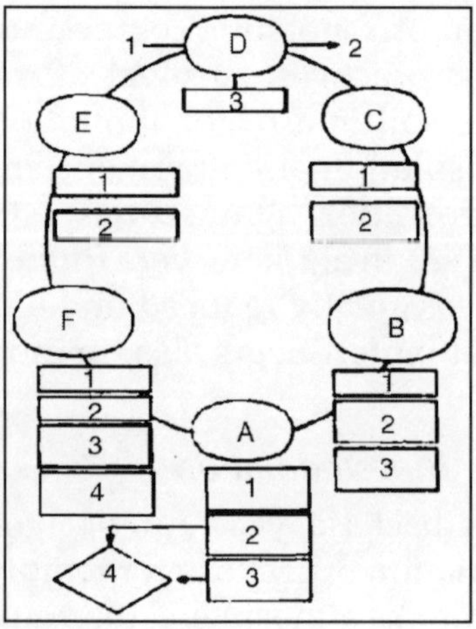

A. Biocatalyst Engineering
   1. Cell Engineering
   2. Process Engineering
   3. Enzyme Engineering
   4. New Reactions
B. Biocatalyst Characterization
   1. Kinetics
   2. Reaction Conditions
   3. Structural Information
C. Biocatalyst Selection
   1. Screening
   2. Enzyme or Cells?
D. Process
   1. Reactants
   2. Products
   3. Economics
E. Product Recovery
   1. Downstream
   2. In Situ Recovery
F. Application
   1. Stability
   2. Immobilization
   3. Cofactor Regeneration
   4. Multiphase Systems

New processes can be developed either due to the availability of a new enzyme which would function with available substrate or due to the identification of new desired products, for which enzyme and suitable substrate are selected. For a conversion involving isomerization or hydrolysis, which do not require regeneration of coenzymes, either pure enzyme or whole cells can be used. But when cofactors are needed, whole cells are preferred, since they enable cofactor regeneration.

Following factors are taken into consideration to optimise the performance of biocatalyst and biocatalytic process:

- If necessary, the enzyme should be engineered for better activity, improved substrate range and enzyme stability (consult next chapter for details);
- The host cell should be selected for solvent resistance, substrate import, product export and elimination of side reactions;
- The biocatalyst (enzyme or cells) may be immobilized and cofactors regenerated for coenzyme-dependent enzymes;
- The reaction medium, consisting of an aqueous phase, an organic phase or a two liquid phase system should be optimized for dissolving substrates and products, while maintaining enzymatic activity.

Industrial biocatalysis is particularly utilized in food and pharmaceutical industries, where reaction selectivity and complex substrate specificity are critical.

Following are three such examples of the use of enzymes in food and pharmaceutical industry:

- Production of high-fructose com syrup, which involves conversion of D-glucose to D-fructose due to xylose isomerase;
- Preparation of semi-synthetic penicillins catalysed by penicilllin amidase;
- Biotransformation of polyunsaturated acids for products with nutritional and biomedical attributes.

Similarly, the following two other examples involve synthesis of speciality chemicals and polymers:

1. Synthesis of phenomic resins, through the use of peroxidases to be used as replacements of conventional phenol formaldehydes
2. Synthesis of acrylamide, through hydration of acrylonitrile, using nitrile hyratase.

Biotransformation of Polyunsaturated Fatty Acids (PUFAs)

An interesting area of biotechnology research is to obtain polyunsaturated fatty acids (PUFAs) and utilize them either directly or for their biotransformation to other bioactive compounds.

Table. Different Polyunsaturated Fatty Acids and their Nomenclature

Fatty Acid (PUFA) m:n (w—x)		PUFA nomenclature m:naz,....nz
Arachidonic Acid (AA)	20:4 (ω–6)	$20:4^{5z, 8z, 11z, 14z}$
a-Linolenic acid (ALA)	18:3 (ω–3)	$18:3^{9z, 12z, 15z}$
Docosahexaenoic aicd (DHA)	22:6 (ω–3)	$22:6^{4z, 7z, 10z, 13z, 16z,z19z}$
Dihomo-g-linolenic acid (DHGLA)	20:3 (ω–6)	$20:3^{8z, 11z, 14z}$
Docosapentaenoic acid (DPA)	22:5 (ω–3)	$22:5^{7z, 10z, 13z, 16z, 19z}$
Eicosapentaenoic acid (EPA)	20:5 (ω–3)	$20:5^{5z, 8z, 11z, 14z, 17z}$
Eicosatrienoic acid (ETA)	20:3 (ω–6)	$20:3^{5z, 11z, 14z}$
Eicosatetraenoic acid (ETRA)	20:4 (ω–3)	$20:4^{5z, 11z, 14z, 17z}$
g-Linolenic acid (GLA)	18:3 (ω–6)	$18:3^{6z, 9z, 12z}$
Linoleic acid (LA)	18:2 (ω–6)	$18:2^{9z, 12z}$

Note: In m:n(ω—x)and m:naz..nz, m and n refer to number of carbon atoms and the double bonds, (ω –x) refers to position, where the final double bond is located, x-carbon atoms (inclusive) away from the terminal carbon atom(ω); in the superscrips az to nz, a and n indicate the positions of double bonds (carbon atoms numbered from carboxyl end).

The biotransformations of PUFAs include two major classes of reactions:

1. Hydrolase-mediated reactions, where PUFA remains unaffected, but overall lipid structure is altered to provide products with desirable nutritional, biomedical or surfactant attributes,
2. Oxidative transformations, which yield molecules with biological activities of relevance to flavour, pharmaceuticals and fine chemical industries.

Hydrolalse-Mediated Biotransformations

In these biotransformations, enriched glycerides, fatty acids and esters are first obtained by lipase-catalysed hydrolysis, alcoholysis, acidolysis, glycerolysis and transesterification of oils obtained from fish, plant seed or microbes. PUFA level can be increased 2-8 fold depending upon lipid substrate, biocatalyst and the reaction employed. Also, EPA, DHA, DPA, and GLA can be incorporated into seed oil (groundnut, rapeseed, soybean, etc.), to enhance their proportion to 40%, so that these oils can be used in food as cheap PUFA fortified oils, Similarly, mixed glycerides with PUFA level approaching 95% can be obtained by esterification of glycerol with acid concentrates; these are used as components of nutritional supplements, infant formulas, and therapeutic agents.

Oxidative Biotransformation (Use of LOXs and COXs)

PUFAs are excellent precursors for a variety of molecules in biological systems. Many of these bioconversions are initiated by the oxidative activation of the double bonds, producing intermediates in the form of hydroperoxide (due to action of lipoxygenases or LOXs) and endoperoxides (due to action of cyclooxygenases or COXs).

Hyperoxyl-PUFAs are transformed due to a cascade of reactions involving cyclases, lyases, dehydrases and peroxygenases, giving rise to:

- Flavour and aroma compounds (feeding attractants, etc.), hormones, defence compounds and other physiologically active compounds (*e.g.* aldehydes, ketoacids, hydroxyacids, eicosanoids). Endoperoxides give rise to cyclic eicosanoids and related compounds, which fulfill regulatory, antimicrobial, reproductive and ecological roles.

Enzyme Engineering

Enzymes can also be specifically engineered to produce forms not available in nature. These will then meet specific needs of commercial application of immobilized enzymes.

They are:

- Protein For Enzyme Engineering
- Engineering For Macromolecules

Chapter 9

Biocatalysis and Enzyme Biotechnology

Biocatalysis involves enzyme catalysis in the living cells and therefore has been the basis of some of the oldest chemical transformations (*e.g.* brewing and fermentation). The foundation of this subject of biocatalysis was laid by distinguished scientists like Liebig, Pasteur and Emil Fisher, whose pioneering work led to the discovery of enzymes (the biological catalysts), which initiate and accelerate thousands of biochemical reactions in living cells.

These enzymes process reactions which are otherwise not possible under normal conditions found in the cell. For instance, although hydrolysis of starch in a test tube requires strong acidic medium and high temperature (boiling), in the alimentary canal it. is hydrolysed and digested under normal conditions of acidity and temperature. This is made possible by starch hydrolysing enzymes available in the stomach. In fact, almost all biochemical reactions in the cell require one or more enzymes for their completion. Non-enzymatic conversions, though are known, but are very few.

Enzymes are proteinaceous in nature, although rarely they can be either RNA enzymes (ribozymes), discovered in 1980s, or DNA enzymes discovered in the year 2000. Catalytic antibodies (abzymes) have also been discovered during 1980-2000. Over 3000 enzymes have already been identified and the number is growing with research in the field of genomics and proteomics. Enzymes can be extracted from living tissues, purified and even crystallized.

Under controlled conditions of isolation, they retain their original level of activity, and in some cases even exhibit an increased activity. Thus, a purified enzyme can be used to carry on a specific biochemical reaction even outside the cell. This property of enzymes has been employed for laboratory experiments as well as for commercial production of several important biochemical compounds, drugs and industrial products. In view of this, enzyme research has become an important area of biotechnology. Some current research activities involving enzymatic synthesis. Different aspects of the techniques utilized in enzyme research and their commercial application will be described in this chapter. Enzymes can also be variously modified through enzyme modelling and gene manipulations for a variety of purposes.

Some Research Activities Involving Enzymatic Synthesis:

- Development of new enzymatic reactions using designed substrates
- Use of organic solvents instead of aqueous media for enzymatic synthesis
- Overcoming substrate and product inhibition
- *Cofactor regeneration*: NAD(P), ATP, CoA, PAPS, sugar nucleotides
- Modification of enzyme activity and' semisynthetic enzymes
- Enzymes immobilization and stabilization
- Use of recombinant DNA for large-scale production of enzymes
- Directed mutagenesis for changes in enzyme properties
- Exploration of entirely new enzymes and the known enzymes from new species
- Scale-up in fine chemical and pharmaceutical synthesis
- Multienzyme systems for in situ and in vivo synthesis
- Antibody catalysis

PROPERTIES OF ENZYMES

Following are the two most important properties of enzymes:

1. Remarkable specificity provided due to enzyme's active site for substrate binding and subsequent chemical transformation,
2. Often a phenomenal acceleration in the rate of reaction, which can reach a level of 10^{10} to $10^{2}3$ times the non-enzymatic rates. In absolute terms, the catalytic events due to an enzyme molecule can be as slow as one event per minute to 10^{5} events per second.

A typical enzyme, having a molecular mass of 50,000 (M_r 50K) is comprised of 450 amino acid residues, of which ~ 400 residues involve 19 chiral-L-amino acids and the remaining ~50 residues consist of glycine (a non-chiral or symmetric amino acid; chiral means asymmetric).

Out of 450 amino acids of an enzyme, it is the 400 chiral amino acid residues, which provide an asymmetric micro-environment for the formation of an active site for substrate binding and subsequent chemical transformation.

Parametres Affecting Enzymatic Reactions

The rate of an enzymatic reaction depends on several factors including the reaction itself, the nature of enzyme involved, and more importantly the concentrations of substrate (S), product (P) and enzyme (E). For instance, if the concentration of the substrate is slowly increased, keeping other things constant, the rate of reaction increases till it reaches a maximum, where further increase in substrate concentration does not lead to any further increase in the rate of reaction. This maximum rate of reaction under a set of constant conditions is described as V_x. Besides substrate concentration, several other parametres have been suggested to describe fully an enzymatic reaction.

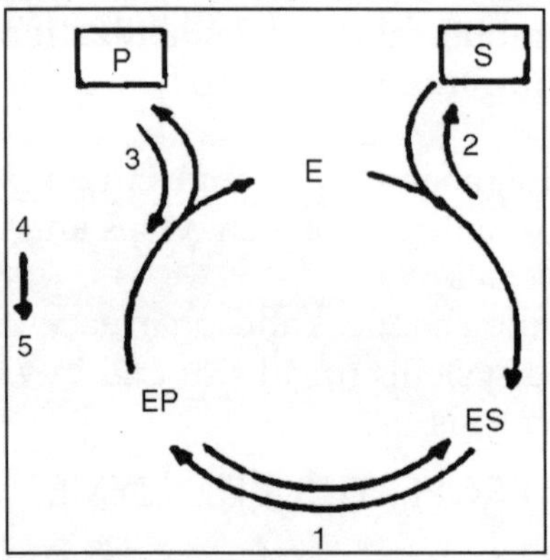

Fig. Different Parametres, Enzymatic Reactions

- K_{cat}: the higher the better
- $K_{cat}/K_M.K_M$ (K_s) Low K_M causes substrate inhibition
- low K_p causes products inhibition
- 1 Umg^{-1} ~ ($K_{cat}{}^2 s^{-1}$)
- 1 mmol product per day

These parametres include the following:

- Dissociation constants of substrate S and product P (K_s and K_p);
- Specific activity or the turnover number, which is quantified by k_{cat} and is obtained, when V_x is divided by. enzyme concentration [E]; this represents the number of substrate molecules processed per enzyme molecule per second and its value is generally 1000, although it can be much greater;
- Michaelis constant (K_m), which represents the substrate concentration, at which the rate of reaction is half its maximum rate V_x described above;

- Specificity or catalytic efficiency of enzyme, determined by maximum specificity constant (k_{cat}/K_m).

Enzyme Kinetics Showing Relationship Between the Rate of Enzyme Reaction and the Substrate Concentration, Km is the Substrate Concentration at which the Reaction Rate is Half Maximal

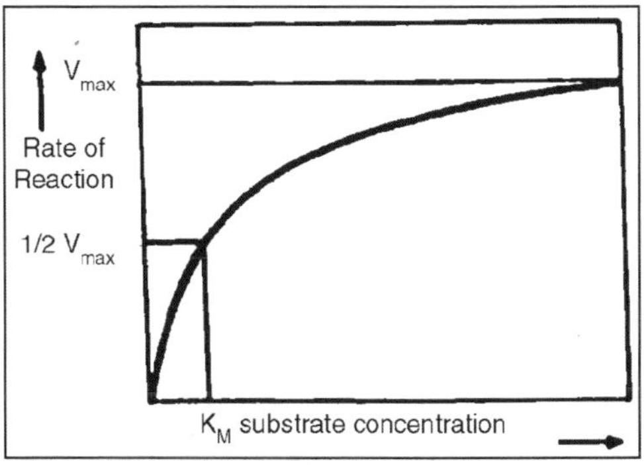

Fig. Enzymatic Reactions

Among the above parametres, the turnover number {k_{cat} (s^{-1})}, which describes the decomposition of enzyme-substrate complex (E-S) is a first order rate constant, and the catalytic efficiency {k_{cat}/K_m ($M^{-1}s^{-1}$)}' which describes the rate of product formation is a second order rate constant. The values of catalytic efficiencies of enzymes lie in the range of 10^5 M^{-1} s^{-1} (*e.g.* aspartate aminotransferase) to 10^9 M^{-1} s^{-1} (*e.g.* fumarase) and cluster around a mean value of ~10^7 s^{-1}, which is common for many hydroxylases. The relationship of these two rate constants and that of a rate constant for the uncatalysed reaction (k_{non}) with the rate of product formation is shown in the following three sets of equations:

1. ES $\rightarrow$ E + S
   $d[P]/dt = k_{cat}$ [ES]
2. E + S $\rightarrow$ E + P
   $d[P]/dt = k_{cat}/k_m$ [E] [S]
3. E $\rightarrow$ P
   $d[P]/dt = k_{non}$ [S]

By comparing the rate constant of an uncatalysed reaction (k_{non}) with that of catalysed reaction (k_{cat}) one can calculate the increase in reaction rate that an enzyme produces; this is described as rate enhancement and is calculated as k_{cat}/k_{non}. Another parametre termed as catalytic proficiency is, calculated as the ratio $(k_{cat}/K_m)/k_{non}$.

Ideally, for the economic' production at the industrial scale, the values of dissociation constants, K_s and K_p should be low, *i.e.*, the enzyme should be subject to minimum substrate and product inhibition. In constrast, the values of turnover number (k_{cat}) and catalytic efficiency (k_{cat}/K_m) should be high to allow high rate of product formation. Specificity of an enzyme for a substrate is also important, because for general-purpose catalysis, enzymes with broader specificity would prove useful, although enzymes with narrow specificity may be more efficient.

Table. Definitions of Some Parametres used for Describing Key Properties of Enzymes

Parametre	Definition
T_m (melting temp)	Temperature at which 50% denaturation of enzyme occurs (determined by fluorescence, circular dichroism (CD) or calorimetry)
T_{opt} (optimum temp)	Temperature for optimum enzyme activity for a given set of reaction conditions
K_s and K_p (dissociation for substrate and product)	Measure of dissociation; low values cause constants inhibition
K_{cat} (turnover number or rate constant)	Maximum number of substrate molecules converted to product per unit time per active site; a function of rate constants for the conversion of enzyme-substrate (ES) complex to enzyme-product complex
K_M (Michaelis constant)	Represents dissociation of the ES complex, indicating substrate binding to active site
K_{cat}/K_M (specificity constant)	Indicates rate of association of enzyme and substrate and also the specificity for competing substrates

Medium for Enzymes Aqueous vs Organic Solvent

Most enzymes function in their natural aqueous media, which limit their use in industry for the production of speciality chemicals and polymers, for which organic solvents are the more suitable reaction media, provided the enzyme can function in such a medium. Normally, the enzymes get denatured in aqueous-organic mixture, leading many to believe that they will do so in pure organic solvents also. However, it was shown during 1985-2000, that various crystalline enzymes retain their structure even in anhydrous solvents. In fact, some enzymatic reactions work better in organic solvents. Sometimes, enzymes may be more stable and specific to the organic solvents, thus providing opportunities for their use in industries, where aqueous solvents become a limitation.

There are several factors, which facilitated the use of organic solvents in enzymatic reactions:

- Transesterification (due to addition of alcohol), aminolysis (due to addition of amines) and thiotransesterification (due to addition of thiols) are suppressed in aqueous solution and are important transformations that are needed in industry.
- Hydrophobic solvents are superior to hydrophilic solvents, because the latter will strip from the enzyme molecules, the tightly bound water (which is essential for catalytic activity).
- No specific pH is needed, while using organic solvents.
- The enzymes in organic solvents can be lyophilized to give enzyme powder without any damage to the enzyme activity. In brief we can say that strategies are being developed to have more efficient enzymes in organic solvents, their activities sometimes being comparable to that in aqueous medium.

Selectivity of Enzymes

One of the important properties of enzymes is to select from a mixture (racemic pairs, prochiral or meso-compounds or diastereometric mixtures), a compound, which is enantiomerically pure (enantiomers are a class of isomers that are mirror images of each other and occur in mixtures).

This is described as enzymatic resolution of substrates and can be achieved by anyone of the following three methods:

1. The enzyme esterifies only a single enantiomer of a racemic substrate and thus allows its separation. Enol esters can be used as transesterification reagents to force the enzymatic process to proceed irreversibly in the desired forward direction.
2. Enzymatic resolution has also been realised by certain hydrolytic enzymes through the use of additional racemization catalysts, which keep on converting the enantiomer that is not a substrate for the enzyme into the enantiomer, which is the substrate. In situ racemization thus gradually increases the overall concentration of the substrate recognized by the enzyme, leading to higher product yield.
3. Enzymatic resolution may involve efficient protecting-group strategies. For instance, some lipases have been shown to catalyse reactions that provide temporary protection of amines. Some examples of 'biocatalytic processes, which involve stereoselectivity of enzymes for enzymatic resolution.

Types of Enzymes

The enzymes that have been exploited for organic synthesis, and the reactions catalysed by each of these types are summarized in. Enzymes have been classified into the following six classes, and a standard nomenclature has been suggested for them by the 'Commission on Biochemical Nomenclature' of the 'International Union of Biochemistry. According to standard nomenclature, each enzyme is assigned' a four number code following the abbreviation EC (for Enzyme Commission). For instance, EC1.1.1.3 means an enzyme number 3 belonging to class 1 (oxidoreductase), subclass 1 and sub-subclass 1.

Oxidoreductases (Class 1)

Oxidoreductases facilitate redox reactions involving transfer of hydrogen or oxygen atoms between molecules. They include oxidases (e^- transfer to oxygen), oxygenases (oxygen atom transfer from molecular oxygen or O_2) prexidases (e^- transfer to peroxides) and dehydrogenases (hydride transfer). Glucose oxidase (EC1.1.3.4) is a popular example of this class of enzymes.

Transferases (Class 2)

Transferases catalyze the transfer of an atom (other than oxygen or hydrogen) or a group of atoms like acyl-, alkyl-, glycosyl- etc. For instance, glycosyl-transferases catalyse transfer of glycosyl group, sulphotransferases catalyse transfer of sulphate group and aminotransferases catalyse transfer of amino group. Aspartate aminotransferase (EC2.6.1.1) is a suitable example of this class of enzymes.

Hydrolases (Class 3)

Hydrolases catalyze hydrolytic reactions or their reverse reactions (condensation) and include esterases, amidases, lipases, proteases, glycosidases, etc. Amidases also help in peptide synthesis (by condensation of acid and amino groups) and esterases also help in the formation of esters (by' condensation between acid and alcohol). Similarly, glycohydrolases are used for the synthesis of glycoproteins. Most of these hydrolytic enzymes are important for' the pharmaceutical industry. Chymosin or rennin (EC3.2.3.4) is a suitable example of this class of enzymes.

Lyases (Class 4)

Lyases catalyze reactions involved in the removal of a group of atoms from the substrate molecule. These enzymes include aldolases, decarboxylases, dehydratases and some pectinases. Histidine ammonia lyase (EC 3.2.1.3) is a suitable example of this class.

Isomerases (Class 5)

Isomerases catalyze the formation f isomers of substrate molecules and therefore, include epimerases, racemases, and intramolecular transferases. Xylose isomerase (EC5.3.1.5) is a suitable example of this class.

Ligases or Synthetases (Class 6)

Ligases catalyze the formation of covalent bonds between two substrate molecules with the help of ATP or GP molecules, which provide the energy for the reaction. Glutathione synthase (EC6.3.2.3) is a suitable example of this class.

Biomimetic Catalysts

Biomimetic catalysts are the synthetic organic or inorganic compounds which perform the same function as the natural enzymes. These biomimetic catalysts may operate under harsher solvent and temperature conditions and may be more robust in terms of lifetime, even if they may be unsuitable for high throughput. Some of the processes that can be catalyzed by biomimetic catalysts include macrocyclization of natural products, production of analogues of hydrogenase or nitrogenase catalysts or the photosynthetic splitting of water.

However, sometimes natural enzymes should be preferred over synthetic reagents and catalysts, as in case of the following:

- Lipases and other hydrolases, have an advantage in kinetic resolutions of intermediates,
- Peniallin acylases are useful in β-lactam antibiotic industry
- Aldolases and transketolases help in the synthesis of oligosaccharides due to aldol condensation.

Table. Enzyme Commonly Used in Organic Synthesis

Enzymes	Reactions
1. Esterases, lipases	Ester hydrolysis, ester formation
2. Amidases (proteases, acylases)	Amide hydrolysis, amide formation
3. Dehydrogenases	Oxidoreduction of alcohols and ketones
4. Oxidases	Oxidation

5. Peroxidases	Oxidation, epoxidation, halohydration
6. Kinases	Phosphorylation (ATP dependent)
7. Aldolases, transketolases	Aldol reaction (C-C bond)
8. Glycosidases, glycosyltransferases	Glycosidic bond formation
9. Phosphorylases, phosphatases	Formation and hydrolysis of phosphate
10. Sulphotransferases	Formation of sulphate esters
11. Transaminases	Amino acid synthesis (C-N band)
12. Hydrolases	Hydrolysis
13. Isomerases, lyases, hydratases	Isomerization, addition, elimination, or replacement of a group

Extremozymes

The use of biocatalysis in chemical reactions is generally restricted to mild conditions of pH, temperature, pressure and aquatic medium. However, enzymes isolated from different living organisms inhabiting ecosystems with extreme conditions of pH, temperature, pressure and solvent, are not constrained with these mild conditions needed for majority of enzymes.

These enzymes, therefore, provide information to improve stability, activity and specificity of many enzymes used for industrial biocatalysis. It is now known that enzymes can function at temperatures as high as 140°C and as low as below freezing point. Organisms living in these conditions are described as extremophiles and the enzymes that function under these conditions are therefore described as extremozymes. Most of these extremophiles belong to archaea group of bacteria.

Extremozymes are broadly classified as follows:

- Extremely psychophilic enzymes, which function at temperatures approaching the freezing point of water; for instance, subtilisin S41 from the psychophile Bacillus S41 differs from other subtilisins in several respects, suggesting structural differences between these psychophilic enzymes and those extracted from mesophiles.
- Extremely halophilic enzymes function better in salt solutions. For instance, aspartate aminotransferase from Haloferas mediterranes is inactivated rapidly at room temperature in low salt solutions, but does not denature even at 78.5°C in 3.3M KCI suggesting that the enzyme is more stable in higher salt concentration. A protease from Halobacterium halobium is also similarly affected.
- Extremely thermophilic enzymes function at high temperatures and several examples of these enzymes are

available now. (DNA polymerase from Thermus aquaticus and Pyrococcus furiosus used for PCR provide one such example).

- Extremely barophilic enzymes have been isolated from microbes living in deep sea which are both barophils and psychophiles.

A number of genes encoding extremozymes (*e.g.* ferredoxin, DNA polymerase, amylase, dehydrogenase, etc.) have been isolated from thermophilic archaea (*P. furiosus, P. woesii* and *Thermococcus littoralis*), cloned and expressed in *E. coli*. Three dimensional structures of some extremozymes have provided opportunities to provide or design more stable versions of enzymes, that can function in extreme salinity, pH, temperatures and non-aqueous media. Selectivity of extremozymes (enantioselectivity, regioselectivity and chemosel-ectivity) can also be adjusted through judicious use of particular enzymes in organic solvents or water/solvent mixtures.

Cofactor-Dependent Enzymes and Cofactor Regeneration

A number of enzymatic reactions require cofactors, some of the common inorganic cofactors being iron/sulphur clusters and metalloporphyrins. These cofactors are costly and therefore can not be used in industry, unless regenerated as a reaction by-product.

Some of the cofactors which can be recycled effectively include the following:

- Hucleoside triphosphates, such as ATP in phosphoryl-transfer reactions,
- Nicotinamide adenine dinucleotide (NAD) and its 3' phosphate (NADP) in oxidoreductions,
- Acetyl CoA in acyl transfer reactions,
- 3'-phosphoadenosine- 5'- phosphosulphate (PAPS) in the formation l of sulphate esters
- Sugar nucleotides in glycosyl transfer reactions. Many cofactor dependent reactions involving regeneration of cofactor have been used for industry. However regeneration of cofactor may require additional enzymes, whose number may be reduced by using bifunctional enzymes.

Modular Enzymes

A modular enzyme is a multicomponent system or an assembly line having several component enzyme domains (also called modules). The order of enzymatic domains or modules in these multicomponent systems specifies which monomer substrates are activated, condensed and elongated. Therefore, the individual domains or components can be interchanged or replaced with functionally distinct analogues from related systems, inorder to obtain products with altered desired structure. The polyketide synthase (PKS) and non-ribosomal peptide synthetase (NRPS) are two such multicomponent systems, which can be modified by gene shuffling to provide designed diversity in complex natural products (*e.g.* antibiotics like penicillins, vancomycin and erythromycin).

Following three general classes of modular biocatalysts can be identified:

1. Enzymes in which catalysis and substrate specifities are separable,
2. Multi substrate enzymes, in which binding sites for individual substrates are modular,
3. Enzymes, where an intermediate is transferred from one active site to another in a metabolic sequence without equilibrating with the bulk fluid. For more details on modular enzymes, the readers are advised to consult an advanced book on enzymes.

Coenzymes

The chemical reactivity and specificity of an enzyme largely depends on its catalytic site, made up of the side chains of chiral amino acids (glycine is the only non-chiral amino acid, the remaining 19 amino acids are chiral in nature). However, sometimes the side chains of chiral amino acids are insufficient and the enzyme may need a non-polypeptide molecule to bind on its surface thus enabling it to function as an enzyme. These non-polypeptide molecules serve as ligands and are called coenzymes. The coenzymes are therefore defined as small molecules that are associated with enzymes, and are essential for the activity of enzymes. Some of these coenzymes are involved in group transfers.

Table. Some Coenzymes Involved in Group Transfer Reactions

Coenzymes	Group transferred
ATP	Phosphate
NADH, NADPH	Hydrogen and electron
Coenzyme A	Hydride ion = H+, acetyl
Biotin	Carboxyl
S – adenosylmethionine	Methyl

Chapter 10

Enzymes in Medical Diagnosis

LACTATE DEHYDROGENASE

Lactate dehydrogenase (LDH or LD) is an enzyme present in a wide variety of organisms, including plants and animals. Lactate dehydrogenases exist in four distinct enzyme classes. Two of them are cytochrome c-dependent enzymes with each acting on either D-lactate or L-lactate. The other two are NAD(P)-dependent enzymes with each acting on either D-lactate or L-lactate. This article is about the NAD(P)-dependent L-lactate dehydrogenase.

REACTIONS

Lactate dehydrogenase catalyzes the interconversion of pyruvate and lactate with concomitant interconversion of NADH and NAD^+. It converts pyruvate, the final product of glycolysis to lactate when oxygen is absent or in short supply, and it performs the reverse reaction during the Cori cycle in the liver. At high concentrations of lactate, the enzyme exhibits feedback inhibition and the rate of conversion of pyruvate to lactate is decreased. It also catalyzes the dehydrogenation of 2-Hydroxybutyrate, but it is a much poorer substrate than lactate. There is little to no activity with beta-hydroxybutyrate.

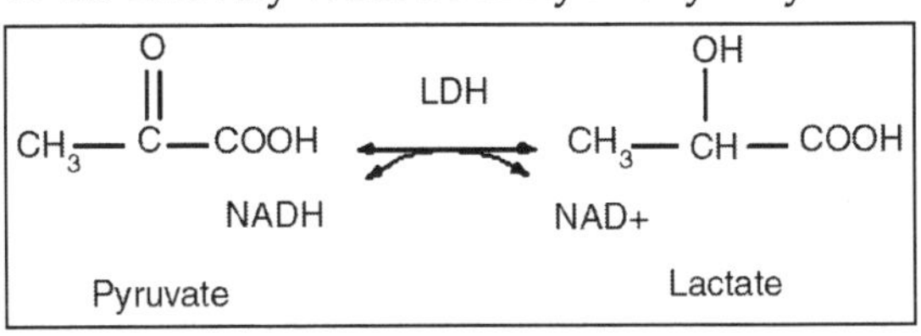

Fig. Catalytic Function of LDH

Enzyme Isoforms

Functional lactate dehydrogenase are homo or hetero tetramers composed of M and H protein subunits encoded by the LDHA *and* LDHB *genes respectively:*

- LDH-1 (4H) - in the heart and RBCs
- LDH-2 (3H1M) - in the reticuloendothelial system
- LDH-3 (2H2M) - in the lungs
- LDH-4 (1H3M) - in the kidneys, placenta and pancreas
- LDH-5 (4M) - in the liver and striated muscle

The five isozymes that are usually described in the literature each contain four subunits. The major isozymes of skeletal muscle and liver, M_4, has four muscle (M) subunits; while H_4 is the main isozymes for heart muscle in most species, containing four heart (H) subunits. The other variants contain both types of subunits.

Usually LDH-2 is the predominant form in the serum. A LDH-1 level higher than the LDH-2 level (a "flipped pattern"), suggests myocardial infarction (damage to heart tissues releases heart LDH, which is rich in LDH-1, into the bloodstream). The use of this phenomenon to diagnose infarction has been largely superseded by the use of Troponin I or T measurement.

Genetics in Humans

The M and H subunits are encoded by three different genes:

- The M subunit is encoded by *LDHA*, located on chromosome 11p15.4 (Online 'Mendelian Inheritance in Man' (OMIM) 150000)
- The H subunit is encoded by *LDHB*, located on chromosome 12p12.2-p12.1 (Online 'Mendelian Inheritance in Man' (OMIM) 150100)
- A third isoform, *LDHC* or *LDHX*, is expressed only in the testis (Online 'Mendelian Inheritance in Man' (OMIM) 150150); its gene is likely a duplicate of *LDHA* and is also located on the eleventh chromosome (11p15.5-p15.3)

Mutations of the M subunit have been linked to the rare disease *exertional myoglobinuria* and mutations of the H subunit have been described but do not appear to lead to disease.

Medical use

Tissue breakdown elevates levels of LDH, and therefore a measure of it indicates, *e.g.*, hemolysis. Other disorders indicated by elevated LDH include cancer, meningitis, encephalitis, acute pancreatitis, and HIV.

Hemolysis

In medicine, LDH is often used as a marker of tissue breakdown as LDH is abundant in red blood cells and can function as a marker for hemolysis. A blood sample that has been handled incorrectly can show false-positively high levels of LDH due to erythrocyte damage.

It can also be used as a marker of myocardial infarction. Following a myocardial infarction, levels of LDH peak at 3-4 days and remain elevated for up to 10 days. In this way, elevated levels of LDH (where the level of LDH1 is higher than that of LDH2) can be useful for determining if a patient has had a myocardial infarction if they come to doctors several days after an episode of chest pain.

Tissue Turnover

Other uses are assessment of tissue breakdown in general; this is possible when there are no other indicators of hemolysis. It is used to follow-up cancer (especially lymphoma) patients, as cancer cells have a high rate of turnover with destroyed cells leading to an elevated LDH activity.

Exudates and Transudates

Measuring LDH in fluid aspirated from a pleural effusion (or pericardial effusion) can help in the distinction between exudates (actively secreted fluid, *e.g.* due to inflammation) or transudates (passively secreted fluid, due to a high hydrostatic pressure or a low oncotic pressure). The usual criterion is that a ratio of fluid LDH versus upper limit of normal serum LDH of more than 0.6 or T! indicates an exudate, while a ratio of less indicates a transudate. Different laboratories have different values for the upper limit of serum LDH, but examples include 200 and 300 IU/L. In empyema, the LDH levels, in general, will exceed 1000 IU/L.

Meningitis and Encephalitis

High levels of lactate dehydrogenase in cerebrospinal fluid are often associated with bacterial meningitis. In the case of viral meningitis, high LDH in general indicates the presence of encephalitis and poor prognosis.

HIV

LDH is often measured in HIV patients as a non-specific marker for pneumonia due to *Pneumocystis jiroveci* (PCP). Elevated LDH in the setting of upper respiratory symptoms in an HIV patient suggests, but is not diagnostic for, PCP. However, in HIV-positive patients with respiratory symptoms, a very high LDH level (>600 IU/L) indicated histoplasmosis (9.33 more likely) in a study of 120 PCP and 30 histoplasmosis patients.

MALATE DEHYDROGENASE

Malate dehydrogenase (EC 1.1.1.37) (MDH) is an enzyme in the citric acid cycle that catalyzes the conversion of malate into oxaloacetate (using NAD^+) and vice versa (this is a reversible reaction). Malate dehydrogenase is not to be confused with malic enzyme, which catalyzes the conversion of malate to pyruvate producing NADPH. Malate dehydrogenase is also involved in gluconeogenesis, the synthesis of glucose from smaller molecules. Pyruvate in the mitochondria is acted upon by pyruvate carboxylase to form oxaloacetate, a citric acid cycle intermediate. In order to get the oxaloacetate out of the mitochondria, malate dehydrogenase reduces it to malate, and it then traverses the inner mitochondrial membrane. Once in the cytosol, the malate is oxidized back to oxaloacetate by cytosolic malate dehydrogenase. Finally, phosphoenol-pyruvate carboxy kinase (PEPCK) converts oxaloacetate to phosphoenol pyruvate.

Isozymes

Several isozymes of malate dehydrogenase exist, depending on where they are localized in the cell and their specific dependence on NAD+ or NADP+ (only in chloroplasts). There are two main isoforms in eukaryotic cells. One is found in the mitochondrial matrix participating as a key enzyme in the citric acid cycle that catalyzes the oxidation of malate. The other is found in the cytoplasm, assisting the malate-aspartate shuttle with exchanging reducing equivalents so that malate can pass through the mitochondrial membrane to be transformed into oxaloacetate for further cellular processes.

Humans and most other mammals express the following two malate dehydrogenases:

Table. Malate Dehydrogenase 1, NAD (Soluble)

Identifiers	
Symbol	MDH1
Entrez	4190
HUGO	6970
OMIM	154200
RefSeq	NM_005917
UniProt	P40925
Other data	
EC number	1.1.1.37
Locus	Chr. 2 p23

Table. Malate Dehydrogenase 2, NAD (Mitochondrial)

Identifiers	
Symbol	MDH2
Entrez	4191
HUGO	6971
OMIM	154100
RefSeq	NM_005918
UniProt	P40926
Other data	
EC number	1.1.1.37
Locus	Chr. 7 cen-q22

Evolution and Structure

In most organisms, malate dehydrogenase exists as a homodimeric molecule and is closely related to lactate dehydrogenase in structure. It is a large protein molecule with subunits weighing between 30 and 35 kDa. Based on the amino acid sequences, MDH has seemingly diverged into two main phylogenetic groups that either closely resembles mitochondrial isozymes or cytoplasmic/chloroplast isozymes. Because the sequence identity of malate dehydrogenase in the mitochondria is more closely related to its prokaryotic ancestors in comparison to the cytoplasmic isozyme, this coincides with the theory that mitochondria and chloroplasts were developed through endosymbiosis. Interestingly, the amino acid sequences of archaeal MDH are more similar to that of LDH than that of MDH of other organisms. This indicates that there is a possible evolutionary linkage between lactate dehydrogenase and malate dehydrogenase.

Each subunit of the malate dehydrogenase dimer has two distinct domains that vary in structure and functionality. A parallel â-sheet structure makes up the domain, while four â-sheets and one á-helix comprise the central NAD+ binding site. The subunits are held together through extensive hydrogen-bonding and hydrophobic interactions.

Mechanism and Activity

The active site of malate dehydrogenase is a hydrophobic cavity within the protein complex that has specific binding sites for the substrate and its coenzyme, NAD+. In its active state, MDH undergoes a conformational change that encloses the substrate to minimize solvent exposure and to position key residues in closer proximity to the substrate. The three residues in particular that comprise a catalytic triad are histidine (His-195) and aspartate (Asp-168), which work together as a proton transfer system, and arginines (Arg-102, Arg-109, Arg-171) which secure the substrate. Kinetic studies show that MDH enzymatic activity is ordered. NAD+/NADH is bound before the substrate.

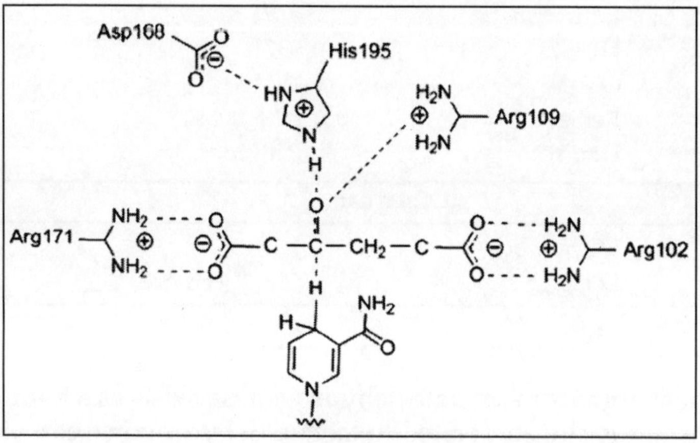

Fig. Active Site of Malate Dehydrogenase

Allosteric Regulation

Because malate dehydrogenase is closely tied to the citric acid cycle, regulation is highly dependent on TCA products. High malate concentrations stimulate MDH activity and conversely, high oxaloacetate concentrations inhibit the enzyme. Citrate can both allosterically activate and inhibit the enzymatic activity of MDH. It inhibits the oxidation of malate when there are low levels of malate and NAD+. However, in the presence of high levels of malate and NAD+, citrate can stimulate the production of oxaloacetate. It is believed that there is an allosteric regulatory site on the enzyme where citrate can bind to and drive the reaction equilibrium in either direction.

CITRIC ACID CYCLE

The citric acid cycle — also known as the tricarboxylic acid cycle (TCA cycle), the Krebs cycle, or the Szent-Györgyi-Krebs cycle, — is a series of enzyme-catalysed chemical reactions, which is of central importance in all living cells that use oxygen as part of cellular respiration. In eukaryotic cells, the citric acid cycle occurs in the matrix of the mitochondrion. The components and reactions of the citric acid cycle were established by seminal work from Albert Szent-Györgyi and Hans Krebs. In aerobic organisms, the citric acid cycle is part of a metabolic pathway involved in the chemical conversion of carbohydrates, fats and proteins into carbon dioxide and water to generate a form of usable energy. Other relevant reactions in the pathway include those in glycolysis and pyruvate oxidation before the citric acid cycle, and oxidative phosphorylation after it. In addition, it provides precursors for many compounds including some amino acids and is therefore functional even in cells performing fermentation.

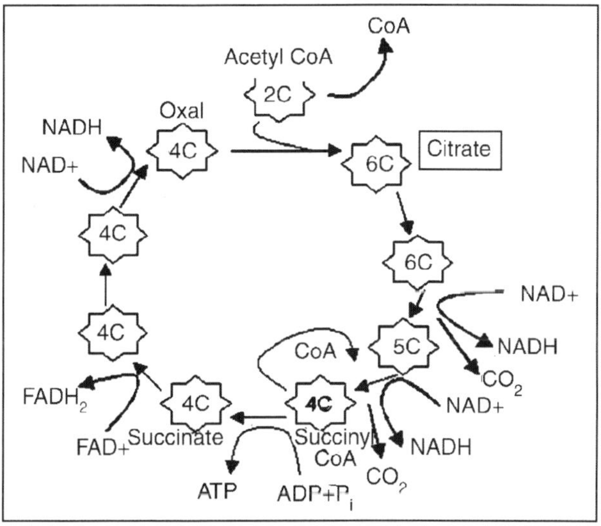

Fig. Citric Acid Cycle

A Simplified View of the Process

- The citric acid cycle begins with the transfer of a two-carbon acetyl group from acetyl-CoA to the four-carbon acceptor compound (oxaloacetate) to form a six-carbon compound (citrate).
- The citrate then goes through a series of chemical transformations, losing two carboxyl groups as CO2. The carbons lost as CO2 originate from what was oxaloacetate, not directly from acetyl-CoA. The carbons donated by acetyl-CoA become part of the oxaloacetate carbon backbone after the first turn of the citric acid cycle. Loss of the acetyl-CoA-donated carbons as CO2 requires several turns of the citric acid cycle. However, because of the role of the citric acid cycle in anabolism, they may not be lost, since many TCA cycle intermediates are also used as precursors for the biosynthesis of other molecules.
- Most of the energy made available by the oxidative steps of the cycle is transferred as energy-rich electrons to NAD+, forming NADH. For each acetyl group that enters the citric acid cycle, three molecules of NADH are produced.

- Electrons are also transferred to the electron acceptor Q, forming QH2.
- At the end of each cycle, the four-carbon oxaloacetate has been regenerated, and the cycle continues.

Steps

Two carbon atoms are oxidized to CO2, the energy from these reactions being transferred to other metabolic processes by GTP (or ATP), and as electrons in NADH and QH2. The NADH generated in the TCA cycle may later donate its electrons in oxidative phosphorylation to drive ATP synthesis; FADH2 is covalently attached to succinate dehydrogenase, an enzyme functioning both in the TCA cycle and the mitochondrial electron transport chain in oxidative phosphorylation.

FADH2, therefore, facilitates transfer of electrons to coenzyme Q, which is the final electron acceptor of the reaction catalyzed by the Succinate:ubiquinone oxidoreductase complex, also acting as an intermediate in the electron transport chain.

The citric acid cycle is continuously supplied with new carbon in the form of acetyl-CoA, entering at step 1 below.

	Substrates	Products	Enzyme	Reaction type	Comment
1	Oxaloacetate + Acetyl CoA + H_2O	Citrate + CoA-SH	Citrate synthase	Aldol condensation	Rate-limiting stage (irreversible), extends the 4C oxaloacetate to a 6C molecule
2	Citrate	*cis*-Aconitate + H_2O	Aconitase	Dehydration	Reversible isomerisation
3	*cis*-Aconitate + H_2O	Isocitrate		Hydration	
4	Isocitrate + NAD^+	Oxalosuccinate + NADH + H^+	Isocitrate dehydrogenase	Oxidation	Generates NADH (equivalent of 2.5 ATP)
5	Oxalosuccinate	α-Ketoglutarate + CO_2		Decarboxylation	Iirreversible stage, generates a 5C molecule
6	α-Ketoglutarate + NAD^+ + CoA-SH	Succinyl-CoA + NADH + H^+ + CO_2	α-Ketoglutarate dehydrogenase	Oxidative decarboxylation	Irreversible stage, generates NADH (equivalent of 2.5 ATP), regenerates the 4C chain (CoA excluded)
7	Succinyl-CoA + GDP + P_i	Succinate + CoA-SH + GTP	Succinyl-CoA synthetase	substrate-level phosphorylation	or ADP$\rightarrow$ATP instead of GDP$\rightarrow$GTP, generates 1 ATP or equivalent
8	Succinate + ubiquinone (Q)	Fumarate + ubiquinol (QH_2)	Succinate dehydrogenase	Oxidation	Uses FAD as a prosthetic group (FAD$\rightarrow FADH_2$ in the first step of the reaction) in the enzyme, generates the equivalent of 1.5 ATP
9	Fumarate + H_2O	*L*-Malate	Fumarase	H_2O addition (*hydration*)	
10	*L*-Malate + NAD^+	Oxaloacetate + NADH + H^+	Malate dehydrogenase	Oxidation	Generates NADH (equivalent of 2.5 ATP)

Mitochondria in animals, including humans, possess two succinyl-CoA synthetases: one that produces GTP from GDP, and another that produces ATP from ADP. Plants have the type that produces ATP (ADP-forming succinyl-CoA synthetase). Several of the enzymes in the cycle may be loosely-associated in a multienzyme protein complex within the mitochondrial matrix.

Alkaline Phosphatise

Alkaline phosphatase [orthophosphoric-monoester phosphohydrolase (alkaline optimum), EC 3.1.3.1] from human placenta is a membrane-bound dimeric protein anchoring to the lipid bilayer through a phosphatidylinositol–glycan moiety. The enzyme catalyses the hydrolysis of the phosphoester linkage of a wide range of substrates. Depending on their primary sequences, kinetic properties and antigenicity differences, the mammalian alkaline phosphatases are divided into four groups: the tissue non-specific (bone-liver-kidney), the intestinal, the placental (Regan), and the placental-like (Nagro) isoenzymes. These isoenzymes have attracted much interest because of their clinical association with many pathological conditions. Less information is available for the mammalian enzymes with respect to structure and catalytic function when compared with the enzyme from *Escherichia coli*.

The three-dimensional structure of *E. coli* alkaline phosphatase was elucidated by X-ray crystallographic analysis more than a decade ago and was achieved to 2.0 Å (10^{-10} m) resolution. The amino-acid sequences of the mammalian enzymes were only 25–30% similar to that of the *E. coli* enzyme. However, computer-modelling results indicate that the mammalian enzymes retain the core of the three-dimensional structure of the *E. coli* enzyme especially at the putative active-site region. The reaction mechanism of all variants of alkaline phosphatase is the same irrespective of tissue source or species.

The reaction catalysed by the *E. coli* enzyme was suggested to follow a dissociative mechanism, called S_N1(P). The mechanism involved an initial dissociative loss of the alcoholic leaving group to generate a putative metaphosphate anion intermediate, which is then captured by a nucleophile, a zinc-activated serine residue, subsequently forming the phosphoryl–enzyme intermediate. Hydrolysis of the phosphoryl–enzyme by water or transfer of the phosphoryl group to an alcoholic acceptor completed the reaction cycle. The rate-limiting step was the hydrolysis of phosphoryl–enzyme intermediate at acidic pH, but, at basic pH, changed to release of enzyme-bound inorganic phosphate. Han and Coleman suggested that the mechanism depends on the pK_a value of the leaving group. When the pK_a value of the leaving group exceeds 15, phosphorylation becomes rate-limiting and the mechanism is primarily associative in character.

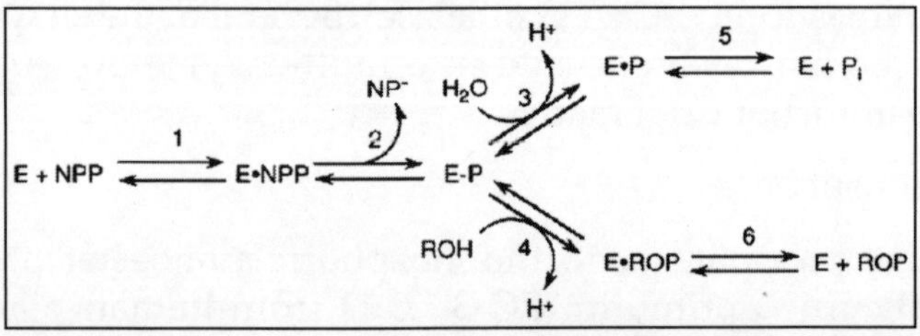

Fig. Reaction Mechanism of Human Placental Alkaline Phosphatase

Abbreviations: E, human placental alkaline phosphatase; NPP, 4-nitrophenyl phosphate; NP^- 4-nitrophenolate anion; P_i, inorganic phosphate; ROH, a phosphoryl-group-accepting alcohol. The non-covalent association is represented by a dot and the covalent chemical bond is represented by a dash.

Step 1: Association of substrate with the enzyme forming a Michaelis complex;

Step 2: Phosphorylation of enzyme serine residue;

Step 3: Hydrolysis of phosphoryl–enzyme;

Step 4: Transfer of phosphoryl group from phosphoryl–enzyme to an alcohol receptor;

Step 5: Dissociation of enzyme-bound phosphate;

Step 6: Dissociation of enzyme-bound phosphorylated alcohol.

Enzyme assay

Assay of the alkaline-phosphatase-catalysed hydrolysis of 4-nitrophenyl phosphate (NPP) in aqueous solution was performed as described previously. The enzyme activity in reverse micelles was assayed by the injection method. All concentrations specified in this study are the final concentration in terms of total volume and not the less definable volume enclosed in reverse-micellar vesicles. The $[H_2O]$/[AOT] ratio (w_o) employed was 22.6, which represented an inclusive volume of approx. 228 nm^3 of the reverse micelles with a core radius of 3.79 nm. This volume is approximately equal to a dimeric enzyme molecule (227 nm^3). The enzyme molecules, therefore, could be regarded as being immobilized in the membranous micellar structure.

The reaction mixture contained 2.5 mM AMP/HCl buffer (pH 8.51–11.76), 20 μM or 2 mM NPP, 66.7 mM AOT and 1.01 μg of alkaline phosphatase in a total volume of 3 ml. The liberation of 4-nitrophenol was monitored continuously at an absorbance of 410 nm and at 30 °C. A molar absorption coefficient of 18.3×10^3 $M^{-1} \cdot cm^{-1}$ for 4-nitrophenol was used in calculations. The dimeric enzyme has an M_r of 115000 and, at A_{280}, the enzyme solution has an absorption coefficient of 115×10^3 $M^{-1}.cm^{-1}$.

Viscosity Effect

Kinetic parametres were determined at 30 °C in water containing various amounts of glycerol (0–35%, w/v) or sucrose (0–36%, w/v) as viscogens. Assays typically contained 50 mM Na_2CO_3/$NaHCO_3$ buffer (pH 9.8) or 2.5 mM AMP/HCl buffer (pH 9.8), 10 μM or 10 mM NPP and 1.01 μg or 2.02 μg of enzyme. The initial enzyme activities measured at high substrate concentrations and v/[S] at low substrate concentrations were used to estimate the k_{cat} and k_{cat}/K_m respectively.

pH Studies

Our previous work has demonstrated that Tris/HCl and carbonate buffers caused quite different effects on k_{cat} or k_{cat}/K_m of the enzyme. To explore further the pH effect on the reaction rate of the enzyme, a phosphoryl-group-accepting buffer (AMP) was used for the entire pH range of our experiments (pH 8.51–11.76, a range over which the enzyme is stable during the experimental period).

All values are means±S.E.M. the hydronation of an ES group with an apparent pK_a value of 9.78±0.6 was detected in aqueous solution. In reverse micelles, however, the data can be optimally fitted to a model that involves two hydronated sites with apparent pK_a values of 9.01±0.15 and 9.86±0.11 respectively. Both groups have to be dehydronated to give maximum reaction rate. The curves show a normal bell-shape at the acidic region. However, the curves level off at greater pH values, thereby conforming to a mechanism having two V_{max} values (V_{max} and V_{max}') because E^nS and $E^{n-1}S$ are both enzymically active where V_{max}' is equal to gV_{max}. The g value was found to be 0.007 in aqueous solution and 0.004 in reverse micelles. The g value could be > 1 with a different buffer system.

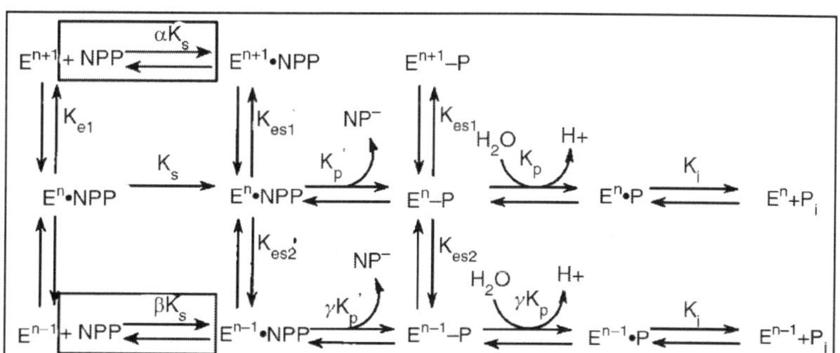

Fig. pH Studies of Enzymes

Scheme 2 Proposed kinetic mechanism of human placental alkaline-phosphatase-catalysed reaction. Assuming that binding of NPP with the enzyme is rapid in comparison with subsequent covalent bond formation or breakdown and that a = b = 1 for rapid equilibrium of substrate and H^+, the binding processes shown in the grey rectangles are not considered.

Equation does not distinguish between the two covalent steps, phosphorylation and dephosphorylation.

Table: Summary of the pK_a Values from the pL Dependence of Kinetic Parametres.

Solvent system	Medium	pL range	Kinetic parametres	Eqn. fitted	pK_a Values and kinetic constants
Aqueous	H_2O	8.51–11.76	V/K	3	pK_{e1} = 9.10±0.52, pK_{e2} = 9.80±0.40, $(V/K)_{lim}$ = 0.2±0.17 $s^{-1}\cdot\mu M^{-1}$
	^2H_2O	8.88–12.42	V/K	3	pK_{e1} = 9.44±0.8, pK_{e2} = 10.62±0.21, DpK_{e1} = 0.34±0.06, DpK_{e2} = 0.82±0.03, $(V/K)_{lim}$ = 0.028±0.003 $s^{-1}\cdot\mu M^{-1}$, $^{2H}(V/K)_{lim}$ = 7.1±3.4
	H_2O	8.51–11.76	V	1	pK_{es1} = 9.78±0.6, V_{lim} = 6.46±0.54 s^{-1}
	^2H_2O	8.88–12.42	V	2	pK_{es1} = 10.27±0.4, DpK_{es1} = 0.49±0.02, V_{lim} = 3.2±0.15 s^{-1}, $^{2H}V_{lim}$ = 2.02±0.13
Reverse micelles	H_2O	8.51–11.76	V/K	3	pK_{e1} = 8.07±0.73, pK_{e2} = 11.23±0.71, $(V/K)_{lim}$ = 1.72±2.48 $s^{-1}\cdot\mu M^{-1}$
	^2H_2O	8.88–12.42	V/K	3	pK_{e1} = 8.78±0.62, pK_{e2} = 12.17±0.61, DpK_{e1} = 0.71±0.06, DpK_{e2} = 0.94±0.05, $(V/K)_{lim}$ = 1.61±2.22 $s^{-1}\cdot\mu M^{-1}$, $^{2H}(V/K)_{lim}$ = 1.07±1.51

H_2O	8.51–11.76	V	1	pK_{es1} = 9.01±0.15, pK_{es2} = 9.86±0.11, V_{lim} = 3.87±0.38 s^{-1}
^2H_2O	8.88–12.42	V	2	pK_{es1} = 9.71±0.22, pK_{es2} = 10.48±0.24, V_{lim} = 4.61±0.21 s^{-1} DpK_{es1} = 0.7±0.01, DpK_{es2} = 0.62±0.01, $^{2H}V_{lim}$ = 0.84±0.06

All values are shown as mean±S.E.M. (n = 15 for H_2O and n = 17 for ^2H_2O).

Solvent Kinetic Isotope Effect

When the experiments described above were performed in deuterium medium, similar results were obtained with apparent pK_a values shifted to the basic region. In aqueous solution, the pK_a value of the amino acid residue involved in catalysis was shifted by 0.49±0.02 unit to give a value of 10.27±0.4.

Two apparent pK_a values were also detected in reverse micelles with DpK_{es1} and DpK_{es2} equal to 0.7±0.01 and 0.62±0.01 respectively. The pH versus log V/K profiles in deuterium medium were similar to those in water.

The DpK_{e1} and DpK_{e2} were shifted by 0.34±0.06 and 0.82±0.03 unit in aqueous solution and 0.71±0.06 and 0.94±0.05 unit in reverse micelles respectively.

Based on the values of V_{lim} and $(V/K)_{lim}$, the pL-independent isotope effect values on V and V/K were calculated to be $^{2H}V_{lim}$ = 2.02±0.13 and $^{2H}(V/K)_{lim}$ = 7.1±3.4 in aqueous solution and $^{2H}V_{lim}$ = 0.84±0.06 and $^{2H}(V/K)_{lim}$ = 1.07±1.51 in reverse micelles.

These values were obtained at high and low substrate concentrations. To examine the detailed isotope effect at various pH values, we performed the substrate saturation experiment at three pL values (9.5, 10.4 and 11.0), which corresponded to regions having a large shift in dissociation constant, minimum effect on V/K and constant V respectively.

At pL 9.5, the isotope effect on V was similar in aqueous solution (3.16±0.09) or in reverse micelles (3.32±0.17). The isotope effect on V/K, on the other hand, was increased from 1.09±0.11 in aqueous solution to 3.43±0.22 in reverse micelles. At pL 10.4, the log V/K value is identical in water or deuterium. The deuterium isotope effect on V/K was almost identical in aqueous solution (1.14±0.04) to that in reverse micelles

(1.16±0.07). These results probably indicate small isotope effect on the phosphorylation step in both solvent systems.

Table: Solvent Kinetic Isotope Effect and Proton Inventory Effect of Human Placental Alkaline-phosphatase-catalysed Hydrolysis of NPP

pL	Solvent kinetic isotope effect $^{2H}(V/K)$	^{2H}V	$^{2H}K_m$	Fractional factor for deuterium (f^T)
		Aqueous system		
9.5	1.09±0.11	3.16±0.09	4.07±1.25	0.26±0.04 (0.53±0.03)
10.4	1.14±0.04	2.52±0.06	2.18±0.12	0.37±0.03 (0.62±0.02)
11.0	1.21±0.03	2.08±0.08	1.68±0.14	0.52±0.04 (0.73±0.03)
		Reverse-micellar system		
9.5	3.43±0.22	3.32±0.17	0.92±0.10	0.15±0.02 (0.44±0.02)
10.4	1.16±0.07	1.52±0.07	1.10±0.19	0.50±0.05 (0.70±0.03)
11.0	0.46±0.04	0.62±0.06	1.68±0.28	1.07±0.16 (1.03±0.08)

All values are means±S.E.M. ($n = 20$ in isotope effect determinations and $n = 9$ in proton inventory experiments at each pL value). Values in parentheses are the results fitting a two-hydrogenic-site model.

If both systems are compared, the log V values plateau only at pL 11. By analogy with the acylation and deacylation in the reaction mechanism of protease, the kinetic parametre k_{cat} has an effective reactant state as the phosphoryl enzyme. The effective transition state is that for the dephosphorylation process. The parametre k_{cat}/K_m has an effective reactant state as the free substrate and free enzyme. The effective transition state is that for the phosphorylation process.

The isotope effect on V/K and V at pL 11.0 thus reflects the effect on the phosphorylation and dephosphorylation steps respectively. Quite different results were obtained in aqueous or reverse micellar solution.

Normal isotope effects of 1.21±0.03 and 2.08±0.08 in an aqueous solution but an inverse isotope effect of 0.46±0.04 and 0.62±0.06 in reverse micelles were observed for $^{2H}(V/K)$ and ^{2H}V respectively.

In an aqueous solution, the K_{mNPP} in water was 80±20, 370±20 and 753±60 μM at pL 9.5, 10.4 and 11.0 respectively. These values changed to 19±7, 170±10 and 449±40 μM in deuterium. In reverse micelles, the K_{mNPP} differences between water and deuterium were always smaller than those in aqueous solution. At pL 9.5, 10.4 and 11.0, the K_{mNPP} values in water were 66±8, 116±22 and 285±60 μM respectively. These values changed to 72±7, 105±17 and 169±20 μM in deuterium.

The isotope effect on K_m was high in an aqueous solution but disappeared in reverse micelles. The dependency of $^{2H}K_m$ on pL was opposite in an aqueous solution and in reverse micelles. At pL 11.0, no difference was observed for $^{2H}K_m$ in these two solvent systems.

Proton Inventory Effect

The reaction rate was determined in H_2O–^2H_2O buffers with various atom fractions of deuterium at the above mentioned three pL values. When $v^2{}_H$ was plotted versus the mole fraction of deuterium (n), linear proton inventory plots were obtained at all pL values. It is not possible to distinguish between one-hydrogenic-site or two-hydrogenic-site models from the data. The deuterium fractionation factors in the transition state (f^T) for the simple one- or two-hydrogenic-site model were estimated. The f^T increases at larger pL values. Taking values of the one-hydrogenic site, the deuterium solvent kinetic isotope effects at pL 9.5, 10.4 and 11.0 were estimated to be 3.85, 2.70 and 1.92 in aqueous solution and 6.7, 2.0 and 0.93 in reverse micelles respectively. The data were also fitted to eqn. (10) to examine the possibility that multiple hydrogenic sites were involved. Large standard errors were obtained for the fitted data. The data were also not compatible with another linear (Gross–Butler) equation, $v^2{}_H = v_H/(1\text{-}n\text{+}nf)$, for the reactant state.

Viscosity Effect

In a study involving the isotope effects on malic enzyme, Karsten *et al.* compared the solvent isotope and viscosity effects and concluded that the inverse solvent isotope effects observed on malic enzyme are the result of the viscosity difference between water and deuterium. We therefore performed the viscosity experiment as a control. The second-order rate constant, k_{cat}/K_m, decreased exponentially and the first-order rate constant, k_{cat}, decreased linearly with increasing solvent viscosity irrespective of whether glycerol or sucrose was used as the viscogen. Above normal viscosity effects correlate with a diffusionally controlled physical step as the rate-limiting step in aqueous solution, similar to that found for the *E. coli* enzyme.

Chemical modification and site-directed mutagenesis results have identified the essential arginine, tyrosine, lysine and histidine residues in the active centre of human placental alkaline phosphatase. The enzyme has been observed to undergo phosphorylation–dephosphorylation during the catalytic cycle. Multiple amino-acid sequence alignments with the *E. coli* enzyme reveal that the putative active-site region of placental alkaline phosphatase is comprised of a conserved Asp^{91}-Ser^{92}-Ala^{93} sequence, the metal triplet (Zn1-Zn2-Mg) and Arg^{166} and other groups nearby. Mammalian alkaline phosphatases are 20- to 30-fold more active than the corresponding bacterial enzymes. Histidine-153 could be responsible for the higher activity of mammalian enzymes. We have constructed several mutant placental alkaline phosphatases and the specific activity of His^{153} Asp and His^{153} Ala is only 7.4% and 0.23%, respectively, of the wild-type value.

In reverse micelles, the magnesium ion induces a conformational change in the enzyme and a curved plot is obtained when the initial rate is measured, thereby complicating the initial rate analysis. For this reason we did not include magnesium ions in the assay mixture and the modulatory role of Mg^{2+} is not considered in the following discussion. If the two-metal-centre reaction mechanism for the *E. coli* enzyme is adopted, the mechanism involves an activation of serine-92 by the zinc ion at the M2 site, which results in a better nucleophile, serine-O^-, with which to attack the phosphorus atom of the substrate molecule, resulting in formation of a phosphoryl–enzyme intermediate. The zinc ion at the M1 site assists the alcoholic product to leave the active centre. Our experimental data on the pH and isotope effects studies may help in elucidating the chemical mechanism of the human placental alkaline-phosphatase-catalysed reaction.

The most significant result is the finding of an inverse isotope effect at pL 11.0 in reverse micelles.

An inverse isotope effect could be augmented by:

- The acid dissociation of a sulphydryl group,
- Solvent isotope effects due to the medium,
- A viscosity effect resulting from viscous deuterium when compared with water
- Dissociation of a metal-bound water molecule.

Of these possibilities, dissociation of sulphydryl group could be ruled out because the DpKa values are markedly larger than that reported for a sulphydryl group (0.2-0.3). The medium effect is characterized by the large number of hydrons involved, with each contributing a small isotope effect. The change in solvent viscosity due to deuterium results in an inverse viscosity effect. However, we observed a normal viscosity effect. Correction for the viscosity effects, if they really exist, will give an even larger inverse isotope effect.

None of the above provides a satisfactory explanation for the observed inverse isotope effects of placental alkaline phosphatase in reverse micelles.

The only explanation left is the dissociation of the metal-chelated water molecule. The fractionation factors for most metal-water complexes range from 0.7 to 1.0, which is very close to the value obtained in the present study. The inverse isotope effect observed in this investigation could be attributed to displacement of a Zn-bound water molecule upon binding of the substrate.

In general, pKa values deduced from the pH-log V/K plot reveal the dissociating groups of the free enzyme or substrate. The dissociation of two hydrons of the phosphate group (pKa1 = 6.8, pKa2 12) may not fall in the pH range 8.51-11.76. In alkaline phosphatase, the active site is a small pocket barely large enough to accommodate a phosphate group, thereby leaving the remainder of the substrate in the solvent region, which accounts for the broad substrate specificities of the enzyme. Two enzyme groups with pKa values of 9.10 and 9.80 were detected in the pH-log V/K plot. The former decreased to 8.07 and the latter increased to 11.23 in reverse micelles. These groups may participate in the phosphate binding. The most likely candidates in these groups are tyrosine-169, arginine-166 and the hydrogen-bonded water-459 and water-457. The latter water molecule, which co-ordinates Mg or hydrogen bonding to lysine-328 in the E. coli enzyme, has been proposed to play a role in hydron transfer. However, in placental alkaline phosphatase a histidine residue is substituted at position 328.

The difference between pKe1 and pKe2 was 0.7 in aqueous solution suggesting a negatively co-operative hydronation process, which indicates that the ionization of a group changes value depending on the ionic state of other groups. Each titrating group is the sum of the fractions of the molecular pKa values. In reverse micelles, the pKe2-pKe1 value increased to 3.16 indicating that the titration constants become effectively equal to the molecular pKa values.

The catalytic group detected in the pH-log V plot has a pKa value of 9.78. The identity of this group is not obvious. Neither Zn1 or Zn2 has an ancillary amino acid nucleophile to aid hydronation and dehydronation steps. However, water-465, which is near the Mg co-ordination sphere, could be involved in the hydronation of serine-92. The pKa value of HOH-Mg2+ is known to ionize with a pKa of 10.6-11.4, whereas the pKa of HOH-Zn is 9.6. The HOH-Zn interaction is strong enough to reduce the pKa from 15.7 to 7 or less. Our data suggest that the pKa 9.78 may contribute to ionization of the water molecule in HOH-Zn.

To obtain a complete separation of solvent isotope effects on pKa and kcat, pL-independent isotope effects on V at pL 11.0 were compared. The linear v2H versus n plot corresponds to generation of the isotope effect by a single transition-state site and is expected for a simple one-site general acid-base catalysis. The kinetic isotope effect can be considered as the ratio of a ground-state and transition-state fractionation factor kH/k2H = fR/fT, on the assumption that a single hydron bond has undergone force-constant changes in the transition state. Vmax/Km and Vmax reflect the conversion of the ES complex to phosphorylated enzyme and the hydrolysis of this intermediate to free enzyme and product respectively. The inverse isotope effect observed on V/K and V signifies a tightening at the hydrogenic site upon formation of the transition state, which then suggests phosphorylation or dephosphorylation as the rate-limiting step.

Another indication of a change in the rate-limiting step in these two solvent systems originated from KmNPP values determined at different pL values. There was a 9.4-fold increase in KmNPP from pH 9.5 to 11.0 in aqueous solution, which is related to the phosphate dissociation as the rate-limiting step at high pH values. In reverse micelles, the increase in KmNPP at high pH is only 4.3-fold, which is not so large when compared with that in aqueous solution. However, since the non-covalent E·Pi complex involves ionic interactions between the negatively-charged Pi and the positively charged arginine-166 and Zn2+ groups, the strength of such interactions might be different in water and deuterium and will make the observed isotope effects more complex. With these caveats in mind, we have interpreted our experimental data as follows.

For *E. coli* alkaline phosphatase, when the pK_a value of the leaving group is greater than 15, phosphorylation of the enzyme molecule becomes rate-limiting and k_{cat} shows a dependence on the pK_a value of the leaving group. More recently we found that the reaction rate of the placental enzyme in reverse micelles depends on the pK_a value of the leaving group. This observation also suggests that the rate-limiting step in reverse micelles is phosphorylation of the active-site serine-92 hydroxy group rather than product dissociation. The pK_a of 4-nitrophenol is 7.15 in aqueous solution, 7.5–7.9 in the water pool of reverse micelles, but 11.5 in the interfacial region. At $w_o = 22.6$, the inclusive volume of the reverse micellar vesicle was presumably barely sufficient to accommodate a dimeric enzyme molecule and thus only a monolayer of bound water at the interface is available for 4-nitrophenol, which, with a pK_a value of 11.5, explains the different rate-limiting step in the reverse micelles.

Based on above discussion, the following chemical mechanism catalysed by human placental alkaline phosphatase is proposed. The concentration of AMP buffer used in this investigation was low, thereby preventing transphosphorylation from competing with the hydrolysis reaction and, thus, only hydrolysis is of concern. In the basic pH region, a dianion of a phosphomonoester is the normal substrate, which binds with the enzyme forming an E·ROP binary complex where one of the ester oxygen co-ordinates with Zn1 and the second oxygen co-ordinates with Zn2, forming a phosphate bridge between Zn1 and Zn2. The other two oxygen atoms form hydrogen bonds with the guanidinium group of arginine-166. Serine-92, activated by Zn2, serves as a nucleophile attacking the phosphorus atom with formation of phosphoryl–enzyme (E–P). A Zn1 co-ordinated water dissociates a hydron to become Zn-$^-$OH, which functions as the nucleophile for hydrolysis of the phosphoryl–enzyme. The non-covalently enzyme-bound phosphate (E·P) forms as the phosphate moves away from serine-92. One of the oxygen atoms of the phosphate again co-ordinates Zn1 to re-establish a phosphate bridge.

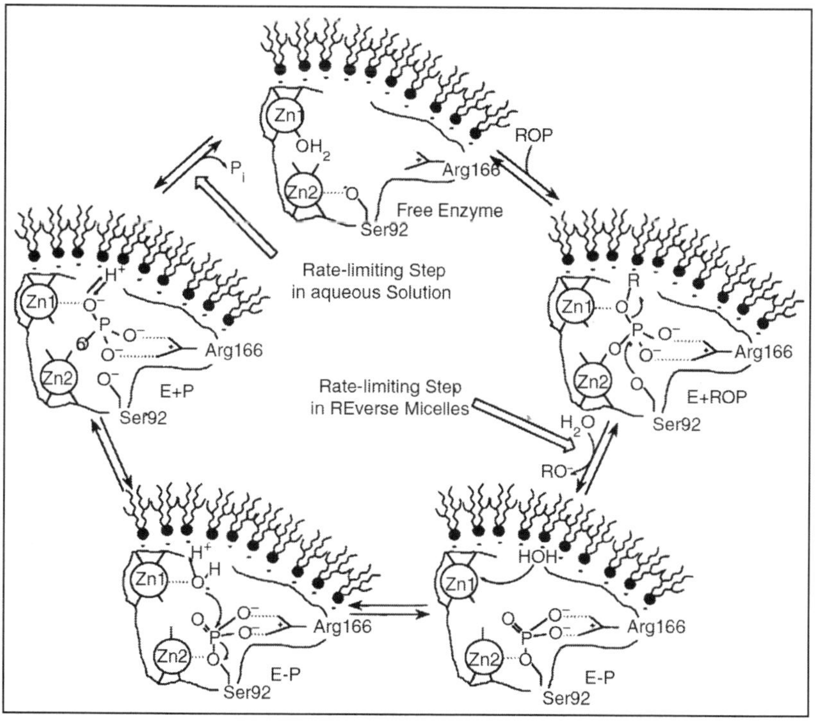

Fig. Proposed Chemical Mechanism of Human Placental Alkaline-Phosphatase-Catalysed Hydrolysis of the Phosphomonoester in Reverse Micelles

The two-metal site catalytic mechanism for *E. coli* alkaline phosphatase was adopted. The rate-limiting step is changed from dissociation of phosphate ion in aqueous solution to phosphorylation step in reverse micelles because of the hindered dissociation of the leaving group 4-nitrophenol. Only part of the reverse micellar vesicle is shown. The active-site region of the enzyme is highly schematic showing only some of the essential parts. In alkali pH, dissociation of non-covalently associated phosphate is the rate-limiting step. In reverse micelles the enzyme is surrounded by membrane-bound water and negatively-charged polar heads of the surfactant, which hinders the ionization of 4-nitrophenol and makes 4-nitrophenolate a poor leaving group. Covalent phosphorylation or dephosphorylation becomes the rate-limiting step.

BACTERIAL

In bacteria, alkaline phosphatase is located in the periplasmic space, external to the cell membrane. Since this space is much more subject to environmental variation than the actual interior of the cell, bacterial alkaline phosphatase is comparatively resistant to inactivation, denaturation, and degradation, and also has a higher rate of activity. Although the actual purpose of the enzyme is still not fully understood, the simple hypothesis, that it is a means for the bacteria to generate free phosphate groups for uptake and use, is supported by the fact that alkaline phosphatase is usually only produced by the bacteria during phosphate starvation and not when phosphate is plentiful. However, other possibilities exist; for instance, the presence of phosphate groups usually prevents organic molecules from passing through the membrane, therefore dephosphorylating them may be important for bacterial uptake of organic compounds in the wild. Some complexities of bacterial regulation and metabolism suggest that other, more subtle, purposes for the enzyme may also play a role for the cell. In the laboratory, however, mutant *Escherichia coli* lacking alkaline phosphatase survive quite well, as do mutants unable to shut off alkaline phosphatase production[*citation needed*].

The optimal pH for the activity of the *E. coli* enzyme is 8.0 while the bovine enzyme optimum pH is slightly higher at 8.5.

Use in Research

Common alkaline phosphatases used in research include:

- For removing phosphate monoester to prevent self ligation
- Shrimp alkaline phosphatase (SAP), from a species of Arctic shrimp (*Pandalus borealis*)
- Calf Intestinal Alkaline Phosphatase (CIP)
- Placental alkaline phosphatase (PALP) and its C terminally truncated version that lacks the last 24 amino acids (constituting the domain that targets for GPI membrane anchoring) - the secreted alkaline phosphatase (SEAP)

Alkaline phosphatase has become a useful tool in molecular biology laboratories, since DNA normally possesses phosphate groups on the 5' end. Removing these phosphates prevents the DNA from ligating (the 5' end attaching to the 3' end), thereby keeping DNA molecules linear until the next step of the process for which they are being prepared; also, removal of the phosphate groups allows radiolabeling (replacement by radioactive phosphate groups) in order to measure the presence of the labeled DNA through further steps in the process or experiment. For these purposes, the alkaline phosphatase from shrimp is the most useful, as it is the easiest to inactivate once it has done its job. Another important use of alkaline phosphatase is as a label for enzyme immunoassays. One common use in the dairy industry is as a marker of pasteurisation in cows' milk. This molecule is denatured by elevated temperatures found during pasteurisation, and can be tested for via colour change of a para-Nitrophenylphosphate substrate in a buffered solution (Aschaffenburg Mullen Test). Raw milk would typically produce a yellow colouration within a couple of minutes, whereas properly pasteurised milk should show no change. There are of course exceptions to this in the case of heat stable alkaline phophatases produced by some bacteria.

Inhibitors

All mammalian alkaline phosphatase isoenzymes except placental (PALP and SEAP) are inhibited by homoarginine and similarly all except the intestinal and placental ones are blocked by levamisole. Heating for ~2 hours at 65°C inactivated most isoenzymes except Placental isoforms (PALP and SEAP).

HUMAN

Physiology

In humans, alkaline phosphatase is present in all tissues throughout the entire body, but is particularly concentrated in liver, bile duct, kidney, bone, and the placenta.

Humans and most other mammals contain the following alkaline phosphatase isozymes:

- ALPI – intestinal
- ALPL – tissue non-specific (liver/bone/kidney)
- ALPP – placental (Regan isozyme)

Diagnostic use

The normal range is 20 to 140 IU/L. High ALP levels can show that the bile ducts are blocked. Levels are significantly higher in children and pregnant women. Also, elevated ALP indicates that there could be active bone formation occurring as ALP is a by product of osteoblast activity (such as the case in Paget's disease of bone). Lowered levels of ALP are less common than elevated levels.

Elevated levels

Main article: Elevated alkaline phosphatase. If it is unclear why alkaline phosphatase is elevated, isoenzyme studies using electrophoresis can confirm the source of the ALP. Heat stability also distinguishes bone and liver isoenzymes ("bone burns, liver lasts"). Placental alkaline phosphatase is elevated in seminomas.

Lowered Levels

The following conditions or diseases may lead to reduced levels of alkaline phosphatase:

- Hypophosphatasia, an autosomal recessive disease
- Postmenopausal women receiving estrogen therapy because of osteoporosis
- Men with recent heart surgery, malnutrition, magnesium deficiency, hypothyroidism or severe anemia
- Children with achondroplasia and cretinism
- Children after a severe episode of enteritis
- Pernicious anemia
- Aplastic anemia
- Chronic myelogenous leukemia

In addition, the following drugs have been demonstrated to reduce alkaline phosphatase:

- Oral contraceptives

Leukocyte Alkaline Phosphatise

Leukocyte alkaline phosphatase (LAP) is found within white blood cells.

White blood cell levels of LAP can help in the diagnosis of certain conditions:

- Higher levels are seen in polycythemia vera (PV), essential thrombocytosis (ET), primary myelofibrosis (PM), and the leukemoid reaction.
- Lower levels are found in chronic myelogenous leukemia (CML) and paroxysmal nocturnal hemoglobinuria (PNH).

GLUCOSE-6-PHOSPHATE DEHYDROGENASE DEFICIENCY

Glucose-6-phosphate dehydrogenase (G6PDH) is expressed in all cells and is responsible for the first reaction of the pentose phosphate pathway in which glucose-6-phosphate is oxidized to 6-phosphogluconolactone with concomitant production of NADPH. he NADPH produced via the pentose phosphate pathway is required for a variety of reductive biosynthetic reactions as well as for the regeneration of the reduced form of glutathione (GSH). GSH is essential for the detoxification of hydrogen peroxide (H_2O_2) and therefore, cellular defence against the oxidizing effects of H_2O_2 is absolutely dependent upon the generation of NADPH. GSH converts H_2O_2 to H_2O via the action of glutathione peroxidase and requires 1 mole of NADPH per mole of H_2O_2. The critical need for production of NADPH via the pentose phosphate pathway is especially true in red blood cells because there are no other NADPH-producing reactions in these cells and they are extremely sensitive to oxidative damage.

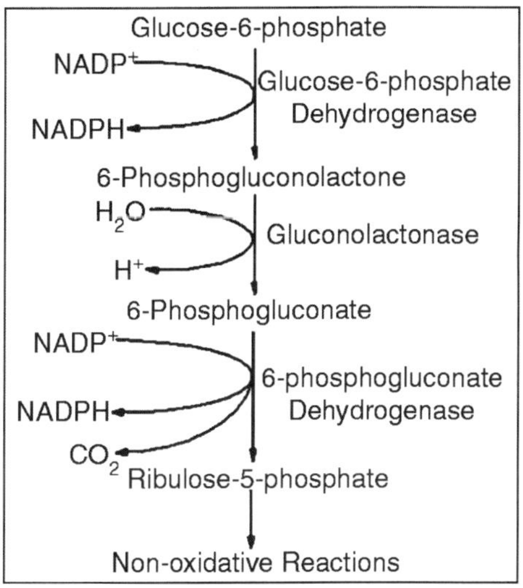

Fig. Oxidative Stage of Pentose Phosphate Pathway

The glucose-6-phosphate dehydrogenase gene (symbol G6PD) is located on the X chromosome (Xq28) about 1 Mb (million base pairs) from the telomeric end. The gene spans 18 kb and is composed of 13 exons encoding a protein of 531 amino acids. The first exon of the gene is a non-coding exon with the translational initiation codon present in exon 2. Post-translational processing G6PDH results in a 515 amino acid functional protein containing an acetylated alanine residue at the N-terminus. Biologically active G6PDH is functional as either a homodimer or a homotetramer and both forms co-exist in equal proportions at neutral pH.

Deficiencies in G6PDH are the most commonly inherited enzyme deficiencies (enzymopathies) world wide with estimates of greater than 400 million affected individuals. The incidence of deficiency approaches 25% of the population of persons of Mediterranean, tropical African, and tropical and subtropical Asian descent. In fact, the incidence of G6PDH deficiency is so high in some populations that the occurrence of homozygous females is not at all rare as would normally be expected for an X-linked disorder. The inheritance of G6PDH deficiency is clinically classified as an X-linked recessive disorder, however, because heterozygous females can develop hemolytic episodes the disorder is not truly recessive in the Mendelian sense. The phenomenon of symptom manifesting heterozygous females is explained by the co-existence of populations of G6PDH positive and negative cells in the same female as a result of X-chromosome inactivation.

Deficiency in G6PDH represents one of the most genetically heterogeneous disorders. There are over 400 different variants of G6PDH defined by their diverse biochemical characteristics. To date a total of 130 different point mutations have been identified in the G6PD gene. Surprisingly, only five in-frame deletions have been identified and no large deletions or insertions have been found. This fact is likely explained by the observation that in mouse models of G6PDH deficiency, complete loss of enzyme activity is associated with embryonic lethality.

G6PDH deficiency cannot be classified by a single mutation but is manifest as a consequence of numerous structural allelic mutants. In addition, G6PDH deficiencies are also classifiable by a variety of physicochemical parametres including chromatographic properties, thermostability, pH dependence, K_M for glucose-6-phosphate, and K_M for NADP. G6PD gene variants are currently divided into five classifications dependent upon enzyme activity and clinical manifestations. In addition, the deficiencies are classified as to whether or not they are sporadic or polymorphic.

The five classifications are:

Class 1: Enzyme deficiency with chronic nonspherocytic hemolytic anemia

Class 2: Severe enzyme deficiency, less than 10% of normal activity

Class 3: Moderate to mild enzyme deficiency, 10-60% of normal activity

Class 4: Very mild or no enzyme deficiency, at least 60% of normal activity

Class 5: Increased enzyme activity.

Mutations in the G6PD gene that result in nonspherocytic

hemolytic anemia have all been found to cluster near the carboxy terminal end of the protein, whereas, the mutations that result in clinically mild symptoms are clustered near the amino terminal end of the protein. Nearly all G6PDH mutations are single nucleotide missense mutations resulting in the substitution of a single amino acid. In most instances of G6PDH mutation resulting in disease it is due to protein instability as a result of the amino acid substitution.

Clinical Features of G6PDH Deficiency

The vast majority of individuals that harbor a mutant form of the G6PD gene go their entire lives without knowing they carry a mutation. The acute hemolysis that is the only common clinical manifestation in G6PD mutations can rapidly be compensated for and thus may remain undetectable. The most common symptoms of G6PDH deficiency are neonatal jaundice and acute hemolytic anemia. In addition, red cell deficiency in G6PDH is the basis of favism, primaquine sensitivity (a drug used in the treatment of malaria similar to quinine) and some other drug-sensitive hemolytic anemias (*e.g.* due to sulfonamides used to treat bacterial infections).

Favism is a term relating to a hemolytic anemia that results from the ingestion of the fava bean (*Vicia faba*) also known as broad beans. In the case of primaquine-induced hemolytic anemia, it was this clinical manifestation, particularly in Blacks, that actually led to the discovery of G6PDH deficiencies. Aside from favism and drug-induction of hemolytic anemia, infection is the most common cause of hemolysis in G6PDH deficient individuals. With respect to favism, the occurrence of acute hemolysis after ingestion of the fava bean was recognized as far back as the time of the Greek mathematician, Pythagoras. Analysis of extracts from fava beans demonstrated that the toxic compounds are the pyrimidine aglycones divicine and isouramil in combination with ascorbic acid.

Index

F

G

H

I

L

M

N

O

P

R